Palgrave Studies in Education and the Environment

Series Editors
Alan Reid
Faculty of Education
Monash University
Melbourne, VIC, Australia

Marcia McKenzie
Graduate School of Education
University of Melbourne
Melbourne, Australia

This series focuses on new developments in the study of education and environment. Promoting theoretically-rich works, contributions include empirical and conceptual studies that advance critical analysis in environmental education and related fields. Concerned with the underlying assumptions and limitations of current educational theories in conceptualizing environmental and sustainability education, the series highlights works of theoretical depth and sophistication, accessibility and applicability, with critical orientations to matters of public concern. It engages interdisciplinary and diverse perspectives as these relate to domains of policy, practice, and research. Studies in the series may span a range of scales from the more micro level of empirical thick description to macro conceptual analyses, highlighting current and upcoming turns in theoretical thought. Tapping into a growing body of theoretical scholarship in this domain, the series provides a venue for examining and expanding theorizations and approaches to the interdisciplinary intersections of environment and education. Its timeliness is clear as education becomes a key mode of response to environmental and sustainability issues internationally. The series will offer fresh perspectives on a range of topics such as:

- curricular responses to contemporary accounts of human-environment relations (e.g., the Anthropocene, nature-culture, animal studies, transdisciplinary studies)
- the power and limits of new materialist perspectives for philosophies of education
- denial and other responses to climate change in education practice and theory
- place-based and land-based orientations to education and scholarship
- postcolonial and intersectional critiques of environmental education and its research
- policy research, horizons, and contexts in environmental and sustainability education

Sara Tolbert • Maria F. G. Wallace
Marc Higgins • Jesse Bazzul
Editors

Reimagining Science Education in the Anthropocene, Volume 2

palgrave
macmillan

Editors
Sara Tolbert
University of Canterbury
Christchurch, New Zealand

Maria F. G. Wallace
University of Southern Mississippi
Hattiesburg, MS, USA

Marc Higgins
University of Alberta
Edmonton, AB, Canada

Jesse Bazzul
University of Regina
Regina, SK, Canada

ISSN 2662-6519   ISSN 2662-6527   (electronic)
Palgrave Studies in Education and the Environment
ISBN 978-3-031-35432-8   ISBN 978-3-031-35430-4   (eBook)
https://doi.org/10.1007/978-3-031-35430-4

ACKNOWLEDGEMENTS

We would like to express deep appreciation to our respective institutions for contributing funding that made it possible for us to publish Volume 1 and Volume 2 of *Reimagining Science Education in the Anthropocene* as open access resources:

Faculty of Education, University of Canterbury

University of Canterbury Library

Centre for Education Research, Collaboration, and Development (CERCD) at the University of Regina

University of Southern Mississippi

Faculty of Education, University of Alberta

Office of the Vice President Research and Innovation, University of Alberta

CONTENTS

Notes on Contributors

Rachel Askew is an Assistant Professor of Education at Freed Hardeman University. She is a former (and forever at heart) middle school science teacher constantly looking for ways to question, challenge, and create within elementary science teacher education. Research interests include rethinking elementary science teacher education and equity in STEM education.

Mahdis Azarmandi is a Senior Lecturer in the School of Educational Studies and Leadership, with expertise in anti-racism and social justice. She is a woman of color who has worked in Germany, Denmark, the United States, and Aotearoa New Zealand and tries to think about race and anti-racism across different contexts.

Laura Barraza holds a PhD on Education from the University of Cambridge, United Kingdom. Currently Director of Education for Sustainability of SACBÉ—Servicios Ambientales, Conservación Biológica y Educación A.C. México—and Advisor for Conservation Education programs in Latin America, Africa, and Australia. For the last 38 years she has dedicated her professional life to work on education in the fields of Indigenous knowledge, environmental values, science education, socio-scientific issues, cultural heritage, and capacity building with women and vulnerable groups.

Jesse Bazzul is an Associate Professor of Science and Environmental Education at the University of Regina. He believes imaginative work in education is needed more than ever to find new ways of living together.

He is the author of *An Intense Calling: How Ethics Is the Essence of Education* (University of Toronto Press).

Aline Carrara is a feminist economic geographer from Brazil, with broad experience in the Pan-Amazonian region. She has worked for over a decade in collaboration with indigenous communities and a variety of institutions on issues related to territorial sovereignty, sustainable supply chains, inclusive conservation, and just futures in the Anthropocene.

Ritodhi Chakraborty is a political ecologist and interdisciplinary social scientist from India. He has worked for the last decade in collaboration with indigenous and agrarian communities, across the Himalaya, to explore pathways of environmental and social justice.

Ryan B. Collis is a doctoral candidate in the Faculty of Education at York University, working at the intersection of autism, expertise, and science fiction. He holds degrees in English (BA), computer science (BScH), education (BEd), and science and technology studies (BScH and MA). He is a high school teacher, co-editor of the *Canadian Journal of Autism Equity*, Member at Large of the Canadian Committee of Graduate Students in Education (CCGSE), and the Graduate Student Officer of the Canadian Educational Researchers' Association (CERA).

Alejandra Frausto Aceves is a science teacher and educational leader currently working on a PhD in learning sciences at Northwestern University. Her research interests include transformative collective (science) learning, intergenerational and community-based co-constructions, imaginations, and praxis, as well as learning and pedagogies towards expansive and agentic present-futures.

Jane Gilbert is a Professor of Education at Auckland University of Technology (in Auckland, New Zealand). She was previously Chief Researcher at the New Zealand Council for Educational Research (NZCER). She has been involved in research and teaching in science education for more than 30 years, focusing in particular on equity issues. Her current work is mainly in the area of educational futures. Recent projects have focused on knowledge's changing meaning, science education's future, complexity thinking, and climate change education. In the last 10–15 years she has published 2 books and 25 refereed journal articles/book chapters on these topics.

Rachel A. Gisewhite is a science educator in the Center for STEM Education at the University of Southern Mississippi. Her research focuses on authentic scientific inquiry experiences in the natural world that increase scientific literacy and provide K–12 students, preservice teachers, and community members with an understanding of their place in the (social and natural) world, the recognition of their responsibility for their collective and individual actions, and a charge to act with generosity to remediate or protect that which they are exploring.

Kelli Gray is an Assistant Professor of English as a Foreign Language in the Department of Foreign Languages in the College of Humanities at the Universidad de Playa Ancha located in Valparaíso, Chile. She is a bilingual, Black woman who was born in DC and raised in Maryland, with expertise in TESOL, critical pedagogy, social justice education, and teacher education. As a teacher educator, Kelli is interested in exploring with teachers how to disrupt the status quo in education and transform classroom curriculum to be more inclusive and justice-oriented through the use of critical pedagogy and other equity pedagogies.

Marc Higgins is an Associate Professor in the Faculty of Education at the University of Alberta and is affiliated with the Aboriginal Teacher Education Program (ATEP). Marc's research investigates the complexities and complications that occur through interfacing Indigenous and Western modern ways-of-knowing and ways-of-being at the intersections of Indigenous education, science education, and creative and critical educational research methodologies in order to unsettle contested curricular concepts (e.g., what "counts" as science) and move towards an ethical otherwise. He is the author of the open-access *Unsettling Responsibility in Science Education: Indigenous Science, Deconstruction, and the Multicultural Science Education Debate* (Palgrave Macmillan).

Amanda Holmes is Kanien'keha:ka (Mohawk) on her mother's side, Highland Scottish on her father's side. She grew up in the Hudson River Valley of New York. She has had her Clan returned to her—she is Turtle Clan. She earned her doctorate in Indigenous language revitalization, education, and decolonizing methodologies in the Department of Language, Reading and Culture in the College of Education at the University of Arizona.

Lars Bang, PhD is an Associate Professor at the Department of Learning and Culture, Aalborg University, Denmark. He has a background as a teacher in special education and holds a master's degree in education (psychology) and a PhD in science. His current research areas are science education, sociology of education, Russian psychology (Leontiev and Vygotsky), and philosophy of science. He is particularly interested in Benedict Spinoza and Gilles Deleuze and how to use their thinking to revisit science education and education in general. He is currently leading two large empirical municipality projects examining the Danish primary school.

Steven Khan is an Assistant Professor of Mathematics Education at Brock University in St. Catharines, Ontario, Canada. Originally from Trinidad, his interests include pre-service teacher education, ethnomathematics, mythopoetic curriculum studies, and more recently the framing of education as being for multispecies flourishing. He enjoys collaborations with graduate students and colleagues from diverse disciplines.

Max Liboiron is an Associate Professor of Geography at Memorial University, where they direct the Civic Laboratory for Environmental Action Research (CLEAR). CLEAR develops feminist and anti-colonial methodologies to study marine plastic pollution. Liboiron is author of *Pollution Is Colonialism* (Duke University Press, 2021) and co-author of *Discard Studies: Systems, Wasting, and Power* (MIT Press, 2022).

Rasheda Likely is an Assistant Professor of Science Education with expertise in culturally sustaining curriculum, assessment, and instruction. She is a Black woman born and raised in Northwest Florida with a previous career as a medical testing scientist and research scientist in Northeast Florida. Her career path shifted after a challenging conversation with seventh-grade Black girls who questioned her identity as a scientist. She is dedicated to expanding and desettling normalized perceptions of science and scientists for K–12 students through incorporating activities such as sports and gaming, DIY hair care product making, and cooking into classroom learning.

Kurt Love is an Associate Professor in the Department of Educational Leadership, Policy, and Instructional Technology at Central Connecticut State University. His work focuses on curriculum and instruction, reframing science education, and integration of sociocultural and ecological contexts.

Susan Naomi Nordstrom is an Associate Professor of Educational Research at the University of Memphis. Her research focuses on qualitative methodological innovations that create a more just world. This focus is built from links between applied and methodological research that are informed by feminist, poststructural and posthumanist philosophies. In her applied work, she merges philosophy, qualitative research and art to create interventions that focus on entanglements in schools and neighborhoods. Her methodological work focuses on developing ways for qualitative researchers to re-consider the work of nonhuman objects in their research practices. Her methodological articles have appeared in leading qualitative research journals.

Huitzilin Ortiz is a science educator and doctoral student at the University of Minnesota. She is a Xicana from metro Los Angeles and raíces in Michoacán and Chihuahua, Mexico. Prior to obtaining her teacher certification, Huitzilin studied microbiology and had a career as a toxicology testing technician. Her current research focuses on the creation and implementation of environmental justice project-based curriculum in the physical sciences. As someone deconstructing her induction into cultural values of whiteness and Western modern science, Huitzilin strongly believes that no student should have to leave their identity at the classroom door.

Katherine Ryker is an Associate Professor of Geoscience Education at the University of South Carolina. Her research focuses on improving student learning and interest as well as post-secondary professional development and teaching beliefs. She is passionate about teaching the Earth in all contexts, having worked as a high school teacher, museum visitor guide, and summer camp counselor prior to becoming a university professor.

Anastasia Sanchez is a PhD candidate in the Learning Sciences and Human Development program in the University of Washington. She is currently working with state-wide organizations, climate justice initiatives, and K–8 teachers to promote antiracist and anticolonial science education. Anastasia's research focuses on the reimagining and reconstruction of science education to better grapple with consequential "wicked problems" in support of collective thriving. This heartwork is inspired by Anastasia's Xicana roots, mOtherhood, and 12 years of experience as a middle school science teacher, language acquisition specialist, and Title 1 curriculum specialist.

Vandana Singh is a Professor of Physics and Environment at Framingham State University in Massachusetts. Although her PhD is in theoretical particle physics, for more than a decade she has been working on reconceptualizing climate change at the intersection of pedagogy, science, society, and justice. Her book *Teaching Climate Change: Toward a Transdisciplinary, Justice-Centered Pedagogy* is forthcoming from Routledge. She is a member of the scientific steering group of My Climate Risk, a project of the World Climate Research Programme, where she facilitates a working group on climate education.

Isabelle Stengers is a Professor Emerita of the Université Libre de Bruxelles. After studies in chemistry and philosophy, she collaborated with Nobel Prize winner Ilya Prigogine to write *Order out of Chaos*, presenting physics as passionate adventure of relevance against its model of rationality and objectivity. An inseparable political and cultural challenge, her approach offers the concept of active ecology, embedding practices in a democratic and empowered, thus demanding, environment. Her practice of speculative, existential constructivism draws from the philosophy of Gilles Deleuze, Alfred North Whitehead, and William James, and the anthropology of Bruno Latour. She has written numerous books.

Sharon Todd is a Professor of Education and member of the Centre for Public Education and Pedagogy at Maynooth University, Ireland. She has published widely in the areas of embodiment, social justice, and ethics in education and is currently engaged in making connections between the climate emergency, art practice, and political aesthetics in education. She is most recently the author of *The Touch of the Present: Educational Encounters, Aesthetics, and the Politics of the Senses* (SUNY Press, 2023).

Sara Tolbert (she/her, tangata tiriti) is an Associate Professor of Science and Environmental Education at the University of Canterbury, with expertise in social and environmental justice, critical pedagogy, and feminist studies. She is a Pākehā/white cisgender woman from the southeastern USA (metro Atlanta), with parental roots in Appalachia and southern Louisiana, and a former bilingual science and environmental educator (Spanish/English), now teacher educator. Her research and teaching focus on science education as/for civic engagement and respons-able relations.

Brittany Tomin is an Assistant Professor of Secondary English in the Faculty of Education at the University of Regina. Her work is situated within curriculum studies and explores how narratives of the future are socially constructed within schools and, contrasting narrow notions of change and progress, how speculative storytelling and pedagogy can help us imagine—and potentially realize—different futures in uncertain times.

Betzabé Torres Olave is a physics teacher and science education researcher currently working as a Postdoctoral Research Fellow at the University of Leeds. Her research interests are around critically debating how neoliberal agendas have shaped (science) education cultures and pedagogies and exploring counternarratives to the neoliberal project as they appear within educators' daily practices. Central to her work in these counterspaces are the concepts of agency, collective praxis, and border crossing.

Maria Wallace is an Assistant Professor of Science Education and Women and Gender Studies Program faculty affiliate in the Center for STEM Education at the University of Southern Mississippi. She earned a PhD in curriculum and instruction with specializations in curriculum theory and science education, and a graduate minor in womens and gender studies from Louisiana State University. Dr. Wallace's research and teaching aim to deterritorialize beginning (science) teachers' subjectivities and practice. Drawing on conventional qualitative research methods and post-qualitative modes of inquiry, Dr. Wallace's research (re)imagines ways beginning science teachers are "known," named, and re/produced.

Matthew Weinstein is a Professor of Education at the University of Washington–Tacoma. He is the author of two books about the public cultures and discourses of science: *Robot World* and *Bodies Out of Control*. His current work focuses on STEM policy and the political economy of science and science education.

Michelle M. Wooten is an Assistant Professor of Physics at the University of Alabama at Birmingham and teaches large enrollment introductory astronomy courses. She is interested in how science teaching at the undergraduate level can become about more than knowledge transmission alone, taking into account students' many strengths, especially to the ends of developing sustainable societies.

LIST OF FIGURES

Introduction: To Be More Relevant the Field of Science Education Needs to Be Less Relevant

Sara Tolbert and Jesse Bazzul

Here we are, nearly three years into the COVID-19 pandemic reflecting on the disparate effects of climate change, as well as ever growing racial injustice, health disparities, and income inequality (all of which were exacerbated by the pandemic). The timing of these two open-access volumes of *Reimagining Science Education in the Anthropocene* seems fitting; nearing two years since its release, the first volume has reached over 110,000 downloads. We remain inspired by these critical and imaginative interventions in the field of science education, specifically by the ways in which the authors of both volumes have made us think, rethink, and feel different horizons in education and life in the Anthropocene(s). Each author has

S. Tolbert (✉)
University of Canterbury, Christchurch, New Zealand
e-mail: sara.tolbert@canterbury.ac.nz

J. Bazzul
University of Regina, Regina, SK, Canada
e-mail: jesse.bazzul@uregina.ca

© The Author(s) 2024
S. Tolbert et al. (eds.), *Reimagining Science Education in the Anthropocene, Volume 2*, Palgrave Studies in Education and the Environment, https://doi.org/10.1007/978-3-031-35430-4_1

created rupture, and together they have made a modest contribution to reconfiguring the landscapes of science and education (like sand bubbler crabs).

For many of us (editors, authors, comrades, compañeras) the field of science education has not always been welcoming. Nor has it always made much sense. As co-editors, we often reflect on our paths in this wild world of academia—and the particular brand of it that is Science Education™. How do we know it's truly worthwhile? How do we do it with integrity and love? Along the way we have found others, kindred spirits, many of whom have been "disciplined" out or simply bewildered by the field and its insidious positionings and strange loyalties. As editors, we set out to make a space for a community of misfits (at least we describe ourselves this way), to write, and thrive, in our peculiarity.

On a very general level the effects of a discipline can be seen multiple ways, and science education researchers and pedagogues should acknowledge that science education operates as a discipline. First, a discipline exercises discipline. This is helpful and productive in some ways. It allows a differentiation of some ways of knowing and being and produces something new in the world. The disciplines of chemistry and wine making are good examples of this. However, there's also a concomitant curtailing of difference, experimentation, and movement. As Shelia Jasanoff has written, "Disciplines cling tightly to their paradigmatic boundaries, reluctant to reflect too deeply on whether they are asking the right questions. ... Disciplinary discourses come with their own tacit claims to sovereignty in the definition of problems and the crafting of solutions" (2021, p. 850).

This discipline of science education operates through specific affordances and limitations, some of which produce something worth having (e.g., justice-focused, inquiry-based learning) and some of which are violent and dangerous (excluding Indigenous ways of knowing and living). While there is much more to be said here, we contend that science education as a discipline has oftentimes veered too close to things violent and dangerous—not with malice, but rather through ignorance, self-interest, and tacit complicity with things like colonization, white supremacy, capitalism, and gender-based violence.

This book attempts to offer some relief by going in the other direction, by offering a transdisciplinary, exploratory, and justice-oriented approach that might open up the (perhaps unavoidable) limiting aspects of a discipline. With the threats posed by the many calamities of the Anthropocene, strict adherence to disciplines today might be seen as deliberately harmful and pathological, in part because "how we acquire and organize our knowledge of the world is always entangled with ideas of how we should govern it" (Jasanoff, 2021, p. 841).

We have been humbled by the responses to the call for chapters for *Reimagining Science Education in the Anthropocene*, not only because the research and scholarship shared in Volumes 1 and 2 comprise a critical transdisciplinary intervention, but also because the overwhelmingly positive response by the education community represents the growing numbers of us who are "going in the other direction." In Volume 2, we share work from around the world that interrogates the status quo of the science education field, while also rendering it "less relevant." It is making the "older" discipline of science education less relevant that is paradoxically important. While most of what has been taught, theorized, and practiced in science education is not so problematic, it is the prescriptive and walled-in aspects of both science and science education that prevent broad pedagogical, methodological, and ethical connection. So, in a sense, the way science education becomes more important, larger if you will, is precisely by diminishing its boundaries and restrictions. There are numerous examples of this in our second volume. From elder practices to science fiction/(re)storying pedagogies, we have encouraged authors to not hold back and/or feel like they need to pay homage to a field of science education that has already described for them—when their inspiration comes from elsewhere (as it should). As in the last volume, we also have a stellar list of interviewees that discuss things as diverse as sorrow, time, colonialism, and what it takes to place matters of care at the heart of science education!

This volume, similar to Volume 1, is divided into five parts. The first is "Kinship, Magic, and the Unthinkable." The chapters in this section focus specifically on topics of kinship and futurity. The chapters falling under this heading in the second volume introduce pedagogical concepts like eroticism and flow in order to have us think about our cultural-geologic location in the times of ecological collapse. Lars Bang Jensen's chapter works against the 'dogmatic image of thought' in science and education: a stifling image that obscures multiplicities, possibilities, and uncertainties. Jane Gilbert argues for a reset of science education based on the undeniable social and ecological transformations of the Anthropocene(s). Rachel Gisewhite's chapter advocates for pedagogies that facilitate intimate encounters with things like oceans and geese and the very sand beneath our feet. Laura Barraza's work explores futurity with students in

Mozambique in order to make questions of uncertainty and precarity a part of everyday pedagogy and research.

The second part is entitled "Anti-colonial Anthropocene(s)," and recognizes that the Anthropocene as an overarching geological label is largely uneven and "plural" in its manifestations. Colonization, extinction, genocides, and systemic racisms all contribute to proliferation of different Anthropocenic worlds. Chapters in this section challenge taken-for-granted pedagogies and their colonial underpinnings and attempt to chart a different anti-colonial path. Carrrara and Chakraborty outline the hegemonic underpinnings of mainstream environmentalism and discuss critical challenges to it from subaltern communities. Mohawk scholar Amanda Holmes reflects on Lakota Elder Rosalie Little Thunder's provocation, "What does it mean to be a good relative?" Holmes draws from Little Thunder's wisdom and Indigenous spiritual geographies to imagine a different way of being in the world that centers connection, relationality, and spiritual interdependences. Anastasia Sanchez articulates a pedagogical framework that brings together presencing with concern and anticolonial critical consciousness to guide liberatory approaches to science and engineering learning. Vandana Singh offers a counterhegemonic approach to climate change education: She discusses how dominant thinking and acting is maintained through stories, which can be challenged by nondominant stories from marginalized communities, stories that act as boundary objects which allow more fluid articulations of possibility in a changing climate, derived from transcending boundaries across disciplines, perspectives, and paradigms.

The third part is called: "Politics and Political Reverberations," which attempts to introduce questions of solidarity and collectivity into science education. Chapters in this section ignite our critical imaginations for a science education driven by hope, love, and collective thriving: Sara Tolbert, Alejandra Frausto Aceves, and Betzabé Torres Olave explore how Freire's notion of 'true generosity' can provide political clarity for science education communities. Mahdis Azarmandi and colleagues share personal experiences to construct a collective and transformative vision for anti-racist science and education. Kurt Love illustrates possibilities for a science education driven not by greed and power but rather by an ideology of thriving, where well-being, sociocultural solidarity, ecological restoration, and societal regeneration are intended outcomes. Susan Nordstrom, reflecting on the Nebraska floods of 2019, reminds us of the importance of affective thinking and everyday caring practices that are so integral to non-human and human ongoing-ness in the Anthropocene.

The fourth section is titled "Science Education for a World Yet to Come," which takes us on speculative journeys into the near and distant future. Chapters in this section illustrate playful transdisciplinary alternatives to the status quo and ways to grow toward justice together. Brittany Tomin's chapter explores speculative pedagogies emerging from science fictions and world building in ways that accentuate the openness of education, but more importantly the various futures students and teachers might create together writ large. Science fiction is also employed in Matthew Weinstein's chapter that examines sociotechnical imaginaries and political possibilities in the face of the climate crisis. Can we indeed arrive at a 'New Deal' with capitalist, colonial powers? Michelle Wooten's chapter takes up the specific need for transdisciplinarity and the way educators move across disciplines and disciplinary knowledges. How is formalized education meant to contend with the perplexing problems of the Anthropocene without epistemic and pedagogical translations that traverse modes of knowing and being? Finally, Rachel Askew's chapter explores what it might mean for teachers to create something less structured and more connective together. How often do students create classes with their teachers based on what they view as pressing and urgent? All the chapters in this section challenge the way we've come to embody our work as science educators; and all of them attempt to make room for pedagogies and ways of being yet to come.

The last section contains interviews the editors conducted over the past two years or so. We are very grateful to have contributions from Sharon Todd, Max Liboiron, Isabel Stengers, and Steven Khan. Overall, we leave it to the reader to choose which sections speak to their needs as educators, students, and scholars. In Volume 2, like Volume 1, chapter authors and interviewees communicate heart-felt, radical, and honest accounts of education, nature, science, living, history, politics, art, and more. And they don't always "agree" with each other. But their nuanced perspectives, experiences, and wisdom all contribute to a vibrant pluralistic vision for science and education in ways that can nourish and sustain us. We hope you feel something, or see room for eccentric movement, when you read these chapters.

With love,

Sara and Jesse

REFERENCES

Al Jazeera. (2022, January 22). The poor die from COVID while the rich get richer, Oxfam warns. https://www.aljazeera.com/economy/2022/1/17/hdlthe-poor-die-from-covid-while-the-rich-get-richer-oxfam-warns

Jasanoff, S. (2021). Humility in the Anthropocene. *Globalizations, 18*(6), 839–853.

PART I

Kinship, Magic, and the Unthinkable

CHAPTER 2

Re-thinking Science Education for the Anthropocene

Jane Gilbert

This chapter argues that the arrival of the Anthropocene era requires a substantial re-set of science education. It makes a case for re-orienting school science education so that meta-level understanding of science is foregrounded over science's "content," its modes of inquiry, and/or its internal social practices. This would be quite unlike the school science curriculum we know today, but, given science's role in the Anthropocene, this is the chapter's main point.

All academic disciplines have four broad features. First, they all have discipline-specific ways of thinking and discipline-specific frameworks for developing and evaluating new knowledge. Second, they generate "products"—concepts, principles, or tools—that although they come from, and may later be changed from within the discipline, can be used by people from outside the discipline. Third, members of the discipline participate in discipline-specific ways of interacting with each other as they work together to generate, evaluate, and distribute new knowledge. Fourth, all

J. Gilbert (✉)
Auckland University of Technology, Auckland, New Zealand
e-mail: jane.gilbert@aut.ac.nz

© The Author(s) 2024
S. Tolbert et al. (eds.), *Reimagining Science Education in the Anthropocene, Volume 2*, Palgrave Studies in Education and the Environment, https://doi.org/10.1007/978-3-031-35430-4_2

disciplines are located in, and influenced by, particular historical and cultural contexts. They exist alongside, and in contrast to, other ways of thinking, doing, and knowing. Although not everyone who is involved in a discipline masters all of these areas equally, the discipline as a whole embodies and it is defined by all four of them.

School curriculum development usually involves selecting knowledge from a "parent" academic discipline and then "re-contextualising" it for educational purposes (Bernstein, 2000; Deng, 2007). These purposes can change over time as education systems respond to developments in the wider social, political, and economic contexts they are enmeshed in. In general, however, each "subject" of the traditional academic curriculum draws on the above four features of its parent discipline. In the curriculum context, these features are "weighted" differently in relation to each other, based on currently prevailing ideas about the educational purposes of that curriculum area. The science curriculum for schools usually emphasises, in different proportions, science's four broad features: first, the *epistemic* aspects of science—the specific forms of intellectual inquiry that scientists use to develop, evaluate, and justify new knowledge; second, the *ontological* aspects of science—its "products" or "content," which can include facts, laws, algorithms, principles, or tools; third, the social and rhetorical strategies scientists use when they *interact with each other as scientists*; and fourth, a *meta-level understanding* of science's role and location in the wider social, cultural, political, and economic context in which it developed.

Over the roughly 150 years or so of science education's existence, science's "products" have been the main component of most school science curricula. Generations of reformers have argued for a more "balanced" curriculum, for greater emphasis on one or more of the other aspects of science, but in general, this has had little impact on classroom practice (DeBoer, 1991). For example, science education researchers have argued strongly that deep understanding of science's "products" is impossible without an understanding of the inquiry processes that produced them (e.g., Newton et al., 1999; Duschl & Osborne, 2002; Osborne, 2014; Kind & Osborne, 2017). Other researchers have advocated pedagogies designed to "socialise" students to think, act, and interact "like scientists" (e.g., Driver, 1983; Driver et al., 1994; Driver & Oldham, 1985; Tobin, 1990). Still others have argued for greater emphasis on developing a meta-understanding of science, through studying its history, philosophy, and sociology (e.g., Matthews, 1994), and there is a large body of work advocating a focus on socio-scientific issues (e.g., Zeidler et al., 2005; Zeidler

& Nichols, 2009) and/or developing scientific "literacy." However, it is fair to say that mainstream science education research, informed as it is by cognitive science, continues to focus largely on addressing the intractable issue of how to support students to achieve "real understanding" of science's "products." Intertwined with all this, influenced by wider trends, there have been calls for science education to be made more "inclusive," "engaging" or "relevant" for more students, and the last couple of decades have seen an increased emphasis on skills, competencies, and what Biesta (2012) calls the "learnification" of education.

This is a complex fruit salad of ideas. However, as a long-term observer of this field, I have found it hard to discern any consensus on the question of the *educational* function/s science is supposed to serve through its inclusion in the school curriculum. A multitude of different purposes for school science are espoused, including providing foundational science knowledge for students headed for science-related careers; providing students with science-related knowledge they might need in everyday life; developing sufficient scientific "literacy" for active citizenship; and empowering students by providing access to "powerful knowledge." These purposes are all very different, and each implies a very different curriculum: however, all are oriented towards acquiring and storing away certain kinds of *knowledge*. This predilection for turning everything into "stuff to be known"[1] seems to be a feature of science education. However, it isn't always clear how acquiring this knowledge is supposed to be *educative*, in the sense meant by Dewey (1938).[2] Having a sense of this is, it seems to me, important for curriculum designers as they decide how to select from—and balance—the four aspects of science outlined above.

None of these issues are new (Gilbert, 2011), but my purpose in rehearsing them here is to suggest that recent events and trends outside education, specifically the coming of the Anthropocene, throw these issues into very sharp relief. In this chapter I want to argue that the "new times" we are now in require us to re-frame science's role in the school curriculum. Substantial change is needed, change that is difficult to even imagine, let alone think about productively and practically. However, for reasons I'll come to shortly, I think there is a moral imperative to attempt this work. In this chapter, I explore whether emphasising the fourth,

---

[1] David Perkins calls this predilection "aboutism" (see Perkins, 2009, p. 5).

[2] Dewey defined "educative" experiences as those that foster ongoing intellectual growth by building the capacity to think in deeper, more complex, more abstract ways.

meta-understanding aspect of science could provide an appropriate frame for the kinds of education we now need. But first, why is change necessary? What exactly *is* this thing called the Anthropocene? And what does it have to do with education?

The Anthropocene is the name now being given to the advent of a new geological epoch, beginning roughly with the Industrial Revolution and the industrial-scale use of fossil fuels, in which human activities came to have a major influence on the earth's physical processes. The term is derived from the Greek: "anthro" meaning "human," and "cene" meaning "new" (geological era) and was coined to signal the termination of the earlier Holocene era. Burning carbon sequestered over hundreds of millions of years by living processes from the atmosphere has vastly increased atmospheric carbon dioxide levels, which has in turn triggered an ongoing increase in mean global temperatures. This is expected to have a major impact on world sea levels, weather systems, and ecosystem stability, which will affect the habitability of the planet for humans and have major implications for human social, political, and economic life. These processes are now well under way (Kress & Stine, 2017; Scranton, 2015; Klein, 2014; McNeill & Engelke, 2014; Hansen, 2009).

As widely discussed elsewhere, collectively we have not managed to put in place measures that could reverse or delay these trends, nor have we developed strategies for adapting to or mitigating their likely effects (Flannery, 2005; Hamilton, 2010; Jamieson, 2014; Oreskes & Conway, 2014). The scientific consensus is that unless we reduce carbon emissions by 45% (from 2010 levels) by 2030, we will exceed 1.5 degrees of global warming, deemed as the upper limit for a habitable planet. According to some scientists, "abrupt" change, that is, change that is so rapid that humans and other natural systems do not have time to adapt, is likely, possibly within the next decade (IPCC, 2018; 2022). It is no longer controversial to say that we are sleepwalking towards disaster, that current practices are destroying the lives of our children, and that anyone who is under sixty years old today is likely to witness the radical de-stabilisation of life on earth. The impacts are likely to be felt first and most by the world's poorest and most marginalised peoples. Most countries are in a state of policy paralysis, at least partly because actually addressing the issue will require major sacrifice, major curtailment of our current economic activities and lifestyles. It isn't at all clear who should bear these costs and/or how they should be distributed. And now, a good thirty years after the science on climate change first became clear, it could well be too late to reverse or delay its effects. If this is the case, then actions additional to those directly related to reducing carbon emissions become important. Any world-improving action—maintaining functioning democracies, functioning legal systems, functioning communities; instituting humane immigration policies; strengthening all human systems, including education—can now be considered climate action.

So what does all this have to do with the school science curriculum? Wouldn't the "topic" of climate action be most appropriately located in the social studies curriculum? Should students be taught the *science* of climate change? Or should climate change action be made a new and distinct curriculum area in its own right, as some are advocating?[3] In what follows, I outline why I think the coming of the Anthropocene requires us to re-set science education. Then I set out why I think a useful place to start this work would be to emphasise meta-level understanding of science over its other three aspects.

Being "in crisis" seems to have been a feature of science education since its inception, (DeBoer, 1991; Aubusson, 2013; Toscano, 2013). Each time a new crisis is identified, reports and new research are commissioned, new approaches to teaching are recommended, and new curricula come into effect. Vocabulary from this work finds its way into policy rhetoric and sometimes classrooms, but usually things continue much as they always have. However, the Anthropocene, because it disrupts fundamental features of the historical period in which science education developed is, it seems to me, the "crisis to end all crises."

Education, science education, and science itself, in their present forms, are products of, and deeply connected to, Western modernity's core assumptions and economic conditions. Modern education was forged in the transition from agriculture-based economies and societies to predominantly urbanised, industrially oriented ways of life. The development of mass schooling was important for its role in producing the human resources—and consumers—modern economies need. The "subjects" of the modern school curriculum, including science, were developed to support the growth of modern capitalist economies/societies. Science is deeply connected to that growth, both in the positive sense of what it has made possible, and in the negative sense of its contribution to the crisis we now find ourselves in (Patel & Moore, 2017). But this period in history, characterised by some as "carboniferous capitalism" because its success has

---

[3] For example, Everth and Bright (2022).

rested on the "cheapening" of nature (Patel & Moore, 2017) and the burning of fossil fuels (Newell & Patterson, 2010), is coming to an end.

As has been well-canvassed elsewhere, this has major implications for the planet and for human social and economic life. But there are also *intellectual* implications, discussion of which is also well under way. For example, some scholars are attempting to set out a new paradigm of "post-carbon" social theory, to re-work "old" (modern) conceptions of society, politics, and the economy for the new times (e.g., Newell & Patterson, 2010; Irwin, 2010; Urry, 2011; Elliott & Turner, 2012; Klein, 2014). Commentators in science-related disciplines talk about the shift to what they refer to as "postnormal times," a world in which things are no longer certain, simple, or stable (if they ever were); instead, uncertainty, complexity, chaos, and contradictions are the "new" normal (e.g., Sardar, 2010; Ravetz, 2011; Slaughter, 2012).

For Bruno Latour, a major figure in the sociology of science, the Anthropocene heralds a major intellectual shift in science itself. In his 2013 Gifford Lectures, he argues that the Anthropocene challenges scientists to think in completely new ways about science—what it *is*, what it is *for*, and what (and who) it should *engage* with (Latour, 2013).[4] He says that scientists need to see nature, not as an "object of enquiry," something we are apart from, or something to be tamed, but rather as something we are deeply *engaged with*, part of, and inextricably entangled with. Rather than investigating nature's "entities" as things-in-themselves, scientists should be exploring what he calls the "crossings," "borders" or "conversations" *between* science and nature. This of course requires completely new ways of thinking: new forms of inquiry, new tools, and new practices. It also requires a new relationship between science and politics (Latour, 2018).

All this, if we accept it, has major implications for education, for science education, and for science itself. If modernity's key concepts no longer apply, then what should education's purpose be? Do we still need (or want) public education? What role, if any, should science play in education? Is it defensible to continue to include science in the school curriculum, given science's contribution to the activities and thinking that produced the Anthropocene? Should we be *reproducing* this kind of thinking? If we think school science could have an educative function in the Anthropocene, what would this look like? How might this differ from

[4] See also: http://www.modesofexistence.org

what it does now? How, if at all, should science intersect with other curriculum areas?

These questions are incredibly difficult to address, mainly because our thinking is structured by a set of conceptual categories that are *part of the problem*. Our thinking is colonised: we can't think outside these categories. All we can do is, to use Derrida's (1991) term, to put them "under erasure," signal that they are problematic, that they may eventually need to be "erased," while at the same time continuing to work with them, because we don't (yet) have an alternative thinking system.

So, given all this, and looking at just one of the above questions, I want to suggest that school science *could* be educative in the Anthropocene context, but only if it is significantly re-framed. However, in considering this re-framing, I don't think we should "throw the baby out with the bathwater": I think we have to work with what we have. In the remainder of this chapter, I attempt to sketch out a curriculum design that foregrounds the meta-understanding aspect of science, maintaining the other three aspects, but in reduced form and with different purposes. This approach doesn't look at all like the school science curricula we are familiar with, which is the central point of this chapter.

Focusing on the epistemic aspects of sciences—that is, its products and/or the way scientists work together as scientists—is useful if our purpose is to *reproduce* these aspects of science, to enculturate or discipline students into the discipline as it is now. These approaches, if they are successful, structure students' thinking in particular ways which, unless they are also exposed to other ways of thinking, make it very hard for them to "see outside" these ways of thinking, and they foster the belief that there is one "right" way of thinking. However, if the aim is to expand students' capacity to think in *different*, more complex ways, or to expand our collective capacity for change, then this isn't such a good strategy.

Change usually doesn't come from the centre of an established discipline (Kuhn, 1970). More often, it comes from the periphery, generated by outsiders who are critical of, but also fluent in, the discipline. Change-makers are often people who are "bi-cultural," people who have participated sufficiently in more than one disciplinary context to see "how things work," what matters, and how the two contexts are similar/different. In other words, they are often people who have developed meta-level understanding. So, if our goal is to build the capacity for change and/or more complex thinking, then fostering meta-understanding, ideally of multiple contexts, seems to me to be a productive strategy.

I'm using the term meta-understanding here to mean a view of science "from above": an understanding of how scientific knowledge has been built in a particular social, political, historical, and cultural context, by particular kinds of people, using particular ways of thinking, particular practices, and drawing on sets of assumptions which differ from those used in other ways of thinking. Advocacy of approaches designed to build meta-understanding isn't new. As mentioned earlier, many reformers have argued for teaching science via a focus on its history, philosophy, and sociology; on socio-scientific issues; and/or on the development of scientific "literacy." And the "nature of science" initiatives that have been added into school science programmes in many countries were intended to encourage critical thinking and meta-understanding of science. However, in practice, none of these initiatives has produced the kind of change expected by their proponents. Perversely, many of these initiatives have generated new and additional sets of propositions students need to "know about"—that is, they have been incorporated into the "aboutism" referred to earlier (Lederman, 2007; Hipkins, 2012). Reformatting these initiatives as yet more knowledge for students to acquire is to (obviously) miss their point, but, importantly for the present purposes, it is unlikely to be *educative* in the Deweyan sense. Because these initiatives haven't worked as intended, and given the present situation, I think something bolder is needed. So, in the remainder of this chapter, I want to make the case for an approach to achieving meta-understanding that uses the concept of deconstruction. This, to people steeped in science and science education, will seem very weird indeed.

The deconstruction concept is commonly used in the humanities and social sciences and sometimes found in education, but for reasons that are probably obvious, it is unfamiliar—and likely to be unwelcome—in science-related contexts. But, as I've tried to show here, science education and science itself are different activities, with different goals. If we follow Dewey's idea of education's purpose as being to foster intellectual growth, to build the capacity to think in ever-deeper, ever more complex, abstract ways, then science's function in the school curriculum is simply to be one of several contexts or "vehicles" educators can use to foster intellectual growth. Science education's primary purpose is to *educate*, not to "communicate" or "to deliver" science (although it may do that). Science education *isn't* science, and, following from this, we wouldn't expect to see a one-on-one mapping between science and science education. So, while deconstruction might not be an appropriate technique to use in science itself, I want to argue here that it could be appropriately used in science *education*.

Deconstruction's purpose is change, particularly in relation to idea-systems, and in situations where these idea-systems are seen to be oppressive. Deconstruction is a process for trying to break out of, and see beyond, the conceptual categories that, at a deep level, structure the way we think. The aim is to look *below the surface* to see how these conceptual categories work together as a system, and how this system becomes possible by excluding or disallowing certain other categories. Deconstruction is different from analysis or critique: its aim is *not* to take apart or knock down the existing categories, but to work *with* them in new ways. Its purpose is to make visible the unacknowledged material that lies between, beyond, and underneath the existing categories, and to then to use this material to think outside these categories (Culler, 1983; Grosz, 1989; Lather, 1991; Davies, 1994).

Deconstructive work is done, not at the level of specific ideas, forms of inquiry, or social practices, but at the level of discursive practices. Discursive practices are systems of thought that emerge from certain sets of ideas, forms of inquiry, and social practices, under certain wider (political and institutional) conditions (Foucault, 1972, 1978). Many different discursive practices exist alongside each other. Each produce "truths" that "work" in the context of that set of discursive practice, within its boundaries, and when its rules are followed. Discursive practices include ideas, forms of inquiry, and social practices, but these are not the focus. What is in focus is the way discursive practices are a medium for wider power relations, and how they work, not to represent reality, but to actively *produce* it. Deconstruction involves exploring how sets of discursive practices "work": it looks at the assumptions they rest on; the practices that define them; the people who participate in them; the political, institutional, and disciplinary structures they are embedded in; the metaphors that organise them; and, importantly, it looks for what *isn't* there, for what or who is excluded, disallowed, or illegitimate.

Science is a discursive practice. It produces particular kinds of knowledge, involves particular practices, rests on particular assumptions, is participated in by people who think in particular ways, and it is part of a wider system of power relations. It produces "truths" that "work" well in this context. Enculturating the students into these discursive practices is useful if the goal is to reproduce them. But, if we accept Latour's argument that the Anthropocene's arrival challenges some of science's deepest assumptions, then this no longer seems defensible.

What might a deconstruction-based approach to building meta-understanding look like in practice? Exploring this fully is outside the scope of this chapter: my purpose here has been to make a case for why it is necessary. However, in other work (many years ago now) I have mapped out in some detail deconstructive approaches to the teaching of first genetics, and then later, animal behaviour, and human evolution for use by high school biology teachers (Gilbert, 1997). I think it is possible to do this work, and I think it could generate the kinds of educative experiences students need to prepare them for life in the Anthropocene. Perhaps now, given today's context, this kind of work might be more palatable to science educators than it was two decades ago. Perhaps.

Science education, it seems to me, has a really important role to play as we transition into the Anthropocene, arguably more so than other curriculum areas. Science and technology are routinely depicted *as* the future, as what will "save" us from the problems we face. But, while technological mitigations for climate change will undoubtedly be developed (Kolbert, 2018), thinking this way sends us down *one* particular pathway to the future. This is dangerous because, while scientific work has identified the Anthropocene's development, scientific ways of thinking and activities have undoubtedly contributed to it. It is important that we are able to think within *and* outside the "science as the future" pathway, to imagine *other* possible pathways to the future. As the Futures Studies scholarly literature tells us, channelling our thinking in particular ways, along particular pathways, effectively closes off other options. Science and technology don't, in themselves, shape our future: developments in science and technology are guided by human values, choices, and actions (Slaughter, 2012). As the futurist Riel Miller puts it, the future isn't something that just "happens" to us: every one of us "*create[s]* the future/s through the choices we make *every day ... starting now*" (Miller, 2006, p. 3). Building on this, the educationist Keri Facer points out that,

> [t]his perspective changes the dominant metaphor for our orientation toward the future. Rather than envisaging ourselves walking forwards into a future in which choices are laid out before us and from which we must choose, carefully selecting paths to avoid risks and fears. Instead we might imagine ourselves walking backwards into an unknowable future, in which possibilities flow out behind us from our actions. (2013, p. 9)

Science education for the Anthropocene should build the intellectual capacities needed to *create* the futures we collectively *want*. It should aim to support the capacity to work *within* current pathways, but also to stand *outside* them. If we can't find ways to see out of the well-worn rut of existing conceptual categories, it is highly likely that we will continue to sleep-walk towards climate catastrophe. We probably have about a decade to wake up and do this work. But, on the other hand, it could well be that it is already too late for the kind of rather abstract strategies proposed here, and that a more productive contribution to the planet's future might be to join the type of activism proposed by the Extinction Rebellion (2019) movement.

References

Aubusson, P. (2013). *Science education futures: Unchaining the beast.* Briefing paper prepared for the Monash University seminar on Science Education Futures 2020–2025. Melbourne, 19 February 2013.

Bernstein, B. (2000). *Pedagogy, symbolic control and identity: Theory, research, critique.* Rowman & Littlefield.

Biesta, G. (2012). Giving teaching back to education: Responding to the disappearance of the teacher. *Phenomenology and Practice, 6*(2), 35–49.

Culler, J. (1983). *On deconstruction: Theory and criticism after structuralism.* Routledge.

Davies, B. (1994). *Poststructuralist theory and classroom practice.* Deakin University Press.

DeBoer, G. (1991). *A history of ideas in science education: Implications for practice.* Teachers College Press.

Deng, Z. (2007). Transforming the subject matter: Examining the intellectual roots of pedagogical content knowledge. *Curriculum Inquiry, 37*(3), 279–295.

Derrida, J. (1991). Of grammatology. In P. Kamuf (Ed.), *A Derrida reader: Between the blinds.* Columbia University Press.

Dewey, J. (1938). *Experience and education.* Collier Macmillan.

Driver, R. (1983). *The pupil as scientist?* Open University Press.

Driver, R., & Oldham, V. (1985). A constructivist approach to curriculum design in science. *Studies in Science Education, 13*, 105–122.

Driver, R., Asoko, H., Leach, J., Mortimer, E., & Scott, P. (1994). Constructing scientific knowledge in the classroom. *Educational Researcher, 23*(7), 5–12.

Duschl, R., & Osborne, J. (2002). Supporting and promoting argumentation discourse in science education. *Studies in Science Education, 38*(1), 39–72.

Elliott, A., & Turner, B. (2012). *Society.* Polity Press.

Everth, T., & Bright, R. (2022). Climate change and the assemblages of school leaderships. *Australian Journal of Environmental Education, 1–20.* https:// doi.org/10.1017/aee.2022.8

Extinction Rebellion. (2019). *This is not a drill: An Extinction Rebellion handbook.* Penguin.

Facer, K. (2013). The problem of the future and the possibilities of the present in education research. *International Journal of Educational Research, 61,* 135–143.

Flannery, T. (2005). *The weather-makers: How man is changing the climate and what it means for life on Earth.* Text Publishing.

Foucault, M. (1972). *The archaeology of knowledge.* Routledge.

Foucault, M. (1978). Politics and the study of discourse. *Ideology and Consciousness, 3,* 7–26.

Gilbert, J. (1997). *Thinking "other-wise": Re-thinking the problem of girls and science education in the post-modern.* Unpublished PhD thesis, University of Waikato.

Gilbert, J. (2011). School science is like wrestling with an octopus. *New Zealand Science Teacher, 126,* 29–31.

Grosz, E. (1989). *Sexual subversions: Three French feminists.* Allen and Unwin.

Hamilton, C. (2010). *Requiem for a species: Why we resist the truth about climate change.* Earthscan.

Hansen, J. (2009). *Storms of my grandchildren: The truth about the coming climate catastrophe and our last chance to save humanity.* Bloomsbury.

Hipkins, R. (2012). *Building a science curriculum with an effective nature of science component. Report prepared for the Ministry of Education.* New Zealand Council for Educational Research. http://www.nzcer.org.nz/research/publications/building-science-curriculum-effective-nature-science-component-0

Intergovernmental Panel on Climate Change (IPCC). (2018). *Summary for policymakers of IPCC special report on global warming of 1.5°C approved by governments.* https://www.ipcc.ch/2018/10/08/summary-for-policymakers-of-ipcc-special-report-on-global-warming-of-1-5c-approved-by-governments/

Intergovernmental Panel on Climate Change (IPCC). (2022). *Climate change 2022: Impacts, adaptation and vulnerability. Summary for policymakers.* https://www.ipcc.ch/report/ar6/wg2/downloads/report/IPCC_AR6_WGII_SummaryForPolicymakers.pdf

Irwin, R. (Ed.). (2010). *Climate change and philosophy: Transformational possibilities.* Continuum.

Jamieson, D. (2014). *Reason in a dark time: Why the struggle against climate change failed and what it means for our future.* Oxford University Press.

Kind, P., & Osborne, J. (2017). Styles of scientific reasoning: A cultural rationale for science education? *Science Education, 101*(1), 8–31.

Klein, N. (2014). *This changes everything: Capitalism vs the climate.* Simon & Schuster.

Kolbert, E. (2018). Climate solutions: Is it feasible to remove enough CO_2 from the air? *Yale Environment 360*, 15 November 2018. https://e360.yale.edu/features/negative-emissions-is-it-feasible-to-remove-co2-from-the-air

Kress, J., & Stine, J. (Eds.). (2017). *Living in the Anthropocene: Earth in the age of humans.* Smithsonian Books.

Kuhn, T. (1970). *The structure of scientific revolutions* (2nd ed.). University of Chicago Press.

Lather, P. (1991). *Getting smart: Feminist research and pedagogy with/in the postmodern.* Routledge.

Latour, B. (2013). *The Anthropocene and the destruction of the image of the globe.* Gifford Lecture No. 4. 25 February 2013. The University of Edinburgh. http://knowledge-ecology.com/2013/03/05/bruno-latours-gifford-lectures-1-6/

Latour, B. (2018). *Down to earth: Politics in the new climatic regime* (C. Porter, Trans.). Polity Press.

Lederman, N. (2007). Nature of science: Past, present and future. In S. Abell & N. Lederman (Eds.), *Handbook of research on science education* (pp. 831–879). Lawrence Erlbaum.

Matthews, M. (1994). *Science teaching: The role of history and philosophy of science.* Routledge.

McNeill, J., & Engelke, P. (2014). *The great acceleration: An environmental history of the Anthropocene since 1945.* Harvard Belknap Press.

Miller, R. (2006). Futures studies, scenarios, and the "possibility space" approach. In *Think scenarios, rethink education* (pp. 93–105). OECD. http://www.oecd.org/site/schoolingfortomorrowknowledgebase/futuresthinking/scenarios/futuresstudiesscenariosandthepossibility-spaceapproach.htm

Newell, P., & Patterson, M. (2010). *Climate capitalism: Global warming and the transformation of the global economy.* Cambridge University Press.

Newton, P., Driver, R., & Osborne, J. (1999). The place of argumentation in the pedagogy of school science. *International Journal of Science Education, 21*(5), 553–576.

Oreskes, N., & Conway, E. (2014). *The collapse of Western civilisation: A view from the future.* Columbia University Press.

Osborne, J. (2014). Teaching critical thinking? New directions in science education. *School Science Review, 352*, 53–62.

Patel, R., & Moore, J. (2017). *A history of the world in seven cheap things: A guide to capitalism, nature and the future of the planet.* University of California Press.

Perkins, D. (2009). *Making learning whole.* Jossey-Bass.

Ravetz, J. (2011). Postnormal science and the maturing of the structural contradictions of modern European science. *Futures, 43*, 142–148.

Sardar, Z. (2010). Welcome to postnormal times. *Futures, 42*, 435–444.

Scranton, R. (2015). *Learning to die in the Anthropocene: Reflections on the end of a civilization.* City Lights Books.

Slaughter, R. (2012). Welcome to the Anthropocene. *Futures, 44,* 119–126.

Tobin, K. (1990). Social constructivist perspectives on the reform of science education. *Australian Science Teachers' Journal, 36*(4), 29–35.

Toscano, M. (2013). *Is there a crisis in science education?* Paper presented to the Annual Conference of the Philosophy of Education Society of Great Britain. New College, Oxford, 22–24 March 2013.

Urry, J. (2011). *Climate change and society.* Polity Press.

Zeidler, D., & Nichols, B. (2009). Socio-scientific issues: Theory and practice. *Journal of Elementary Science Education, 21*(2), 49–58.

Zeidler, D., Sadler, T., Simmons, M., & Howes, E. (2005). Beyond STS: A research-based framework for socio-scientific issues education. *Science Education, 89*(3), 357–377.

CHAPTER 3

What Future Do Young Mozambicans Envision in a Time of Humanitarian and Environmental Crisis?

Laura Barraza

Introduction

What does it mean to talk about the future in vulnerable communities? What future is visualized by teenagers from a poor country that has been devastated by a natural disaster? Talking about the future in vulnerable communities means discussing concepts such as resilience, social vulnerability and cultural identity as aspects that allow us to understand the mobility and relationship dynamics that are generated in these communities. It is also critical to discuss changes in cultural values and development as vital elements of a future increasingly affected by climate change. Using feminist standpoint theory, which aims to empower the oppressed, and considering that standpoint is about "histori-cally shared, group-based experiences" (Collins, 2017), in this chapter, I look at the vision that a

L. Barraza (✉)
SACBÉ -Servicios Ambientales, Conservación Biológica y Educación A.C., Mexico City, Mexico

© The Author(s) 2024   23
S. Tolbert et al. (eds.), *Reimagining Science Education in the Anthropocene, Volume 2*, Palgrave Studies in Education and the Environment, https://doi.org/10.1007/978-3-031-35430-4_3

community of young Mozambicans have towards the future in light of the impact of Cyclone Idai in March 2019.

Identifying adolescents' concerns and views about the world and the environmental crisis now can help build a more resilient society able to face the problems of the future with better tools. Education for the future provides the opportunity for empowerment, so that individuals can work towards their chosen future (Barraza, 2001). Therefore, images of the future play a crucial role in relation to human behaviour and present-day actions, on both the personal and societal levels (Hicks & Holden, 1995). Images that are seen daily affect an individual's imaginary to project the future they want.

The future has become an emotionally charged matter. We live in a world of uncertainty and, given that, we have to adapt and take responsibility for ourselves (Figueroa-Diaz, 2018). It's about the uncertainty linked to what Ulrich Beck calls a risk society: "the development of modernity embodies contradictions between the advances that the era brings with it and the risks and uncertainties that implies" (Beck, 2013, p. 34). According to Levitas (2017), our images of the present do not identify agencies and processes of change. This is because the images that show today's world are images of extreme violence, suffering, decay and desolation. This can only change when we have better analyses of the present which identify possible points of intervention, paths and agents of change (Levitas, 2017). Our actions in the present inevitably help to determine what kind of future will emerge (Levitas, 2017). Therefore, it is not difficult to produce imaginary maps of the future but to produce adequate maps of the present which permit images of a connected but transformed future (Levitas, 2005). To visualize the future we want, it is necessary to work on our present actions and work consciously on what we can do today to avoid a daunting future. To speak of the future is to speak of the possibilities of development, growth and action, but also of contradictions, inequities and inequalities (Figueroa-Diaz, 2018).

It is also fundamental to discuss changes in cultural values and in development models as critical elements of a future increasingly affected by climate change. The sustainable development model incorporates future generations in the responsibility that humans of today should assume in environmental terms. This fact has resulted in looking at a reality linked to different expectations about the future, and that in some specific groups there is a concern for values and topics such as the environment, biodiversity, gender equality, inclusion, multiculturalism, cultural diversity and austerity in the consumption, among others, under the premise that sustainability must be as much environmental as well as social (Figueroa-Diaz, 2018). Ecological pressures suggest that human survival may require more than gradual ameliorative adjustments to our present way of life (Levitas, 2017). The questions Levitas (2017) suggests we should be asking are: what kind of a society can enable us to prosper and thrive in a way that is genuinely sustainable both ecologically and socially? How do we collectively think about the problems this presents? And how might we move in the direction of appropriate change?

In 2000, world community leaders established a shared vision of development based on the fundamental principles of freedom, equality, solidarity, tolerance, respect for nature and co-responsibility, which resulted in the Millennium Declaration adopted by the Assembly General of the United Nations. These Millennium Development Goals (MDGs) were replaced in 2015 by the 2030 Agenda for Sustainable Development. This new agenda includes 17 Sustainable Development Goals (SDGs) that will stimulate action in five critical areas: people, planet, prosperity, peace and partnership. The declaration states,

> We resolve, between now and 2030, to end poverty and hunger everywhere; to combat inequalities within and among countries; to build peaceful, just and inclusive societies; to protect human rights and promote gender equality and the empowerment of women and girls; and to ensure the lasting protection of the planet and its natural resources. We resolve also to create conditions for sustainable, inclusive and sustained economic growth, shared prosperity and decent work for all, taking into account different levels of national development and capacities. (United Nations, 2015a)

Eight years after this new declaration, there is an increasingly intense dehumanization that takes us away from the harmony and peace we want for our future. Inequalities within and among countries and enormous disparities of opportunity, wealth and power are rising in the world today. Why is this happening? It is worth distinguishing between the international aspirations declared by the United Nations and the realities of the neoliberal agenda. On the one hand, the SDGs as stated by the UN clearly indicate the criteria by which each country must commit to fulfil them. The problem is that in many countries the interests of those in power with neoliberal policies are counter to the Sustainable Development Goals, and hegemonic governments are increasingly moving away from ending poverty and hunger and not doing what is needed to ensure a healthy environment. Neoliberalism has profound consequences in the way we make use of the natural world, since many economic sectors depend directly on the environment and its resources, and in many cases, economic interests put ecosystems at risk (Durand, 2014). One of the biggest neoliberal conservation problems is that it is founded on an abstract idea of society, in which economic, political and cultural disparities are ignored and the dynamics of power remain invisible (Durand, 2014). Consequently, achieving the 17 SDGs requires a profound transformation in the way we live, think and act.

YOUTH RESILIENCE AND NATURAL DISASTERS

Young people play a fundamental role in the future they want to build. It is they who start looking for new ways to relate to themselves, to others and to the environment. Their voice begins to be heard more loudly; they are now clamouring for a greater say in how their societies are being configured. The young have also the potential to propel sustainable development more widely and urgently (UNESCO, 2014). Greta Thunberg has been an example of this and is certainly mobilizing young minds to act.

> Young people are doing this so that you adults wake up. Young people are doing this so that you put your differences aside and begin to act as you would in a crisis. Young people are doing this because we want to recover our hopes and our dreams. (Thunberg, 2019, quoted in Kettley, 2019)

In 2014, UNESCO launched the *Roadmap for Implementing the Global Action Programme on Education for Sustainable Development* (ESD). One of the priority action areas of this roadmap is the empowerment and mobilization of young people. Youths between 15 and 24, now more than one billion individuals, make up the largest group ever to be in the process of transitioning to adulthood (UNESCO, 2019). Youth have a high stake in shaping a better future for themselves and generations after. However, populations of youth in extreme poverty are often the victims of calamitous development and natural disasters. They are affected much more directly by environmental degradation and the lack of economic and social sustainability (UNESCO, 2014). It is estimated that half of those who are affected in natural disasters are children and youth (UNICEF,

2015), and they often have been understudied in disaster research (Fletcher et al., 2016).

In 2013 the Global Alliance for Disaster Risk Reduction and Resilience in the Education Sector (GADRRRES) was established with the main purposes to strengthen global coordination, increase knowledge and advocate on risk reduction education and safety in the education sector. The work of the Global Alliance ultimately contributes to a global culture of safety and resilience through education and knowledge (GADRRRES, 2017).

Mozambique: A Vulnerable Country

The vulnerability index indicates how vulnerable a country is. This index includes social and physical factors. Physical factors refer to the "natural" vulnerability that countries face, such as historical frequency and average strength of natural disasters (e.g. deforestation, reduction in coastal mangrove density and destruction of wetlands all remove natural barriers to storm surge and wind damage). Social vulnerability, on the other hand, refers to poverty, education, health, inequality and other relevant demographics of a society (Pelling & Utto, 2001; Adger et al., 2004; Birkmann, 2006). According to these two vulnerability factors Mozambique is a highly vulnerable country. The analysis of the IPM-MZ 2014/15 indicates that 53% of its population is multidimensionally poor. There is a big difference between rural and urban areas: in rural areas, 70% of the population is multidimensionally poor, while in urban areas, 17% is in this condition (Zavaleta & Moreno, 2018). Multidimensional measurement allows a country to see things that income measurement does not show.

Mozambique is also highly vulnerable to extreme climatic conditions, with two out of three people living in coastal areas vulnerable to rapid-onset natural hazards such as cyclones, storms and flash floods (OCHA, 2019). In March 2019, five provinces in Mozambique were nearly destroyed by the Cyclone Idai. This weather system brought destruction and damage to 1.85 million people in Mozambique alone (Humanitarian Response Team, 2019). According to the United Nations (2019), this devastating cyclone could be considered the worst disaster the southern hemisphere has suffered, leaving almost 3 million people affected.

Some social-ecological systems are more resilient than others (Folke, 2006). Vulnerable countries are often more resilient to adapt to different adversities—economic crises, acts of terrorism, mounting crime and violence, the HIV pandemic and other disease outbreaks, food shortages, failing education systems and natural disasters—due to their ability to constantly fight back for survival. According to Folke (2006), the ability to cope with extreme stress and resume normal function is thus an important component of resilience, but learning, reorganizing and changing over time are also key. In social-ecological systems, resilience-building practices include adaptive governance, ecosystem management and disaster risk management (Pinchoff & Hardee, 2018). The term resilience is often used in conjunction with other terms such as adaptation and transformation. Resilience can be distinguished from adaptation by considering resilience a trait—the ability to bounce back from adversity—while adaptation can be considered the actions taken to react to shocks (Pinchoff & Hardee, 2018).

Resilience can be defined as "the ability of a social or ecological system to absorb disturbances while retaining the same basic structure and ways of functioning, the capacity for self-organisation, and the capacity to adapt to stress and change" (IPCC, 2012, p. 586). Social resilience is understood as having three properties comprising aspects of how people respond to disasters: resistance, recovery and creativity (Kimhi & Shamai, 2004). A community that is highly resilient has the capacity to demonstrate each of these properties. In a resilient community, cultural identity plays an important role since the social context of a community, its organization, values, traditions and their sense of togetherness gives them a sense of unity in solving problems together. The resilience of an individual over the course of their development depends on the function of complex adaptive systems that are continually interacting and transforming. These adaptive systems are shaped by biological and cultural evolution (Masten, 2014). As a result, the resilience of a person is always changing, and the capacity for adaptation of an individual will be distributed across interacting systems (Masten, 2014).

STUDY SITE

Penvenne and Sheldon describe Mozambique in the following way:

> Mozambique is a scenic country in south-eastern Africa. It is rich in natural resources, is biologically and culturally diverse and has a tropical climate. Its extensive coastline, fronting the Mozambique Channel, which separates mainland Africa from the island of Madagascar, offers some of Africa's best natural harbours. These have afforded Mozambique an important role in

the maritime economy of the Indian Ocean, while the country's white sand beaches are an important attraction for the growing tourism industry. Fertile soils in the northern and central areas of Mozambique yield a varied and abundant agriculture, and the great Zambezi River has provided ample water for irrigation and the basis for a regionally important hydroelectric power industry. (Penvenne & Sheldon, 2019)

This study was carried out in the city of Chimoio, at the International School, founded in 1998 by a group of parents with the aim of providing their children a better-quality education using English as the language of instruction. It is a small school with only 150 students, of which 90% are Mozambicans. In addition to following the national curriculum, the school offers extracurricular activities such as art, music, sports and Portuguese language.

Chimoio is the capital of Manica Province and the fifth largest city in Mozambique. It is located 164 kms northwest of the port city of Beira which was 80% destroyed by Cyclone Idai. The city of Chimoio was also largely affected by the devastating cyclone. "Centrally located, it is also a commercial and industrial centre. The Chicamba Real hydroelectric-power plant on the nearby Revuè River provides power for the city's cotton, steel, and saw mills and for the manufacture of coarse textiles and processing of other agricultural and mineral products. Chimoio is connected by road and railway southeast to the port of Beira" (Penvenne & Sheldon, 2019).

METHODS

This study was conducted just ten days after the devastating Cyclone Idai. It is a qualitative research study in which what interests the author is to have a better understanding of the ideas that students have about their future, their culture and their values in a time of a humanitarian and environmental crisis. The study focused on the ideas that young Mozambicans have towards their future, after being severely affected by a cyclone, rather than their perceptions of the cyclone itself.

Two aspects of feminist standpoint theory were explored in this study: first, to enable students to identify any oppression they may have seen or experienced and then to see what their understanding was about this issue, so they could think on ways to improve their conditions for a better future. Gender norms influence the terms on which men and women communicate (Kalbfleisch, 1995), so in this study the ideas that boys and girls have towards the future were taken into account.

Participants were 43 students from grades 7–11, between 12 and 17 years in age. Girls comprised 53.5% of the participants and 46.5% were boys. Data was collected with each grade level dedicating 90 minutes of questions and dialogue for each group, for a total of five sessions. There was no discussion before the session except to introduce the activity. Students were asked to pretend they were aliens from another planet sent on a special mission to visit Mozambique in 2069. Students had to describe, in writing, what life in Mozambique would be like in 50 years' time.

Content analysis was used to review adolescents' vision of Mozambique's future. All narratives were then used in the construction of thematic categories, for example, infrastructure, development, health, education and technological innovation.

With a group of 15 students, representing 55% of those who had a positive vision towards the future, I conducted a focus group; this session lasted two hours. In the focus group, aspects discussed were related to how humans should behave in 50 years' time. What habits and attitudes should be changed? And what would be the values of humans by 2069? Additionally, with nine students from grade 8, there was a discussion about what they thought would happen to the culture of Mozambique in 50 years. These questions provided evidence of students' ability to identify some examples of oppression in their culture and speak of their major concerns as well as ways to improve their conditions.

RESULTS AND DISCUSSION

When discussing how Mozambican adolescents visualize the future in 50 years' time, the author found that 67.4% of the students had a positive vision, 23.25% had a negative image of their future and 9.3% said that there would be good things happening, but there would also be negative and bad things occurring in the world.

The ideas that prevailed in the optimistic mind of the students towards a promising future were focused on six main areas: education, health, infrastructure, poverty reduction, a clean environment and technological development. They mentioned that Mozambique would be the capital of a more united Africa, where peace would prevail throughout the continent. Mozambique would be a developed country in which poverty would cease to exist, the death rate would fall and the population would live in better conditions with greater access to quality education. Health services would be improved; there would be more hospitals and access to free medicines for the whole population. In general, there would be better infrastructure, the roads would connect the whole country and the country would go from being a developing country to being an industrialized nation.

It is interesting to mention that the students of seventh and eighth grade, aged between 12 and 14 years, gave simple answers without having a deeper discussion, their level of information and the issues they addressed were also more limited, and they visualized more images of robots, more buildings and flying cars. The ideas of these students focused more on the areas of infrastructure, development and technological innovation.

> I think after 50 years that Mozambique will be a beautiful and a peaceful place to live. People in Mozambique will build very big and nice houses. (13-year-old girl)

> After 50 years many things will change, nature, roads, schools and more new things will appear like flying cars, flying motorbikes, and everything will be clean and neat. (13-year-old boy)

> Mozambique in the future will be a paradise. There will be a lot of improvements, skyscrapers, houses, schools, hospitals, playgrounds. I will feel very proud to see Mozambique in 50 years' time. (12-year-old boy)

The conceptualization of the ideas of a positive future by students from 15 to 17 years showed not only greater diversity in topics, but they also connected and related some of the current problems with possible solutions in the future. During the discussion they handled a good level of information on various topics, as the following quotations show:

> I think Mozambique will be more developed in 50 years' time because precautionary measures have been taken. Mozambique will no longer be seen as a less developed country in which leaders were poor and hungry and they ended up stealing the money and using it for themselves instead of making a development for the country they were ruling. Mozambique will be viewed as a more economically developed country. (15-year-old girl)

> Today Mozambique is facing some difficulties as a nation, we are currently trying to survive after the devastating impact that Cyclone Idai had on us.

We are working hard to try and help our brothers and sisters who lost everything. That is the current situation in Mozambique, but what about 50 years from now? Hopefully we will have recovered from the natural disaster, but there are a lot of other very important things to take into consideration, for instance our forests. People have been cutting down trees for their own personal benefit, the forest is having to heal itself naturally and that takes a lot of time. I believe that if people open their eyes and see that they slowly are damaging our country and help it heal instead of further destroying it, then this land could surely go back to being the beautiful place it used to be in 50 years. (15-year-old boy)

For these students, their vision of the future is strongly related to the current reality. Bourdieu (1998) points out that individuals orient their present actions and orientations towards the future in relation to objective potentialities in the present and in relation to a probable future. The testimonies and ideas of the adolescents mentioned above highlight a reality that they live day to day and are a part of and therefore want to change. The visions of a positive future are based on the problems that young people perceive from a present they don't like. These positive images respond to a hopeful feeling of young people who visualize that despite the problems they experience in the present, they can take actions which will improve their present and help create a better future.

With respect to the images of a negative future, adolescents showed great concern for nature, the loss of species of animals and plants, pollution and war. They also fear that there would be more natural disasters and that this would bring destruction to the country, volcanoes would cover the country with lava, earthquakes would separate the countries of Africa, and the city of Beira would be covered by water and become part of the ocean. They fear that foreigners from rich countries would continue to extract minerals and precious stones to take them to their countries, leaving Mozambique poorer.

In the narration given by one of the students shown below, we can see the level of information and the depth of her ideas expressed in a very critical way. She clearly shows her genuine and realistic concern for a better future.

I don't think Mozambique will continue being as rich as it is in 50 years' time. Most of the endangered animals would have been killed for their skin, horns, fur, skeletons to be made in many different products. The sea might become harmful for us humans and other animals as toxic waste from many industries will be released in the oceans. I once read something that Vladimir Putin, the president of Russia said: "When an African man is rich, his bank accounts are in Switzerland, he invest in Germany, his children study in France, he goes to Canada, U.S.A., Japan for tourism, but when he dies, he is buried in his native country, Africa is just a cemetery for Africans", how can a cemetery become developed? Africans are not proud of their culture, nor their background. Africa will only grow when the people start to appreciate it more. (15-year-old girl)

This study found that 23.25% of the students fear that the world will be worse than today. Such a result accurately reflects the prevailing cultural pessimism of the time, but it also reflects the real situation that these young people live in. Similar results have been found in other studies regarding the fears expressed by young people for the future that lies ahead (Barraza, 1999; King, 1995; Hicks & Holden, 1995). The context and the situation of the moment lived by the devastation of the Cyclone Idai strongly influences the images of fear, pain and anguish that were revealed by the students in their answers. Environmental problems affect the way individuals see life (Barraza, 1999), and the environmental problems affecting the world are devastating, for example, the amount of rubbish that is generated every day, air and water pollution, almost half of the forests no longer exist, underground water sources and fish stocks are reducing rapidly, and land degradation and acidification of oceans are getting worse (Vos & Jahan, 2012). This reality is clearly affecting the population and students in this study. They are very aware, concerned, afraid and informed about it. Bourdieu, in Rawolle and Lingard (2013), sees all social phenomena in relation to their location in a given field and in relation to others in the field. To change the scenario of a negative future, it is necessary to work on the fears and expectations that young people have in the present (Hicks & Holden, 1995). We must work on our present actions to have the future we want.

Of the adolescent participants, 9.3% felt that positive things would happen in 50 years, such as technological development and with it some improvement in the aspects of education and health, but there would be a greater increase in the population and with it a higher crime rate. In addition, they predict there would be serious problems arising from pollution and climate change.

MOZAMBICAN CULTURE IN THE FUTURE: VALUES AND ATTITUDES

One of the aspects discussed in the focus group session was how students think humans needed to behave to live that positive future that they envisioned 50 years from now. What are the values that Mozambican culture will have? What habits and attitudes should be changed?

Regarding the first question—how should humans behave in 50 years' time—everyone agreed that humans must behave differently from the way they are today. Students said that the essence of the human being must change to live in the future they have imagined. They point out that a person must stop thinking about himself or herself and instead develop a sense of community. They say that everyone should work together, in collaboration, to solve problems together. Additionally, they commented that they should be more respectful of nature and the environment in general. Other qualities that humans must have in 50 years are that they need to be kind, clever, hardworking and healthy, and need to take care of themselves. They should help each other in times of crisis, and they should contribute for better development of the world. Humans need to listen to each other's ideas with respect and tolerance; they need to know how to forgive and accept apologies. They also need to be more aware of their surroundings and become more open minded and learn to adapt to different conditions and contexts. They should be generous and loving individuals.

> Humans will understand that cooperation is the only way forward for mankind. (15-year-old boy)

Regarding the second question about what habits and attitudes should be changed, students mentioned that the habits and attitudes that must be changed to live in a positive future are aggression, violence and conflict in general. There must be habits of equality and respect for both women and men. Everyone must be offered the same opportunities. At the moment there are still many children who do not go to school, and everyone must go. They should change to be less greedy and lazy, and instead, they must work with passion and interest to contribute to a better society. They must stop being selfish and rude and instead have a big heart; they must give up the practice of forcing underage girls to marry older men. We need to change the habit of littering the streets and create an attitude of respect for nature and the environment.

> The mindset of killing and being self-centred should be changed. We should learn to have a good relationship with everyone around us and not view ourselves as superior. We must start viewing education as a very valuable asset. (14-year-old boy)

Young Mozambicans have shown a clear vision about the changes that humans must undergo in order to live in a better future. They have listed a number of behaviours and values reflected in 10 of the 17 SDGs declared by the United Nations for the new agenda 2030 (SDGs 1, 4, 5, 8, 10, 11, 12, 14, 15 and 16). The shared goals include eliminating poverty; reducing violence between countries; achieving gender equality and empowering women, children and vulnerable groups; having access to good quality education; promoting economic growth; providing health services for everyone and protecting all ecosystems to ensure a healthy environment.

The ideas expressed by the students as to how humans should behave in 50 years show profound aspects related to the ontology of being; that is, the changes they suggest are not physical changes, but changes that require a higher level of consciousness. They are changes that reveal the desire to be a new being, where new habits and attitudes will have to respond to a new culture. According to the United Nations (2015b), the vision of development for the future must be centred on the 17 Sustainable Development Goals (SDGs) of the new agenda for 2030, considering human rights, equality and sustainability in five priority areas. These SDGs have been expressed by the students and prevail in the attitudes and values that need to remain as an intrinsic part of the human being in order to live a better future. According to Jickling et al. (2018), in this new geological epoch of transition, "the Anthropocene, or the age of human impact," what must be changed is the relationship that humans have with Earth.

This relationship is reflected in how developed countries have lost their knowledge and respect for living well with place, that is, living with care and compassion for other beings, and to live with wonder on Earth itself (Jickling et al., 2018). The ideas and concerns of adolescents in this study reflect that although their culture has been based on having a relationship of respect for Mother Earth, the Western system has disqualified and marginalized their way of seeing and understanding their life. Changing relationships with Earth and its other beings will require learning through active engagement with the natural world (Jickling et al., 2018). New ways of teaching in schools, and new visions and paradigms that help recreate a new human being, are needed if we want to live a better future.

Regarding the third question—what will the values of humans be in 2069?—there was a good discussion about values. Students connected how a person should be in 50 years' time with the habits and attitudes necessary to live a harmonious and respectful future. Many of the ideas that were discussed focused on being much more thoughtful and careful with each other, but mainly being careful, honest and humble with ourselves. The values would focus more towards the care of nature, our environment and every living being. The main value would be respect for life, and therefore, there would be no early marriages, and girls as well as boys could take up any profession they choose. The value of peace would prevail in the homes of the entire population.

> Humans will be less tolerant of corruption and dictatorship. Most humans will think about how their actions affect their neighbours before doing something. Discrimination will no longer exist, there will be no black, white, Arabs or coloured, we will all be just people. (17-year-old girl)

Feminist standpoint theory postulates that those marginalized in social or political power relations will rise to challenge the social order within which they find themselves (Harding, 2004). In this study, young Mozambicans identified power relations as mechanisms of oppression of vulnerable groups, particularly of women and children. This theory was presented as a way of empowering oppressed groups to value their experiences and to point towards a way of developing "an oppositional consciousness" (Harding, 2004, pp. 1–3). Awakening this oppositional consciousness is what allows marginalized groups to see their state of oppression and ideally motivates them to wake up. "Standpoint theories claim, in different ways, that it is important to account for the social positioning of social agents" (Collins, 2017, p. 119).

Images of inequality in the treatment and marginalization that women live are reflected in the values that young Mozambicans want to change to live in a fairer future. The girls in this study indicated that there are still many practices of domination and power on the part of men towards women. They point out that there is a lot of domestic violence and abuse by men towards women in their culture. They have seen how their mothers suffer abuse, and they wish that this abuse no longer existed. They also mentioned that underage girls in the rural communities of the country are offered to older men to marry in exchange for a few meticais (Mozambican currency); this means that girls cannot have a school education. School conditions also discriminate on the basis of gender, where girls leave school with cultural pressure, to take care of younger siblings or sick family members, or get married (Benque, 2020). Girls in this study hope that in the future there will be respect for women and that they will be allowed to perform professionally with the same opportunities as men. What is interesting is that both girls and boys in this study pointed out that there should be no difference between men and women, that both should have the same opportunities and that everyone should be treated as equals. According to Harding (2004), standpoint projects must be part of critical theory, revealing the ideological strategies used to design and justify the sex-gender system and its intersections with other system of oppression. To do this we must work with different groups—both those who are the marginalized and those who are the oppressors. This will help to identify aspects in their behaviours that can be worked on by the different groups.

This study shows the ideas of concern that young Mozambicans have for these practices, and everyone agrees that the only thing that will make Mozambique prosper will be the equal treatment between men and women and the termination of premature marriages of older men with younger girls. But there is also concern about the power relations exercised by developed countries over Mozambique. Students in this study believe that their culture will cease to exist in 50 years due to the intervention of foreigners and that policies of other countries will dominate their ideas and traditions.

Hebdige (2005) suggests that "the various ways in which different futures are imagined will themselves be something we have to begin thinking seriously about. We shall have to establish how particular discursive strategies open out or close down particular lines of possibility; how they invite or inhibit particular identifications for particular social fractions at particular moments" (p. 275).

The issues that most stood out in the discussion from the focus group carried out with students from grade 8 about what they think will happen to their Mozambican culture in 50 years, were related to specific aspects of life in this country: its land, food, family, traditions and ceremonies. Young Mozambicans in this study pointed out that, without a doubt, the culture of a country is what gives it identity and what gives pride to its people. Although they recognized that there are many good things about their culture, such as the family being a social nucleus and the support that exists between the families, they also said that there are practices in their culture that must change, such as premature marriages, the lack of support for children in general to go to school and, particularly, that education for girls must be encouraged. Some students fear that the culture of Mozambique will no longer exist in 50 years and in response to the human selfishness that prevails today. Children will not listen to adults, and respect for elders will be lost. Society itself will have an absence of values without there being a clear direction of where to go to improve things. Students believe that the loss of values that will happen in the Mozambican culture is due to a process of displacement of their own traditions due to the influence of foreign countries that have come to Mozambique with their own cultures.

Educational and Philosophical Touch Points for the Anthropocene

The adult literacy rate in Mozambique is around 47% (Dezanii, 2017). Teachers have very limited knowledge and are poorly qualified (Benque, 2020). From a survey done in seven African countries, Mozambique has the worst qualified teachers (Club of Mozambique, 2015). It is a national priority to work with teachers and raise the quality of education in the country. For this, we need new teaching tools and methodologies that respond to the interests and needs of the communities.

To create a significant change in the educational practice, it is necessary to offer programmes with new methodologies that will help teachers not only to manage groups in a better way but to guide the individual to become an independent and responsible individual in their own learning process (Barraza & Franquesa-Soler, in preparation). Teachers need to see themselves as a conscious subject in a process of critical reflection on their practice to identify patterns of change. They need to be teaching students how to think, how to solve problems, how to be resilient and how to relate to each other and with the environment in order to be able to adapt to one of the worst humanitarian crises ever faced (Barraza & Ruiz-Mallen, 2017). We need new forms of teaching to guarantee effective place-based learning and create problem solvers for this critical situation. Hence, it is crucial to re-think, recreate and generate methodologies outside the traditional conventions that help to promote open minded, critical and sensitive consciences, capable of respecting all forms of life. We need different and novel paradigms that can respond to the interests, needs and work demands of the world today (Barraza & Franquesa-Soler, in preparation). We need to start from the premises:

- What should every citizen be learning in school?
- What should be taught?
- What kind of education do we need to promote in the current context of uncertainty?
- What skills do we need to promote?

CONCLUSIONS

Young Mozambicans in this study have shown their sensitivity and a high level of concern as a vulnerable community towards the environmental and social problems suffered by their country. Their familiarity with and knowledge about the environmental and social problems that their country is experiencing is clearly shown. They have also identified aspects of oppression towards vulnerable groups such as women and children, showing an understanding of those problems and providing some ideas on how they can reduce and remedy these problems. Access to a good education is clearly a solution they all agree on. They say they hope to recover soon from the devastation they suffer from Cyclone Idai, so that they can continue advancing the resolution of more severe problems such as those associated with climate change.

A majority of the participants in this study visualize a positive future, but it implies a change in human behaviour. People need to reconnect with nature with respect and a great sense of community in which the fundamental value will be collective work and support among all. For this, a change in the paradigms of education and in educational practice is urgent.

Their ideas and images of the future respond to the social, environmental and political problems of their country. They demonstrated a broad and critical knowledge about the problems they are currently experiencing. Their ideas and visions clearly show a deep connection to current problems. Both girls and boys in this study evidenced great concern about the marginalization that Mozambique and the rest of Africa has been suffering.

Despite the initiatives taken by the Global Alliance for Disaster, Risk, Reduction and Resilience in the Education Sector (GADRRRES), there is still a need to find a place for children and youth in Mozambique on the disaster research agenda. It is essential to create secure conditions in schools in Mozambique and to expand knowledge about this important sector of the population. Such knowledge would provide a more complete understanding of the impact of hazards and disasters on vulnerable societies in Africa.

Acknowledgements The author is grateful to Epifania Phiri, the school principal of the International School of Chimoio, who kindly opened the doors to make this study possible. A big thank you to all the students that participated in this study and to Elisangela Carvalho for her support contacting the school. Thanks also to Isabel Ruiz-Mallen and Adriana Otero Arnaiz for their valuable comments on this chapter. A special thanks to Molly Siple for the English revision of this chapter.

References

Adger, W. N., Brooks, N., Bentham, G., Agnew, M., & Eriksen, S. (2004). *New indicators of vulnerability and adaptive capacity.* Tyndall Centre for Climate Change Research.

Barraza, L. (1999). Children's drawings about the environment. *Environmental Education Research, 5*(1), 49–66.

Barraza, L. (2001). Perceptions of social and environmental problems by English and Mexican children. *Canadian Journal of Environmental Education, 6*, 139–157.

Barraza, L., & Franquesa-Soler, M. (in preparation). *Teachers of hope: A pedagogical proposal to develop agents of change.*

Barraza, L., & Ruiz-Mallen, I. (2017). The 4D's: A pedagogical model to enhance reasoning and action for environmental and socio scientific issues. In P. B. Corcoran, A. E. J. Wals, & J. P. Weakland (Eds.), *Envisioning futures for environmental and sustainability education.* Wageningen Academic Publishers.

Beck, U. (2013). *La sociedad del riesgo: Hacia una nueva modernidad.* Paidós España.

Benque, L. (2020). Realizing children's rights in Mozambique. https://www.humanium.org/en/mozambique/

Birkmann, J. (Ed.). (2006). *Measuring vulnerability to natural hazards: Towards disaster resilient societies.* University Press.

Bourdieu, P. (1998). *Practical reason.* Stanford University Press.

Club of Mozambique. (2015). Review of Mozambique's education system. https://norainmozambique.wordpress.com/2015/04/23/review-of-mozambiques-education-system/

Collins, P. (2017). Intersectionality and epistemic injustice. In I. J. Kidd, J. Medina, & G. Pohlhaus (Eds.), *The Routledge handbook of epistemic injustice* (Kindle ed., p. 119). Taylor and Francis.

Dezanii, L. (2017). 7 Things to know about education in Mozambique. https://borgenproject.org/7-things-education-in-mozambique/

Durand, L. (2014). Does everybody win? Neoliberalism, nature, and conservation in Mexico. *Sociológica, 29*(82), 183–223.

Figueroa-Diaz, M. (2018). El futuro como dispositivo: La mirada de algunos estudiantes universitarios. *Política y Cultura, 50,* 177–201.

Fletcher, S., Cox, R., Scannell, L., Heykoop, C., Tobin-Gurley, & Peek, L. (2016). Youth Creating Disaster Recovery and Resilience: A Multi-Site Art-Based Youth Engagement Research Project. *Children, Youth and Environments, 26*(1),148–163.

Folke, C. (2006). Resilience: The emergence of a perspective for social–ecological systems analyses. *Global Environmental Change, 16,* 253–267.

GADRRRES. (2017). Marco integral de seguridad escolar. http://gadrrres.net/resources/comprehensive-school-safety-framework

Harding, S. (Ed.). (2004). *The feminist standpoint theory reader: Intellectual and political controversies.* Routledge.

Hebdige, D. (2005). Training some thoughts on the future. In J. Bird, B. Curtis, T. Putnam, & L. Tickner (Eds.), *Mapping the futures: Local cultures, global change* (pp. 274–284). Routledge.

Hicks, D., & Holden, C. (1995). *Visions of the future: Why we need to teach for tomorrow.* Trentham Books.

Humanitarian Response Team and the United Nations Resident Coordinator's Office in Mozambique. (2019). 2018–2019 Humanitarian Response Plan. Mozambique. Revised following Cyclones Idai and Kenneth, May 2019. United Nations Office for the Coordination of Humanitarian Affairs. https://reliefweb.int/sites/reliefweb.int/files/resources/ROSEA_20190525_MozambiqueFlashAppeal.pdf

Intergovernmental Panel on Climate Change (IPCC). (2012). *Managing the risk of extreme events and disasters to advance climate change adaptation.* Special report of the Intergovernmental Panel on Climate Change (IPCC). Cambridge University Press.

Jickling, B., Blenkinsop, S., Timmerman, N., & De Danann Sitks-Sage, M. (2018). *Wild pedagogies: Touchstones for re-negotiating education and the environment in the Anthropocene.* Palgrave Macmillan.

Kalbfleisch, P. (Ed.). (1995). *Gender, power, and communication in human relationships.* Erlbaum.

Kettley, S. (2019). Greta Thunberg speech in full: Read the climate activist's damning message to the UN. Express https://www.express.co.uk/news/science/1183377/Greta-Thunberg-speech-full-read-climate-change-UN-speech-transcribed-United-Nations

Kimhi, S., & Shamai, M. (2004). Community resilience and the impact of stress: Adult response to Israel's withdrawal from Lebanon. *Journal of Community Psychology, 32*(4), 439–451.

King, L. D. (1995). *Doing their share to save the planet: Children and environmental crisis.* Rutgers University Press.

Levitas, R. (2005). The future of thinking about the future. In J. Bird, B. Curtis, T. Putnam, & L. Tickner (Eds.), *Mapping the futures: Local cultures, global change* (pp. 260–270). Routledge.

Levitas, R. (2017). Where there is no vision, the people perish: A utopian ethic for a transformed future. Centre for the Understanding of Sustainable Prosperity. https://cusp.ac.uk/themes/m/m1-5/

Masten, A. S. (2014). Global perspective on resilience in children and youth. *Child Development, 85*(1), 6–20.

Office for the Coordination of Humanitarian Affairs (OCHA). (2019). Mozambique, East Africa. https://www.unocha.org/southern-and-eastern-africa-rosea/mozambique

Pelling, M., & Utto, J. I. (2001). Small Island developing states: Natural disasters vulnerability and global change. *Global Environmental Change Part B: Environmental Hazards, 3*(2), 49–62.

Penvenne, J. M., & Sheldon, K. E. (2019). Mozambique. Encyclopedia Britannica. https://www.britannica.com/place/Mozambique

Pinchoff, J., & Hardee, K. (2018). *Building resilience in communities most vulnerable to environmental stressors.* The Population Council.

Rawolle, S., & Lingard, B. (2013). Bourdieu and educational research: Thinking tools, relational thinking, beyond epistemological innocence. In M. Murphy (Ed.), *Social theory and education research understanding Foucault, Habermas, Bourdieu and Derrida* (pp. 117–137). Routledge.

UNESCO. (2014). *Roadmap for implementing the global action programme on education for sustainable development.* United Nations Educational.

UNESCO. (2019). *Framework for the implementation of education for sustainable development beyond 2019.* United Nations Educational, Scientific and Cultural Organization. https://unesdoc.unesco.org/ark:/48223/pf0000370215.page=7

UNICEF. (2015). *Fondo de las Naciones Unidas para la Infancia.* Informe Anual 2014. Nuestra Historia.

United Nations. (2015a). The Millennium Development Goals report. https://www.un.org/millenniumgoals/2015_MDG_Report/pdf/MDG%202015%20Summary%20web_english.pdf

United Nations. (2015b). Transforming our world: The 2030 agenda for sustainable development. https://sustainabledevelopment.un.org/post2015/transformingourworld/publication.

Vos, R., & Jahan, S. (2012). *El Futuro que queremos para todos: Informe para el Secretario General.* UNDP UN-DESA. http://dev.un.org/millenniumgoals/beyond2015.shtml

Zavaleta, D., & Moreno, C. (Eds.). (2018). Mozambique: Primer país africano con IPM nacional oficial. Red de Pobreza Multidimensional (MPPN). *Dimensiones, 5*(1), 9–12.

CHAPTER 4

How a Phenomenology of Place in Science Education Can Grant Erotic Generosities for the Ocean

Rachel A. Gisewhite

I am writing this chapter in the thick of the COVID-19 pandemic. My spouse is essential military personnel and continues to go to base for work, while I work remotely, and my children attend school from the comfort of a makeshift learning center in our living room. Though important, social distancing for my family has had its challenges. Like for many, there is much we miss about physical social interactions, and we are worn down from the changed schedule and the excessive time spent in our home. More and more we inherently seek natural environments to clear our heads and work our bodies. We can often be found in our yard gardening, bird watching, or throwing the Frisbee to our dog. Occasionally I catch my children sitting perfectly still on the back deck, eyes closed, feeling the breeze or the warm sun's rays as if this instinct is reminding their bodies of a wildness they have nearly forgotten as they stare at computer screens

R. A. Gisewhite (✉)
Tucson, AZ, USA
e-mail: skennen7@gmail.com

© The Author(s) 2024   45
S. Tolbert et al. (eds.), *Reimagining Science Education in the Anthropocene, Volume 2*, Palgrave Studies in Education and the Environment, https://doi.org/10.1007/978-3-031-35430-4_4

and interact through video to complete their schooling. On weekends though, we like to explore the water.

This last weekend we went to Deer Island, a small island, now a coastal preserve, off the coast of Biloxi, Mississippi. As my children and I walked along the shoreline, pointing out shells and wading through tide pools, their usual curious questioning while investigating a new place turned to questions about pollution and how it ended up there. The amount of garbage littering the beach was overwhelming. Once inhabited by fewer than twenty homes, the impact of hurricanes and time had returned Deer Island to a natural, unoccupied state save the wildlife, which includes a diverse seabird population and several endangered species. Now it is considered a crucial coastal wetlands habitat, and the state of Mississippi has made restoration efforts through prescribed burning and planting trees and seagrass. But the Mississippi Department of Marine Resources isn't the only one playing with fire. As I walk with my sons along the water's edge, dodging broken glass and picking up various plastic pieces, it *is* important to question how it got there but equally important to consider what we need to do next.

Though approximately 40% of the US (NOAA, 2014) population lives within sixty miles of a coastline, few of us have an intimate connection with the ocean and what lies beneath, often as a result of decreased value placed on spending time outdoors and increasingly busy schedules that come with the drive to stay afloat in the economy. The goal of this chapter is to outline the impact of the Anthropocene on the ocean-Other and describe how a phenomenology of place in science education can enhance the lives of our youth to be meaningful within the marine environment so that they can become ocean literate and capable of making the kinds of decisions that benefit, not harm, marine and aquatic environment.

THE ANTHROPOCENE AND THE OCEAN

In the grand scheme of human impact on the ocean, marine debris is only the tip of the metaphoric iceberg. Hundreds of millions of dollars were spent in the 1970s and 1980s on the deep-sea mining of manganese nodules. These nodules are formed when dissolved metals in the water column or sediment precipitate around some nucleus on the seafloor, causing the metals to build up over time. The manganese nodules are primarily composed of manganese but also contain other metals like cobalt, nickel, and copper, making them a fiscal curiosity. The feat promised so much economically that the Law of the Sea negotiations in the 1980s were stalled to determine which country was worthy of claiming such a reward. None wanted to give up the potential capital. The Law of the Sea Treaty of 1994 later included policies to protect against deep-sea mining, but the wording was vague enough to exclude copper, gold, nickel, cobalt, and silver because of the potential economic possibilities. The process of deep-sea mining is taxing on many levels, from the amount of energy spent on the endeavor, to the interruption of the natural habitats and disruption of the seabed, the loss of biodiversity, and the possible contamination and mortality that occurs from transporting such large quantities of metals.

Since the 1950s the demand for oil as a fuel source has exponentially increased, increasing with it the plastics, pharmaceuticals, pesticides and fertilizers, cars, airplanes, and so forth that pollute our natural systems and our bodies. Moreover, the increasing oil demand has quickly burned through several millions of years' worth of fossilized forests and microbes of ancient oceans, quickly diminishing these sources and requiring more deep-ocean exploration, drilling, and pipelining. This, in turn, also leads to the interruption of the seafloor and natural habitats, health risks, and marine organism mortality. These negative ramifications do not even include the very serious effects of oil spills, which are as small as the oil left under our cars, traveling through the groundwater back to the ocean, to the very large spills resulting from such events as the *Exxon Valdez* tanker spill and the *Deepwater Horizon* pipeline explosion. The oil issue extends even further through the creation of exclusive drilling rights. Fishing access and water supply have also been privatized in response to fears over scarcity and degradation. Marine organisms are kept in aquariums, training centers, or swim-with-dolphin programs, unable to live naturally in their ocean habitat. Our consumer culture is driven by the demand for the latest good or service, where such demand triggers greater resource extraction, production, packaging, and distribution. From there, the consumer uses energy to obtain the product, discards the packaging, uses the product, and eventually discards that as well. Though much of this disposal occurs on land, the waste still makes its way to the ocean, resulting in nearly 80% of the plastic debris in the ocean worldwide (Li et al., 2016). The ocean is seen as a commodity, possibly in part due to the myth sustained in its vastness—that it is just so large that it can never be irreparably damaged. We know this idea is not accurate. In 2016, the United Nations Food and Agriculture Organization determined that nearly 80% of monitored fish stocks were fully exploited, overexploited, or depleted (FAO,

2018). Asia could run out of exploited fish for seafood by 2048, with global fish stocks also in decline if we continue to deplete our fisheries at current rates (IPBES, 2018). Overfishing, pollution, and other environmental influences are negatively affecting species populations worldwide. These factors make it difficult for species to reproduce and resist disease.

Marine policy has been most influenced by economic development of which commercial fishing, marine tourism, and offshore oil exploration are only a few examples from a long list. In recent decades there has been a push for increased environmental protection measures for the ocean, including marine biosphere reserves and endangered species legislation, though such measures are influenced by economic development through ecotourism and sustainable development. The anthropocentric viewpoint reigns in this conversation of economy, and policy choices are influenced by values. For example, policymakers push to devise fishing practices that ensure a healthy future population but only because they will benefit the human community. This links global overfishing to the property rights of fisheries. In consideration of global fisheries, for example, it is claimed that the privatization of fishing originated to prevent the collapse of fish stocks (Costello et al., 2008). The ocean is one of our greatest commons, and yet we are headed for a tragedy of the commons.

ENCLOSING THE COMMONS

Resource privatization is a problematic solution for more than environmental issues, as it raises concern over the inequalities resulting from the privatization of access rights. Enclosures are the privatization of those things that were previously considered to be part of the commons—non-monetized natural and diverse cultural systems (i.e., cultural knowledge, intellectual skills, narratives, habitats, or even digital worlds) (Mueller, 2008). There are plenty of examples of ocean-related enclosures that threaten both cultural and environmental commons, including the right to own beachfront property and the allocation of property for aquaculture, both of which drive out family fishing practices and local fisheries people. Major commercial fishing companies capitalize on wealth while jeopardizing place-based livelihood, especially for low-income and small-scale fishers and fisher people in small rural communities. Augustina Adusah-Karikari (2015) describes another example of enclosure in the ocean commons through oil production in the coastal communities of West Ghana:

When ordinary people and their environments become victims of disruptive economic expansion without adequate protection or provision of alternative means to improve their livelihoods, they remain vulnerable. The women of the coastal communities become vulnerable to the political and economic power of the oil companies and the government. Clearly, the strategic economic interests of these power structures took precedence over the community welfare and these women's livelihoods ... since the production of oil commenced ... there are already visible signs of abject poverty, economic deprivation, lack of social amenities, destitution and unemployment in these oil communities. (p. 30)

The result is that the commodification of the ocean encloses and marginalizes many communities, making ways of understanding the ocean, lifestyles, and rights vulnerable. Not surprising, those marginalized are often most in need of the ability to perform the duties of their livelihood.

Consider, for example, the people of Ecuador who face enclosure because of mangrove depletion (Hamilton, 2020). Mangrove habitats stabilize bottom sediment and protect against storm surges. They are an important filter for runoff from inland regions and a nursery, shelter, and source of food for many marine organisms. The mangrove ecosystem has provided for the people of Ecuador in a variety of ways: Ecuadoreans find sustenance in the fish, mollusks, and crustaceans that live in the mangrove habitat and use their wood for charcoal, construction, and fuelwood. Adults and children collect mangrove cockles from mangrove roots to sell at the market for family income. Yet, mangrove habitats in Ecuador are depleted for the space they occupy, which can be used for aquaculture to meet the ever-growing global demand for cheap seafood. If mangrove deforestation continues, then these communities face increased risk of food insecurity, loss of livelihoods, and issues related to soil erosion. As youth make meaning of ocean phenomena through a phenomenology of place, they will learn to break down and balance the tensions or barriers of rapidly increasing enclosures and learn to protect and sustain the commons for the future.

OCEAN LITERACY AND AN EROTIC ETHIC

We are living in a time of major social, political, and economic changes. Our knowledge and ways of understanding change too with growing technological advances, globalization, and the subsequent generation and organization of information. In consideration of Hodson's (2011) argument that scientific literacy is necessary to help students cope with an uncertain and constantly changing world, I argue that ocean literacy is essential in helping students tackle constantly evolving and changing ocean-related issues and understandings for the health of themselves, their community, and the ocean. Ocean literacy provides a space to utilize knowledge of the ocean to act for the resolution of specific issues relating to ocean science or for the betterment of a community because of issues relating to the ocean and its resources. This knowledge does not strictly have to be from formal education and can include knowledge from one's home, culture, community, or knowledge from some other domain. The ocean's relevancy to human lives is timeless and extraordinary, though the demand for more ocean-literate students has not reflected this relationship as intensely as is needed in the schools due to ocean sciences not being explicitly noted in the national standards and benchmarks. However, educators should use marine education to promote ocean literacy as a way of enhancing the lives of our students to be meaningful within the marine environment. Students cannot know the marine ecosystem, and therefore how to protect it, without exposure or involvement that allow for the creation of personal meaning through these lived experiences. As students make meaning of ocean phenomena through a phenomenology of place, they can uncover the value of human and nonhuman Others. This revelation provides an opening for students to experience eroticism that encourages them to act generously for the ecojustice of these enclosures.

An erotic relationship (Luther, 2013) is one that fosters erotic generosities between parties (e.g., students and the ocean) or a giving of oneself for the sake of the Other (e.g., the ocean and its inhabitants) because of the relationship. A central tenant in an erotic relationship is ambiguity, as humans are ambiguous by nature. We are simultaneously our bodies while also not our bodies, both a subject and an object, no longer part of the past or yet part of the future. Therefore, it requires an embodiment because the erotic dimension is exposed as consciousness is coupled to the body. As embodied beings, we are passionate and thoughtful. We are influenced by the push and pull between the natural world and society, and we make choices based on these influences. We use our bodies to act upon decisions based on our passions and our emotions. Oppressive institutions make it difficult to recognize the freedom or need for such in the Other but embodied people have erotic intentionalities, which provide a space for responsibility and generosity with lived bodily experiences. An erotic ethic also does not focus on who or what is to blame, or what is specifically right or wrong, but rather that we work harder to become more responsible, compassionate, ethically acting people. An erotic ethic can then provide a pathway between an Anthropocene of negative human impact on our natural systems and one of responsibility, care, and generosity.

PHENOMENOLOGY OF PLACE

Edmund Husserl (1970) explains that a phenomenological reduction occurs when one peels away or "brackets" the assumptions and presuppositions of culture from a phenomena, anything that can objectify it, like peeling away the layers of an onion. In doing so, all inessential details are disregarded, revealing only the immediate level of consciousness, where the phenomena or entity can "speak for itself." Through phenomenological reduction, we can experience the things as they are, free of prejudice and presumption. Maurice Merleau-Ponty (1964) reinterprets Husserl's description of phenomenological reduction to mean that human consciousness returns to the "perceptual pre-conceptual experience of the child" (Moran, 2000, p. 402). In doing so, we can return to a level of being and way of knowing that we once had but lost through experience and age. A child, for example, begins to perceive from birth, before it is even capable of speech. At such an unadulterated stage of perception of the world, children are full of the sense of wonder at the world within which they live. They don't immediately know the role of tools until they are demonstrated for them, nor do they have judgments for the natural world beyond what their senses reveal. Indeed, they are very sensual beings. As a toddler, for example, my oldest son woke each morning anxious to go outside and play. He loved searching for the neighbor's cat, finding spiders in the mailbox, digging in and examining dirt, and collecting acorns and leaves. For the most part, I let him explore our yard, a nearby park, or other outdoor area uninhibited. When we were outside, he wasn't afraid to walk or climb anywhere, as he had no understanding or experience yet with such things as poison ivy, burrows, or moss-covered rocks. On a trip to the State Botanical Gardens of Georgia, we were walking through a pile of leaves, when he bent down to pick up something that had caught his eye—the partial carcass of a deceased metallic green Japanese beetle. He stared in awe as he turned it over and over in his hands, until finally noticing its shiny body was just hollow enough that he could slip his little finger through to wear it almost like a ring. His perception of the Japanese beetle was not of a disgusting bug, a pest, nor did he probably understand that it was once alive. His perception of the beetle was what the beetle presented itself to be, simply, and what his senses gauged and used the sensuality of the thing before him to make mindful meaning. There was no need to understand the beetle, just to allow the phenomena to reveal itself. If we could all become more like children, we would come closer to the idea of the phenomenological reduction. As elders revert back to their youthful understandings of the world, they too serve as an example. These things are also experienced culturally, for instance, in indigenous connections with the Earth.

An erotic ethic is fitting for a phenomenology of place in science education, which allows youth to tap into their childish conceptions and reveal the most basic phenomena of their place in relation to the ocean. As students engage their senses and make meaning of the phenomena, they can determine what is valuable and worthy of protection and care. This idea is diametrically opposed to consumer-driven curricula and lifestyle, where the ocean-Other is not seen phenomenologically, as valuable or beneficial in its own right, but instead as a commodity for exploitation. Therefore, utilizing a phenomenology of place is a valuable way to teach science content while promoting moral and ethical thinking such that students can afford nature-equivalent moral considerations in their community and gain the tools necessary to tackle ecojustice issues in their community and the local environment.

Eroticism for the Ocean

Globally, we have an unsustainable system, where the oceans' resources will continue to dwindle and degrade if humans worldwide continue on our current path of consumption. The ocean may provide an anecdote to the pollution of our lives if we respect and protect the eroticism, or intimate connection, inherent in and fostered with the ocean that leads to gifted reciprocity and generosity instead of seeking to harness the powers of the ocean to fulfill our desires. Those that benefit from this system are content so long as they are benefiting, but continually push for more, faster, and better. The beneficiaries feel as though the system fails them during times of natural disasters, war, or depression—all things the beneficiaries categorize as not of our own making. How is this logical? As erotic embodied beings, we are responsible for our actions. We have failed the natural system through consumerism, overspending, waging wars, and polluting our air and waterways. This disregard for the value of nature is embedded in Western society unconsciously through the influence of root metaphors and the consequences of rationalist culture, which not only distances humans from nature but promotes domination and constructive thinking and behaviors (Luther, 2013).

Val Plumwood (2002) argues that rationalism, or emphasizing reason as knowledge, distorts contemporary thinking under the influence of capitalism. Plumwood explains that the dualism of reason associated with men and nature associated with women is a recipe for oppression or a justification for domination. The focus on capitalism in the reason-centered Western culture has distorted how nature is perceived, allowing for the domination of nature and Other(s), including other cultures and marginalized people, resulting in a commodification of the world. *Other* historically refers to "lesser beings" that are oppositional to Western rationality, culture, and philosophy (Plumwood, 2002), and is associated with the characteristics of women in patriarchal society: weak, voiceless, passive, and valued based on the potential for production (Beauvoir, 1948, 2011). Because it is considered inferior and feminized through patriarchal constructions, the Other is generally characterized as having less possibility. Subsequently, the Other is more vulnerable to objectification and oppression. In consideration of this notion of Other, the ocean and its inhabitants are an example of an Other—the ocean-Other. The ocean is not a single living entity in the sense that it cannot be sexed or gendered. However, as the ocean as Other becomes ocean-Other, the ocean takes on personification. For the phenomenologist, the personification of the ocean for the sake of valuing its Otherness is appropriate and significant (c.f., Abram, 1996).

Given this, how is it possible to ensure that the ocean is protected and respected? Perhaps when we first consider that the ocean is more than the greatest commons, it is the greatest unifier. The ocean touches every continent. Oceans reach into every continent through the connection with inland water systems. Though nations may be physically, economically, socially, politically, and culturally different, the one thing they share is the shoreline. All people are reliant on the ocean in the same way. The ocean is a source of security, as it provides water, medicine, food, energy, and planetary governance. It is an essential and unifying system that connects us all. The value of the ocean is so great that it can sustain us far into the future for energy, protein, and water. This can only happen, however, if we grant erotic generosities to the ocean. In the face of our best attempts to objectify the ocean, when we stand at the ocean shoreline and watch the crashing waves, witness a seabird dip beneath the water's surface to catch its dinner, or feel the pelting of rain on our skin from a tropical storm, we know in our bodies that dominating the ocean is not in the realm of possibility. This mindful emotional and intuitive response to the condition of the ocean-Other may be the opening for the establishment of an erotic relationship within us.

Embracing an erotic ethic in consideration of the ocean, students realize the breadth of their responsibilities to the ocean-Other. As they engage in activities that promote a better understanding of the ocean, they are able to more clearly see how the ocean influences their lives and what effects their action has on the ocean. Further, through the development of an erotic relationship and authentic practices, they are better prepared to act more compassionately and ethically for the sake of the ocean-Other. There are innumerous ways to grant erotic generosities to the ocean-Other, including, for example, cleaning debris in local waterways, reducing our use of oil-based fuels, raising public awareness of marine-related issues, and purchasing seafood that is harvested sustainably or fished locally. Our growing erotic relationship with the ocean yields mutual reciprocity that is joyfully sustaining. It is easier to develop this type of relationship through intimate experiences with the sea, but how do people without these experiences or exposure to the sea develop erotic relationships with the ocean? Moral value can be assigned when we consider something to be worthy of our respect, often associated with those embodied experiences that allow intellectual and emotional appreciation to blend. If we cannot conjure an emotional connection or valuable memory, then how can we assign moral value? Utilizing a phenomenology of place in science education provides an opportunity for students to engage in activities in or centered on the marine environment that promote an understanding of the ocean that enables them to more clearly see the influence the ocean has on their lives and what effects their actions have on the ocean.

IMPLICATIONS IN SCIENCE EDUCATION

Scientific knowledge leads to provocation for action against social and environmental injustice (Aikenhead, 1985; Kolstø, 2001). It can bring the people of communities together to improve their local conditions.

Scientific knowledge and thinking scientifically can essentially provide a framework for people to be better citizens. Following this logic, as youth gain an understanding of the marine environment, they are more likely to care for it and take action to protect the marine environment. Science teachers ought to strive to implement a model of erotic marine science education that provides students with the knowledge to provoke action for the betterment of their community and their local aquatic environments, which are linked inextricably to marine and freshwater environments worldwide.

An important component of what I am advancing here is that youth use their content knowledge, phenomenological experiences, and erotic relationship with the ocean to work together and socially construct or re-envision what the future might look like based on their proposed solutions through cultural, environmental, and virtual heuristic considerations. If youth perceive themselves as capable of doing this through this process of making changes in their environment for the betterment of the natural world and their community, then they are more likely to realize, understand, and work toward reaching their potential as responsible citizens—a reality made evident by young people like Greta Thunberg. Through an erotic marine science education, students gain the knowledge necessary to act as citizen scientists. As citizen scientists, students then share the responsibility for issues in the community and relating to the marine environment by participating more fully in democratic discourse.

Our current curriculum focuses on preparing students for active citizenship through ethnocentric and nationalist practices, where students are not fully able to make meaning of civic education (Ladson-Billings, 2004). This type of civic education has issues, like a lack of meaningful content and training in thinking and process skills, focus on passive learning, avoidance of controversial topics, a low-quality curriculum for underrepresented students, and a lack of attention to global issues (Cotton, 1996). Until our students see models of active citizens in their schools and classrooms, they will be unable to make the connections needed to learn and engage in active citizenship. Therefore, for an ocean-literate person to take action through erotic generosities, they need to know how to act. Action through erotic generosity is a critical component of an erotic ethic because it is in action that we can recognize and demonstrate that the strangeness of another is valuable and worthy of care. Through erotic generosities, we grant freedom and assume our responsibility.

How can science educators prepare science teachers to meet the needs of our students and demonstrate erotic generosities for them, particularly if students need to see physical, human examples of active citizenship? One possibility is through a humanist perspective ideology, which "promotes practical utility, human values, and a connectedness with societal events to achieve inclusiveness and a student orientation" (Aikenhead, 2006, p. 22). It is important, however, to amend this definition to include ecological consideration—an *ecohumanist* perspective, which allows for the valuation of the natural environment and its resources to human interests (Mikulak, 2007). According to Mikulak, an ecohumanist perspective considers Heidegger's (1962) philosophical understanding that we cannot separate ourselves from our environment and that in killing part of our environment, we are killing part of ourselves. This neglect is in stark contrast with the traditional ideology of science education, which often focuses on creating the next generation of scientists through mental training and scientific orientation. Zimmerman (1994) explains that the traditional ideology is an inauthentic existence that "seeks to protect and complete itself by dominating other people and by devouring the planet" (p. 111). Heidegger, on the other hand, posits that through an ecohumanist perspective, students "dwell authentically and in tune with [their] surroundings in a way that allows things 'to be,' through a movement towards a more holistic, interdependent model of understanding [their] relationship with the environment" (Mikulak, 2007, p. 20). In other words, if science teachers help their students shift perspective to a more ecohumanist perspective, the students may strive to dwell authentically in their erotic relationship with the ocean and gain a more holistic understanding of the ocean-Other to act for its freedom.

To achieve this, science educators need to move beyond the goals of traditional Western science education to include what is relevant to students, "usually determined by students' cultural self-identities, students' future contributions to society as citizens, and students' interest in making personal utilitarian meaning out of various kinds of sciences—Western, citizen, or indigenous" (Aikenhead 2006, p. 23). Science educators can use this ecohumanist perspective to promote a science curriculum that gets at the very basic understanding of phenomena. Through a connection to the community, science educators can prepare science teachers to hone an erotic ethic in the classroom by demonstrating that situations provide opportunity and possibility, rather than limitations. Water percolates from the surface to the groundwater, which is an essential process for sustainable groundwater management. It recharges the water table and replenishes aquifers. My claim aligns with an ecohumanist perspective, as fostering an erotic ethic can develop citizenship if we imagine our students as water percolating through the water table of their community. As they establish and develop erotic relationships within the community, they allow their "water," or their passion, sensuality, generosity, and care for the Other, to flow through the community, recharging and revitalizing it. With a basic understanding of the phenomena in their place—their community, local waterways, the ocean, natural environments—students are better able to think more clearly and meaningfully about issues affecting the phenomena while drawing connections back to the community and their actions.

EROTIC GENEROSITIES IN SCIENCE EDUCATION

Service-Learning

One way for science teacher preparation programs to utilize an ecohumanist perspective is in preparing science teachers to use service-learning strategies comprised of erotic ethics. Specifically, the services provided in this type of service-learning should be erotic generosities bestowed upon the Other for its freedom and because of its moral worth. In science education programs, service-learning provides opportunities for pre-service teachers to develop a multicultural science teaching practice, which allows them to make meaningful connections with community members and authenticates the kind of science they do (Barton, 2000). Students engaging in service-learning also benefit, including, through increased academic achievement, improved personal and social skills, developed citizenship, and improvement in school–community relationships (Kielsmeier et al., 2004). Moreover, as science students participate in community-based service-learning activities, they learn science authentically, which prepares them for lifelong learning and active participation in society (Handa et al., 2008).

Though the integration of service-learning activities is nothing new to science teacher education and science education, what I propose is different—service-learning based on erotic generosities. This kind of service-learning would include the traditional components (Barton, 2000; Phillipson-Mower & Adams, 2010), but it also capitalizes on the erotic relationships students will have (un)knowingly developed with their community and the ocean-Other. This focus is significant because students learn how to grant erotic generosities to the Other through their erotic relationship when explicitly explored. Service-learning based on an erotic ethic allows students to get back to the basic essence of the link between people, their community, and the ocean-Other, but this work needs to be done on the front end. Science teacher preparation programs need to teach science teachers how to help students bracket out the inessential details of their intimate relationships with Others to reveal the pure state. Ultimately, this allows students to live more authentically and generously with Other(s). One way science teachers can help students get to a phenomenon is by directing them to connect with members of the community, like community elders, to glean from their intergenerational, scientific, and cultural knowledge. These interactions help students to begin stripping away inessential and irrelevant layers to get at the basic connection they have to this knowledge. The early understanding community members convey can highlight how best to serve or act generously for the marine environment, because it helps students to uncover only those details that are pertinent. Through the understanding that comes from this phenomenological reduction, teachers can then help students feel better prepared to serve the ocean in a way that protects and sustains the basic integrity of the natural environment.

Once students begin to grapple with ocean phenomena, what might they learn about the phenomena by going to the sea or another aquatic environment, where they can take off their shoes and let their feet explore the hot sand? What might they learn about science and the natural world through service-learning activities, as the sea breeze whips their hair around their faces, seaweed washes ashore, and signs prevent them from trampling the dunes? Consider the fifty senior high school students engaging in water quality testing in Santa Rosa Sound through NOAA's Watershed Education and Training program. The students focus specifically on baseline testing to track future changes in the water quality to determine steps for the health of the environment. Through the program, the students go into the field and collect water samples for data analysis, then eventually communicate their findings to local officials and 1500 fifth graders (Escobedo, 2019). There are other examples, such as South Carolina high school students that paddle through the marsh on kayaks collecting water samples to investigate nutrient abundance, sedimentation, and types of pollution (Hedelt, 2019). The students simultaneously learn about local culture and history, wanting to make connections between the science content they were learning in class and their community, or establishing a sense of place. Through these service-learning projects, students learn important science content, but they are also more likely to act as lifelong stewards through phenomenological experiencing and nurturing erotic relationships. Community involvement might also be more sustainable, as the community has a moral responsibility to act generously to achieve transcendence for self and ocean-Other. Service-learning is one major example of erotically based marine science projects that can be utilized in the science classroom to promote erotic generosities and erotic relationships, but there are other examples of pathways that stimulate co-evolution of students' erotic relationships and the care and conservation of the marine environment.

Projects and Activities to Stimulate Co-evolution

Some scientists and marine resource managers are concerned for the ecology of the ocean because of its own inherent value, rather than for the remediation or conservation measures of marine ecology for human utility. Hale and Dilling (2010) argue that we exercise the precautionary principle and stop using marine resources arbitrarily because we are not able to control the results of human activities on the ocean. Through these considerations, marine resources would be distributed broadly and equitably among present and future generations. Moreover, this type of care might lead to erotically based, sustainable environmental management, rather than a focus on already dwindling resources. Perhaps what Hale and Dilling are defending is a co-evolution with our natural environments, where we recognize the transformations occurring in these environments through our erotic relationship and experiences with them. We in turn adapt to meet the needs and changes of the natural environment and learn to live sustainably within our limits. If we co-evolve with our natural environments, not only we are more capable of adapting by way of erotic thinking, but we also focus on and strive for an erotic relationship with our environment that is based on reciprocity, where it is mutually sustaining and conserving. We need science education programs that prepare science teachers to engage students in projects and activities that stimulate this co-evolution through the development of meaningful erotic relationships.

To determine how science education programs bolster this sort of curriculum, we might consider the work of Rachel Carson, who had a fierce erotic relationship with the sea and spent her life granting erotic generosities to the ocean because of this relationship. What about her erotic ethic is meaningful for science education? Carson sparked the interest of public and government officials alike; her passionate writing became the ignition for interest and action in environmental conservation efforts around the world. She called for critical thinking on scientific issues, action for scientific learning, and growth toward ecojustice. She established the significance for children to always have a sense of wonder about the natural world. Carson was deeply embedded and actively engaged in inquiry for her research because of the love she had for the sea. These qualities are all necessary to move the field of science education and marine science education forward, just as she was able to do with her environmental conservation efforts. It's time for our science educators to focus on fostering the kind of curriculum in science education that aligns with Carson's lived experiences, where authentic inquiry is key to developing or maintaining a sense of wonder about the natural world. Moreover, authentic inquiry activities, where students investigate legitimate issues that concern them, promote a co-evolution with the marine environment to build and strengthen our erotic relationships.

Through their lived experiences, students actively engage in the world over. According to Beauvoir (1948), we actively engage in the world to experience freedom. As something acted out, freedom derived from an erotic ethic should inspire action of value for the Other. This action will require "a pedagogy whereby educators explicitly connect student experience to the subject of study in the present moment in such a way that the past and the future are open, emerging, and in process" (Slattery & Morris, 1999, p. 30). Freedom must be based on ambiguity, not certainty. Henriksen wrote of the erotic as "open and opening, not closed and closing" (2010, p. 225). Just as plunging into the deep unknown of the ocean, "descent into the depths of consciousness necessitates a fluid and changing self, the dissolving of solidity and form into new energies for life, an openness to mysteries both within the self and beyond" (Victorin-Vangerud, 2001, p. 175). In science classes, students should be encouraged to embrace ambiguity and plunge into the unknown in order to experience freedom and open themselves to new experiences for spontaneity and action. This freedom should be particularly true when engaging in laboratory activities or through interactions with nature and the ocean, so as not to confine the results, nature, or ocean, thereby ensuring the possibilities of their ambiguity. Experiencing their science class in this way may allow the consciousness of the students to expand as well, as they would themselves be open to the possibilities of their ambiguity.

Other Pathways for an Erotic Ocean Science Education

Two additional possible routes to investigating an erotic ethic for ocean science education would be through the exploration of digital commons and within the science classroom itself. It could be argued that the digital world is a natural environment. If this is true, inland teachers can explore the uses of technology to effectively reveal the eroticism of the ocean (consider Google tools) when the marine environment is not readily available or necessarily familiar to them and their students. Therefore, it might be worthwhile to consider the use of online social networking such as Facebook to connect with others across the country or even worldwide. For example, what are the implications of inland classrooms setting up their own Facebook group that they could use to discuss marine-related topics with other classrooms along the coastline or to pose questions for their community regarding marine-related community issues? Students could even engage in social media gaming such as Coral Greef or Ocean Sweeper, in which players clean the world's oceans as part of a larger curriculum. Similarly, they can use other gaming options like Beyond Blue, which teaches about lesser-known aspects of the deep ocean environment, and NASA's NeMO-Net, which teaches users how to classify coral reefs and then allows them to classify real images that NASA's supercomputers use to improve their automated classification coding. Virtual gaming can be an excellent teaching opportunity, particularly when used appropriately. Beyond social networking and gaming, teachers could set up a classroom blog to discuss or advocate for pertinent marine-related community matters. Video conferencing technology provides a perfect venue for connecting with coastal classrooms, knowledgeable others, or oceanographers, as evidenced by the success of Skype a Scientist. Depending on internet accessibility, students could even video conference with students or oceanographers at or on the ocean to increase exposure. The appropriate use of technology can enhance the students' experiences with the ocean when they cannot physically be near it and can connect inland students with their community and coastal students and marine scientists to gain a better understanding of the ocean. Given these benefits, it would be interesting to consider the possibilities of a digital world enhancing or developing an erotic relationship with the natural world.

The ocean is peaceful, healing, sensuous, comforting, and embracing. Perhaps an erotic ethic can bring those qualities of the ocean to students, whether the ocean is present or not. An erotic relationship with the ocean can extend into science classrooms, regardless of proximity to the ocean. Perhaps in instances when even authentic inquiry is not always a possibility, teachers can create an "ocean" in their classroom using an erotic ethic. This ocean can serve as a model for students to practice freedom and ambiguity, lessons they will need when they transition into the adult world. Teachers can model mutual reciprocity of freedom in the classroom by viewing their students as free, not as objects in their classrooms. Through this model, students can learn the value of freedom not only for the ocean, but for all Others, and begin to understand how to grant this freedom themselves. As freedom is granted, students protect the Other from undue harm.

Because in many patriarchal societies, nature is gendered as female, teachers can analyze the language they use in the classroom and be conscious of the use of gendered pronouns and patriarchal metaphors about the ocean. It is not uncommon to hear metaphors like "the rape of mother earth" when referencing the human exploitation of natural environments. Ships and research vessels are said to "penetrate" the female sea. These root metaphors often originate in creation stories, in which the moist and cold properties of water symbolize the life-giving womb, coupled with the color blue that symbolizes female creativity. Water is seen as seductive and transformative, aspects of women that men historically fear (Pararas-Carayannis & Laoupi, 2007). In ungendering the ocean, the teacher acknowledges the ambiguous ocean. This ambiguous ocean can be erotic and subjective, not vulnerable to oppression or domination. As students recognize the ambiguity of the ocean, they see the ocean-Other as having possibilities for spontaneity. In this recognition, the student can view the ocean as free so that its end is freedom. The students can realize through this freedom that they must act with the ocean and Others to ensure the ocean continues to have such possibilities and reject any desires that would harm or negate the ocean-Other's freedom.

CODA

Adopting an erotic ethic in consideration of the ocean is essential for conservation and freedom. Through the development of an erotic relationship with the ocean-Other, we recognize the possibilities of self and the ocean-Other that allow us to consider the ocean-Other as free and worthy of care. We become more open to granting erotic generosities for the ocean-Other's security in recognition of its moral worth. An erotic relationship with the ocean-Other is necessary for all people across the country, regardless of the distance to the sea, age, or social class. This becomes evident when we consider how our actions are directly linked to the health of the ocean, which is in turn linked to the health of our bodies. With this knowledge, in part discovered through relationships with (O)thers, and preparedness to act, we can work against seeing the ocean as a commodity and act to sustain and protect its resources. Revealing the eroticism of the ocean and developing an erotic relationship with the ocean can promote the use of erotic generosities for the ocean-Other and provide a framework for the inclusion of an erotic ethic in environmental and science education. Through the inclusion of erotically based marine science curricula in science education, students can foster their own erotic relationship with the ocean, become aware of how intimately they are linked to the ocean, and gain the tools and knowledge necessary to make decisions to act in a way that protects and preserves the ocean-Other through erotic generosities.

References

Abram, D. (1996). *The spell of the sensuous*. Vintage Books.

Adusah-Karikari, A. (2015). Black gold in Ghana: Changing livelihoods for women in communities affected by oil production. *The Extractive Industries and Society, 2*(1), 24–32. https://doi.org/10.1016/j.exis.2014.10.006

Aikenhead, G. S. (1985). Collective decision making in the social context of science. *Science Education, 69*(4), 453–475.

Aikenhead, G. S. (2006). *Science education for everyday life: Evidence-based practice*. Teachers College Press.

Barton, A. C. (2000). Crafting multicultural science education with preservice teachers through service-learning. *Journal of Curriculum Studies, 32*(6), 797–820.

Beauvoir, S. de. (2011). *The second sex* (C. Borde & S. Malovany-Chevallier, Trans.). First Vintage Books. (Original work published 1949).

Beauvoir, S. de. (1948). *The ethics of ambiguity* (B. Frenchtman, Trans.). Philosophical Library.

Costello, C. S., Gaines, J. L., & Lynham, J. (2008). Can catch shares prevent fisheries collapse? *Science, 321*(5896), 1678–1681.

Cotton, K. (1996). *School size, school climate, and student performance*. School Improvement Research Series. Northwest Regional Educational Laboratory.

Escobedo, D. (2019, November 18). Students conduct Santa Rosa Sound water quality test. *The Destin Log*. https://www.thedestinlog.com/news/20191118/students-conduct-santa-rosa-sound-water-quality-testing

Food and Agriculture Organization of the United Nations (FAO). (2018). *The state of world fisheries and aquaculture 2018: Meeting the sustainable development goals*. Licence: CC BY-NC-SA 3.0 IGO.

Hale, B., & Dilling, L. (2010). Geoengineering, ocean fertilization, and the problem of permissible pollution. *Science, Technology, & Human Values, 36*(2), 1–23.

Hamilton, S. E. (2020). *Mangroves and aquaculture* (Vol. 33). Coastal Research Library: Springer.

Handa, V., Tippins, D., Thomson, N., Bilbao, P., Morano, L., Hallar, B., & Miller, K. (2008). A dialogue of life: Integrating service learning in a community-immersion model of preservice science-teacher preparation. *Journal of College Science Teaching, 37*(6), 14–20.

Hedelt, R. (2019). Caroline High students culminate watershed education by paddling, testing water at Port Royal. Fredricksburg.com. https://www.fredericksburg.com/news/local/caroline-high-students-culminate-watershed-education-by-paddling-testing-water-at-port-royal/article_9e5e8dc8-7075-5899-91d7-9cbf49c3a228.amp.html

Heidegger, M. (1962). *Being and time* (J. Macquarrie & E. Robinson, Trans.). Harper and Row.

Henriksen, J.-O. (2010). Eros and/as desire—A theological affirmation: Paul Tillich read in the light of Jean-Luc Marion's *The Erotic Phenomenon*. *Modern Theology, 26*(2), 220–242.

Hodson, D. (2011). *Looking to the future: Building a curriculum for social activism*. Sense Publishers.

Husserl, E. (1970). *Logical investigations* (J. N. Findlay, Trans). Humanities Press.

IPBES. (2018, March 23). Biodiversity and nature's contributions continue dangerous decline, scientists warn [Media release]. https://gallery.mailchimp.com/5da0fed71c7e4399fb28ab549/files/867bde03-d799-4ba9-b577-630e0fa32eda/20180323_IPBES6_Media_Release_Regional_Assessments.pdf

Kielsmeier, J. C., Scales, P. C., Roehlkepartain, E. C., & Neal, M. (2004). Community service and service-learning in public schools. *Reclaiming Children and Youth, 13*(3), 138–143.

Kolstø, S. D. (2001). Scientific literacy for citizenship: Tools for dealing with the science dimension of controversial socioscientific issues. *Science Education, 85*(3), 291–310.

Ladson-Billings, G. (2004). Differing conceptions of citizenship. In N. Noddings (Ed.), *Educating citizens for global awareness* (pp. 69–80). Teachers College Press.

Li, W. C., Tse, H. F., & Fok, L. (2016). Plastic waste in the marine environment: A review of sources, occurrence and effects. *Science of the Total Environment, 566*, 333–349.

Luther, R. A. (2013). Fostering eroticism in science education to promote erotic generosities for the ocean-other. *Educational Studies, 49*(5), 409–429.

Merleau-Ponty, M. (1964). *Sense and non-sense* (H. Dreygus, Trans.). Northwestern University Press.

Mikulak, M. (2007). Cross-pollinating Marxism and deep ecology: Towards a post-humanist eco-humanism. *Cultural Logic, 10*(1), 1–25.

Moran, D. (2000). *Introduction to phenomenology.* Routledge.

Mueller, M. (2008). Ecojustice as ecological literacy is much more than being "green!". *Educational Studies, 44*(2), 155–166.

National Oceanic and Atmospheric Administration (NOAA). (2014). What percentage of the American population lives near the coast? https://oceanservice.noaa.gov/facts/population.html

Pararas-Carayannis G., & Laoupi, A. (2007). *The role and social significance of women and youth in ancient Mediterranean cultures in sustaining development through connection with the sea as revealed by implied symbolisms of marine mythology.* Theme 2: Involvement of women and youth within the Millennium Development Goals strategies. International Ocean Institute: University of Malta.

Phillipson-Mower, T., & Adams, A. D. (2010). Environmental education service-learning in science teacher education. In A. Bodzin, B. Shiner Klein, & S. Weaver (Eds.), *The inclusion of environmental education in science teacher education.* Springer.

Plumwood, V. (2002). *Environmental culture: The ecological crisis of reason.* Routledge.

Slattery, P., & Morris, M. (1999). Simone de Beauvoir's ethics and postmodern ambiguity: The assertion of freedom in the face of the absurd. *Educational Theory, 49*(1), 21–36.

Victorin-Vangerud, N. M. (2001). The sacred edge: Women, sea and spirit. *Sea Changes: The Journal of Women Scholars of Religion and Theology, 1.* http://www.wsrt/com.au/oldseachanges/volume1/victorin-vangerud.html

Zimmerman, M. E. (1994). *Contesting Earth's future: Radical ecology and postmodernity.* University of California Publishing.

CHAPTER 5

The Ghost of Laplace's Demon: Revisiting the Anthropocene

Lars Bang

OVERTURNING THE DOGMATIC IMAGE OF THOUGHT IN SCIENCE EDUCATION

Science education and its various actualizations, and education in general, are haunted by a dogmatic image of thought (Bang, 2017; Bang & Valero, 2014; Bazzul et al., 2018; Deleuze, 1994), a ghost of inadequate representation. As science education propels forward with new sustainable UNESCO goals for the twenty-first century, Baby PISA tests designed by the OECD and similar practices of international measurement in the higher education arms race, the problematic of the dogmatic image of thought continues to be repeated and reproduced ad nauseam. Science education in 2020 need concepts based upon an adequate understanding of humanity's place in nature. Before outlining a reconceptualization of the Anthropocene in science education it is thus necessary to address the dogmatic image of thought and how that potentially limits the connection between *learning* about nature and *being* in and of nature, or more

L. Bang (✉)
Department of Culture & Learning, Aalborg University, Aalborg, Denmark
e-mail: lbje@ikl.aau.dk

© The Author(s) 2024   67
S. Tolbert et al. (eds.), *Reimagining Science Education in the
Anthropocene, Volume 2*, Palgrave Studies in Education and the
Environment, https://doi.org/10.1007/978-3-031-35430-4_5

broadly, between the activity of the mind and the activity of the body. In science education, this is actualized both in general subject matter and specifically in regard to the Anthropocene, in subject matter such as sustainability, ecological footprint, carbon footprint, climate change, and so forth.

To address the dogmatic image of thought is to "overturn" it, to dramatize and unfold the idea, or perhaps more accurately to retrace the idea (Deleuze, 2004). The specific dramatization invoked here utilizes two historical cases related to contemporary science education and the philosophy of science. Firstly, the work and "event" of Pierre-Simon Laplace (1749–1827), more specifically his germinal work *A Philosophical Essay on Probabilities* from 1825. Secondly, the "event" and work of Richard P. Feynman (1918–1988), a lauded educator of physics and Nobel Prize winner, and his lectures published in *The Feynman Lectures on Physics* (1963), specifically the lectures on probability and the theory of gravity (Feynman et al., 1963). Pierre-Simon Laplace and Richard P. Feynman are, in Deleuze and Guattari's lens, "conceptual persona" (Deleuze & Guattari, 1994), placeholders or envelopes for specific ideas on the plane of immanence.

Laplace's essay is usually seen as a manifesto for determinism or as a work which captures the scientific zeitgeist of the era, building upon the scientific tradition of the French Enlightenment. The scientific tradition in the French Enlightenment is not a monolithic entity but rather the actualization of many different ideas combining both philosophical and ethical/moralistic ideas regarding humanity and reason (see for instance Michel Foucault's essay on *What is Enlightenment?* for a reading of Kant's enlightenment [Foucault, 1984]) and new scientific practices and breakthroughs in the French academies. In other words, in the French Enlightenment, we see a fusion of ideas of reason and progress with a specific moral character as belonging to the scientist, or as I have argued elsewhere, the rationality of the Man of Science (Bang, 2017). Rereading Laplace's work, setting the rationality of the Man of Science aside for a spell, using Deleuze's method of dramatization, one can see the contours of a new idea, of a science based upon a Leibnizian worldview incorporating a principle of sufficient reason (Van Strien, 2014). The Deleuzian monstrous reading forwarded here is thus to reread Laplace with a Spinozist, rather than Leibnizian, approach to utilize a new outline of the concept of the Anthropocene.

Feynman is quite literally a Man of Science and in many ways captures the new American scientific spirit of the 1960s; his lectures became famous for their ingenuity and clarity, representing the "best practice" of physics education. Additionally, Feynman became one of the first science celebrities and was greatly used by the media. A close "monstrous" reading of Feynman's lectures points, similarly to Laplace's, though the way to thinking of nature in terms of flows; the physics concept of gravitation and force is especially impossible to understand, without a concept of flow. Feynman's moon example literally consists of flows and gravitational force and thus become a fulcrum, where Spinoza's theory of learning becomes clear and the Anthropocene can be rearticulated.

Before the monstrous historical readings, it is necessary to map and enunciate the problematic concept and representation of the Anthropocene.

MAPPING PREVAILING REMARKS AROUND THE ANTHROPOCENE

The concept of the Anthropocene was first forwarded by Paul Crutzen and Eugene F. Stoermer in 2000 in their entry in the *Global Change Newsletter* (Crutzen & Stoermer, 2000). They wanted a concept and representation encapsulating the modern geological age and how mankind since the late eighteenth century (beginning with James Watt's invention of the steam engine in 1784), have wrought geological, measurable changes upon the world. Despite their noble intentions of bringing the problematic activities of mankind and the impact of industrialization to the forefront of the climate debate with the new concept (or more adequately representation) of "the Anthropocene" (Crutzen & Stoermer, 2000), it is unfortunately in many ways an inadequate and flawed conceptual representation, especially when viewed through the lens of Deleuze's philosophy and Spinoza's metaphysics. The conceptual representation of the Anthropocene is useful though as an ideological rallying cry to address climate change and global warming, and my critique of the conceptual representation in light of science education is thus not a critique of activism against global warming or climate change, but merely aimed at arriving at a better concept to understand the role of humans on this planet and a concept of the Anthropocene cleansed of a dogmatic image of thought.

Claire Colebrook similarly points toward a certain ambivalence when talking and writing about the Anthropocene in light of Deleuze and Guattari's philosophy:

> It seems the only response to the vogue for Anthropocene thinking is ambivalence: yes we are finally—perhaps—thinking beyond our own time and interests, but we are doing so by way of a parochial concept of the species ("anthrops"), accompanied by a resurgence of seemingly counter-humanist rhetorics that are all too human. (Colebrook, 2016)

Colebrook's solution is to relegate and de-universalize the Anthropocene to a narrative similar to the other narratives or strata that Deleuze and Guattari outlined in *A Thousand Plateaus* (1987) and *Anti-Oedipus* (1983), while simultaneously retaining an inclusive disjunction retaining the ambivalence in the Anthropocene strata, as both productive and problematic. In term of Deleuze and Guattari's philosophy and especially their philosophical works on capitalism and schizophrenia (*A Thousand Plateaus* and *Anti-Oedipus*) Colebrook's solution, "the Anthropocene as strata in an inclusive disjunction," is seemingly adequate. The problem with Colebrook's ambivalence and the either/or and both/and argument concerning the Anthropocene is that it becomes a detached representation, a flawed concept and an abstraction at best. Spinoza's metaphysics, which is a cornerstone in Deleuze's philosophical oeuvre, simply can't accept such a "fuzziness," when it comes to concepts and learning. In short, without connecting the Anthropocene amply to the body it never becomes an adequate concept, which is exactly the connection Deleuze and Guattari drew in their work *Anti-Oedipus*, where they connected capitalism with the body (bodily flows, desire, and the oedipal structure). After this mapping and re-articulation of the problematic of the Anthropocene, it is necessary to readdress the dogmatic image of thought before demonstrating the interplay between these two issues.

THE NATURE OF THE DOGMATIC IMAGE OF THOUGHT

Deleuze writes in *Difference and Repetition*, "we do not speak of this or that image of thought, variable according to the philosophy in question, but of a single Image in general which constitutes the subjective presupposition of philosophy as a whole" (Deleuze, 1994, p. 167).

This sickness of representation, the dogmatic Image of thought and Deleuze's "patient-zero" in the history of philosophy, has many fathers, and can be summarized as *the subjective presupposition*, which the above quotation also hints at. In other words, the subjective, anthropomorphic center of (hu)man who thinks thoughts and makes representations ad nauseam, or perhaps with a Nietzschean flavor—the tragic pretentious subject. To overturn and struggle against this dogmatic image of thought in its various actualizations (science education being the actual case here) is to arrive at a fresh philosophy, a new way of thinking education, which in Deleuze's words would be:

> the condition of a philosophy which would be without any kind of presuppositions appear more clearly: instead of being supported by a moral Image of thought, it would take as its point of departure a radical critique of this Image and the "postulates" it implies. It would find its difference or its true beginning, not in agreement with the *pre-philosophical* Image but in a rigorous struggle against this Image, which it would denounce as *non-philosophical*. (1994, p. 167)

Deleuze connects several postulates to the dogmatic image of thought, and in relation to the Anthropocene "overturning" outlined here the first postulate related to recognition and representation is especially critical in regard to the Anthropocene, which is why Colebrook was right in mentioning the other representations, such as the Capitalocene, the Corporatocene and similar versions of the concept of the human age (Colebrook, 2016). Following Deleuze's method of dramatization in regard to science education to overturn the Anthropocene is thus not to ask, "what is it," but instead "what can it do" and to locate the active affirmative part of the dogmatic image of thought within the Anthropocene and link it in turn to Deleuze and Guattari's notion of individuation and becoming. Science educators should didactically pose the active question, "What can the Anthropocene do?" in the largest sense of the word, linking the thinking about the Anthropocene with bodily activity. One such instance of "Anthropocenic" didactics, which recently has sprung up in science education in Denmark is the activity of picking up garbage near the coastline (or around the schools in general). Such an endeavor seems to both highlight the futility of the activity, while combining it with the larger question "What can I do?," potentially opening up the issue that "I" is both the problem here and potential solution, connecting local bodily activity to global bodily activity. For the concept of the Anthropocene to be adequate in Spinozist and Deleuzian terms, it needs to be connected to processes of becoming instead of an "I," or perhaps more aptly an fractured "I" showing the composition of processes of becoming, before the "we" can be articulated adequately.

THE ANTHROPOCENE AND BECOMING-ANIMAL

Climate change and global warming are starring in Hollywood movie after Hollywood movie depicting the end of days due to their vicious effects— now recast as the great villain of the twenty-first century. Climate change is a vicious monster and who other than the radical activist can combat it? However, the villainizing of climate change is an effect of the dogmatic image of thought—it becomes a person, a global international representation: "the evil corporate American president Donald Trump, who ignores global warming" or "the evil militaristic president Jair Bolsonaro, who ignores the consequences of burning the Amazon" and so forth. But make no mistake, modern capitalism and its various presidents and spokesmen are THE overall problematic of the twenty-first century (just to stipulate that this is not a defense or negligence of Donald Trump and his Twitter regime of stupidity or Jair Bolsonaro's politics), and its effects are terrible, the personification and villainizing does not help us understand the Anthropocene and let us arrive at a point where it becomes clear that "man [sic] follows nature," to paraphrase Spinoza. Only by connecting the drastic changes of the world to the multiplicity and the constant changes within us can one arrive at an understanding. Seeing the reason to change is a principle of necessity, not as a villain to fight, but simply as living a life.

Deleuze and Guattari drew upon the writings of H.P. Lovecraft to explain becoming-animal and our fascination and horror with the multiplicity within us and as an "Outsider." Lovecraft writes in his novel *The Call of Cthulhu* about this condition of humanity:

> The most merciful thing in the world, I think, is the inability of the human mind to correlate all its contents. We live on a placid island of ignorance in the midst of black seas of infinity, and it was not meant we should voyage far. The sciences, each straining in its own direction, have hitherto harmed us little; but some day the piecing together of dissociated knowledge will open up such terrifying vistas of reality, and of our frightful position therein, that we shall either go mad from the revelation or flee from the deadly light into the peace and safety of a new dark age. (2002, p. 139)

Exactly the *abject* nature of the Anthropocene posited as something inevitable, something foreign and coming from beyond our existence and ultimately something outside of our scope, reason, and influence is the dogmatic image of thought conjured—a conjuration which installs a concept of the Anthropocene, as an inhumanity outside of our control, whose intent is to devour and dissolve us. In other words, to adequately understand the Anthropocene is to link it to Deleuze's notion of becoming-animal (and other related becomings). Deleuze and Guattari drew upon the writings of H.P. Lovecraft to explain becoming-animal and our fascination and horror with the multiplicity within us and as an "Outsider." Deleuze and Guattari's concept of becoming is on one hand drastically simple and easy to understand, as the myriad of biological processes within us, ever-changing and the engines behind our individuation, but following solely this vein of thinking leads us to Michael DeLanda's somewhat "scientific" interpretation of Deleuze (see for instance [DeLanda, 2013], which in my perspective drastically reduces Deleuze and Guattari's concept of becoming). On the other hand, the connection between the biological processes of becoming and our psyche is harder to explain without unpacking Deleuze and Guattari's particular take on Lacanian psychoanalysis and psychology in general (and how literature, the arts and so forth shows us / open up to processes of becoming). The various becomings (animal, other and so forth) are the realization, the crack in the surface of the "I," where individuation slips through and the potential appears for the abandonment of subjectivity. To summarize Leonard Lawlor's interpretation of Deleuze and Guattari's concept of becoming:

> In Deleuze and Guattari, becoming is never a process of imitating, yet the one who becomes finds himself before another who ends up being in oneself. With the other in me, however, I am not substituting myself for another; the structure of becoming is not reciprocal. It is a zigzag in which I become other so that the other may become something else, but this becoming something else is possible only if a work (oeuvre) is produced. (2008, p. 170)

The drawing called *Dredging up the Arcadian Dawn* by C. Bang, Fig. 5.1 below, depicts the problematic of the Anthropocene and the conceptual representation offered by the dogmatic image of thought, as a veritable Lovecraftian monstrous swamp thing rising from the undifferentiated ground and feeding when the stars are right. But the monster is dredged up "within," it is always and already inside us. We are all Deep Ones from

Fig. 5.1   Drawing by C. Bang

the Cthulhu Mythos, or as Deleuze and Guattari wrote in *Anti-Oedipus*, "We are all schizos" (1983, p. 87). To adequately capture the linkage between the Anthropocene and becoming, one needs a heuristic vehicle, actualized in a conceptual persona, which takes us by the hand and shows us how the determinism, causality, and inhumanness of the world are linked to our understanding, our ratio. In other words, we turn to Laplace's writings and Feynman's lecture to dramatize the idea of the Anthropocene.

OVERTURNING THE FIRST TRAIT OF THE ANTHROPOCENE: THE ORDERED UNIVERSE OR GOD AS A HYPOTHESIS

In the last section, we saw the monstrous conceptual representation of the Anthropocene, as a veritable Lovecraftian monster. This heuristic vehicle for understanding the Anthropocene in science education has (at least) two traits, which allows us to crack it open, to dramatize it and potentially revisit the concept and overturn it through connecting it to processes of becoming. The first trait is the otherness, the abject nature of the Anthropocene, as something which happens outside us, outside our zone of control, something which escapes reason and our ability to do something about it. It becomes geological, inevitable, the churning of the age and epochs slowly moving toward humanity's doom. This trait is a monstrous form of determinism which we trace and dramatize here through Laplace. The second trait is the doomed hero and antagonist in Lovecraft's stories, the agent of reason, the Man of Science set up to fail in the uncovering of the monstrous. There is no salvation of reason in Lovecraft's oeuvre, only the potential to go mad if you dare to look close enough. Lovecraft's actualized Man of Science is the "shadow" or perhaps the unconscious of the Man of Science seen in Feynman's lectures, which doesn't allow for the monstrous, which sees science as an apt focus of reason, but fails to aim at a totality or synthesis between human and nature. In other words, the wild side of Niels Bohr's complementarity principles and Albert Einstein's theory of relativity have been filtered out in favor of the zeitgeist of American reason of the 1960s where the Russians are winning in the space race.

Laplace was a French mathematician and astronomer and mainly preoccupied with solving Newton's problem of proving how the trajectories of the planets were stable and didn't need God's hand to uphold the machinery. Laplace used differential calculus to solve this problem, and his main work in five volumes is called *Mécanique celeste*. Newton's *Principia*, the last volume of the *Mécanique celeste*, was published in 1825 (Rouse Ball, 1960). Rouse Ball refers to Laplace's connections with Napoleon and repeats the famous conversations they anecdotally had, although other biographers are less convinced about the content of this alleged conversation, in which he famously retorts to Napoleon "I didn't need this hypothesis (God)" (Rouse Ball, 1960).

Laplace writes famously in his *Philosophical Essay on Probabilities* from 1825:

We ought then to consider the present state of the universe as the effect of its previous state and as the cause of that which is to follow. An intelligence that, at a given instant, could comprehend all the forces by which nature is animated and the respective situation of the beings that make it up, if moreover it were vast enough to submit these data to analysis, would encompass in the same formula the movements of the greatest bodies of the universe and those of the lightest atoms. For such an intelligence nothing would be uncertain, and the future, like the past, would be open to its eyes. The human mind affords, in the perfection that it has been able to give to astronomy, a feeble likeness of this intelligence. (1995, p. 2)

This account signals a certain kind of determinism, but as recently pointed out by Marij Van Strien (2014) there is more to Laplace's above passage then a simple manifesto for causal determinism in a ordered universe. This casual determinism we often see repeated in science education as Newtonian mechanics. The intelligence Laplace refers to is God, and he conceptualizes our human mind, and the reason behind his/our discoveries, as having a resemblance, "a feeble likeness," to this intelligence. Here we see a similarity to Spinoza's statement *deus sive natura* (God or nature) and his conceptualization of human mind (Lord, 2010). The above quotation has also been termed Laplace's demon, where the intellect referred to is a demon instead.

In other words, both in Laplace and in Spinoza's physics and metaphysics we have a conceptualization of causality at work. To overturn the dogmatic image of thought in the Anthropocene and re-actualize it in a new science education requires a revision of causality, linking becoming-animal with the climate change (perhaps becoming-nature in its various forms), and bringing together both Laplace's determinism and his ordered causal universe with Spinoza's principle of sufficient reason, thus linking thinking (the dogmatic image of thought revisited) with matter in Spinoza's parallelism. To reconceptualize the Anthropocene is to thus both to understand causality as a parallel movement consisting of (1) an expanded mechanistic determinism, a necessitarian principle and (2) the human activity of thinking. The movement toward understanding thus becomes a movement toward doing and changing according to the laws of nature, which we are governed by, or in Deleuze and Guattari's term, understanding is becoming (Deleuze & Guattari, 1987). When humans do this, nature reacts as follows, or, as Spinoza says, in the *Ethics* regarding the causal nature of the Affects:

Most of those who have written about the affects, and men's [sic] way of living, seem to treat, not of natural things, which follow the common laws of Nature, but of things which are outside Nature. Indeed they seem to conceive man in Nature as a dominion within a dominion. For they believe that man disturbs, rather than follows, the order of Nature, that he has absolute power over his actions, and that he is determined only by himself. [...] Therefore, I shall treat the nature and powers of the affects, and the power of the mind over them, by the same method by which, in the preceding parts, I treated God and the mind, and I shall consider human actions and appetites just as if it were a question of lines, planes, and bodies. (EIIIPref)[1]

Our conceptual persona, Laplace, already had a principle of sufficient reason embedded in his philosophy (Van Strien, 2014). The task at hand with regard to the insertion into the Anthropocene is thus to slightly revisit it as a Spinozist principle of sufficient reason.

Overturning the First Trait of the Anthropocene: Spinoza's Principle of Sufficient Reason and Laplace

Earlier we saw that the concept of the Anthropocene had two traits, which needed a revision. In the overturning, we propose that a connection has to be made between the concept of the Anthropocene and our human becoming. The first trait we attempted to overturn was a specific kind of determinism, and to overturn this I fertilized Laplace determinism with Spinoza's metaphysics. A crucial part of Spinoza's determinism, or more aptly his principle of necessity, is his principle of sufficient reason, which I shortly will unpack.

I agree with Michael Della Rocca's (2003) interpretation of Spinoza's principle of sufficient reason and how that in many ways is the overarching rationalism in Spinoza's philosophy. The principle of sufficient reason is the "principle that each fact has an explanation or, equivalently, that there are no brute facts" (Della Rocca, 2003, p. 75). Or as we see in Spinoza's main work, *Ethics* (1996), "For each thing there must be assigned a cause for its existence and non-existence" (EIp11d2). This reading of Spinoza is more rationalistic than the more usual readings deployed by new

---

[1] I am using Edwin Curley's Spinoza nomenclature, as seen in his translation of the collected works of Spinoza. Curley, E. (Ed.). (1985). *The Collected Works of Spinoza. Volume 1* (Vol. 1). Princeton University Press. (Originally published 1632–1677).

materialists and similar contemporary readers as seen, for example, in Elizabeth Grosz's work (1994). The necessary link between causation and conceivability points backs to both Laplace and Spinoza's great "hope" for the human mind: that we are part of the system and can thus potentially conceive it, or order things by way of the common order of nature, which is parallel to the order of the intellect. Laplace and Spinoza's "hope" is similar to a hope for science education, that through adequate concept, experimental didactics and so forth humanity can "become" more rational and wiser. This positivity of Spinoza's philosophy is highly affirmed by Deleuze and is thus in sharp contrast to arguments more pessimist and fatalistic regarding the human race. In Laplace, we can see the principle of sufficient reason directly quoted just before the statement on the intellect/demon:

> The connection between present and preceding events is based on the evident principle that a thing cannot come into existence without there being a cause to produce it. This axiom, known as the principle of sufficient reason, extends even to actions between which one is indifferent. The freest will is unable to give rise to them without a specific reason; for if, all circumstances of two situations being exactly the same, it (ie. the will) were acting in the one but not in the other, its choice would be an effect without cause; it would then, says Leibniz, be the blind chance of the Epicureans. The contrary opinion is an illusion of the mind that, losing sight of the fleeting reasons for the choice of the will in matters between which we are indifferent, persuades itself that it (ie. the choice) is determined of its own accord and without motives. (1995, p. 2)

In the above essay, Laplace's account of the principle of sufficient reason refers to causes instead of reason, unlike Leibniz's version of this principle. Van Strien rightly points out Laplace's determinism is more in line with Spinoza's principle of sufficient reason than Leibniz's. We see causes in Laplace's determinism, but not a notion of final causation, which again is similar to Spinoza's refutation of this. I have argued that the principle of sufficient reason is a crucial rationalist strand in Spinoza's philosophy and in Laplace's physics and philosophy, and that this principle becomes a crucial pivot, which breaks the dogmatic image of thought open, potentially overturning by the mere inclusion of this principle together with the other major Spinozist principle—the principle of necessity or Spinoza's necessitarianism.

Spinoza's necessitarianism is linked to his principle of sufficient reason. Here I agree with Della Rocca's conclusion that Spinoza's dual commitment to explaining existence and causation leads to a necessitarian principle or, as he says, "to insist that existence not be treated as a mystery" (Della Rocca, 2003, p. 90). I think this insistence is the core of Spinoza's rationalism and why you, for instance, in Spinoza's work *Theological-Political Treatise* (Curley, 2016) encounter one of the first historical criticisms of the Bible and the "miracles" reported within. In other words, Spinoza's philosophy is both a new naturalism and a new rationalism. Both principles articulated above, together with Laplace's determinism, can be encultured, or perhaps injected like a virus, in the concept of the Anthropocene to be utilized in science education. First, when you have reiterated what a concept is (and can do) in Deleuze and Guattari's philosophy, we can fully understand the ramifications of how the Anthropocene can be actualized in science education. Before we can return to this enculturation of the Anthropocene, the second trait of the dogmatic image of thought in the Anthropocene needs to be briefly unpacked.

OVERTURNING THE SECOND TRAIT OF THE ANTHROPOCENE: RICHARD P. FEYNMAN'S MAN OF SCIENCE

I stated earlier that the second trait of the Anthropocene was linked to a specific way of thinking by the scientific hero or the stereotypical scientist. Similar to our dramatization of Laplace, we are not presenting a critique of Feynman's lectures here, rather we are searching for the instances where the lectures (or conceptual persona of Feynman) points toward a link between thinking in flows and forces, ultimately paving the way of seeing a connection between human nature and the forces of the cosmos. In short, we are searching for a new, more monstrous blueprint of the scientific hero. Feynman is a very good example of the Man of Science, and the lectures are another example of the zeitgeist of the 1960s, where he constantly links his teachings of space and the cosmos to the Russian success in this endeavor (this is 1963), thus politically highlighting issues regarding the state of science education in America and the higher education arms race. In his lecture on the theory of gravity (preempted by his lecture on probability) we see though how he opens up for a way of thinking science, of how to formulate an active participation in regard to understanding in his remarks "What else can we understand when we understand gravity?" and "What else can you do with the law of gravitation?" (Feynman et al., 1963). Earlier in the lecture, he used the example of the moon and the tides, which both shows the potential confusion of the human mind in regard to the phenomena, and also how the understanding of gravitation solves our "common sense" issues. Finally, he points toward the need to establish a connection between theories of different forces and remarks how closely electrical forces resemble gravitational forces. Feynman's lecture thus overall becomes an example of the necessity of connecting our various ways of understanding, how everything fits together and how our understanding increases in terms of flows, seeing flows and systems between various forces. The revisited scientific hero or "Human of Science" sees in terms of flows and forces, constantly linking the cosmos and nature to the nature and thinking "within." Feynman's statement below aptly summarizes this particular form of rationalism and hope of understanding the universe and resonates clearly with Spinoza's hope for humanity and our rationality:

> It is hard to exaggerate the importance of the effect on the history of science produced by this great success of the theory of gravitation. Compare the confusion, the lack of confidence, the incomplete knowledge that prevailed in the earlier ages, when there were endless debates and paradoxes, with the clarity and simplicity of this law—this fact that all the moons and planets and stars have such a simple rule to govern them, and further that man could understand it and deduce how the planets should move! This is the reason for the success of the sciences in the following years, for it gave hope that the other phenomena of the world might also have such beautifully simple laws. (Feynman et al., 1963)

THE GRAVITY OF THE ANTHROPOCENE

After having unpacked and dramatized the two traits of the Anthropocene and how to overturn the concept of the Anthropocene, I return to science education and outline how this can be adequately linked.

I thus insert both Laplace and Spinoza's principles and determinism in the concept of the Anthropocene as such together with Feynman's new scientific hero, showing an outline of how science education can use the Anthropocene in its various practices. Deleuze and Guattari's distinction between functives and concepts in their last work *What is Philosophy* (1994) helps us to didactically understand the dogmatic image of thought in regard to the Anthropocene and the task for educators. Deleuze and Guattari write:

> The object of science is not concepts but rather functions that are presented as propositions in discursive systems. The elements of functions are called functives. A scientific notion is defined not by concepts but by functions or propositions. (1994, p. 117)

If we take the law of gravitation $F = G\dfrac{mm'}{r^2}$, which Feynman famously dubbed "one of the most far-reaching generalizations of the human mind" (Feynman et al., 1963, pp. 7–11), we see an example of one of the functions Deleuze and Guattari refer to. This function "discursively states" how every object in the universe attracts every other object with a force which for any two bodies (m and m') is proportional to the mass of each and varies inversely as the square of the distance between them (Feynman et al., 1963).

This function has no relation to philosophy as such but is used by scientists to reflect and communicate (inserted in discursive systems and so forth). But as Deleuze and Guattari continue: "when an object—a geometrical space, for example—is scientifically constructed by functions, its philosophical concept, which is by no means given in the function, must still be discovered" (1994, p. 117). Gravitation is thus simultaneously a philosophical concept, a concept related to bodily individuations, multiplicities and becomings. For example, consider a child in the early stages of life experimenting with gravitation, with object-permanency and so forth. A life is by itself, experimentation with gravitation. We get up, we fall, we see objects falling, and so on. In other words, we are surrounded by forces of gravitation.

The function often referred to as mechanistic determinism in physics is $\dfrac{d^2r}{dt^2} = F(r)$, a function which discursively states that if we have all the information of all the positions, velocities, and forces present, we can predict the future and past states of the system in question (Van Strien, 2014) is often attributed to Laplace. This function though has nothing to do with the philosophical principle of sufficient reason and isn't derived from it. I agree with Van Strien's conclusion (which corresponds to Deleuze and Guattari's distinctions quoted above) that Laplace's determinism stems from his philosophy and the law of continuation, not from his mechanics (Van Strien, 2014). When inserting Laplace's determinism in the Anthropocene it is a philosophical determinism we graft onto the Anthropocene, not specific functions. It is the Anthropocene as a philosophical concept we are interested in, wrested from a dogmatic image of thought. The didactical task becomes then a matter of connecting the understanding of how gravitation works in nature with how gravitation plays out in our body and daily life.

An Outline of the Revisited Anthropocene

A science education, or perhaps more accurately a science didactics, based upon Spinozist principles can only be outlined here but is necessarily linked to experimentation of bodies and of sensation as Aislinn O'Donnell points out (2018), while simultaneously being an activity of thinking, of ratio, and of linking functions to phenomena in an expanded notion of scientific inquiry (Bang, 2018). To understand the Anthropocene in terms of the principle of sufficient reason, Spinoza's necessitarianism, and Laplace's determinism is to understand climate change as a cause, the natural reaction to human capitalism and the folly of human being conceptualized as a "subject." It is an insistence to understand and link the multiplicity and change outside us, to the multiplicity and change inside us, our various becomings. Understanding climate change and the Anthropocene is similar to understanding necessary bodily cycles like the menstrual cycle, bowel movements, human aging, and so forth. Linking the inevitability of the processes of the body with the inevitability of the geological processes. In other words, if we return to the example of garbage collection near the Danish coastline, children collect garbage in science education not to improve the environment as such but to understand the processes of waste in themselves and how they affect the world. Only through these didactics of connections through our ratio can we begin to have a practice in science education, which connects our bodies to the changes in the world bodies, linking weather systems to hormone cycles, starvation to fossil fuels, and, to paraphrase Nietzsche, everything is a matter of health especially our morals and collective living. Everything is devoid of notions of ecological morality, or perhaps more accurately we can glimpse a new ethics. "Thou shall not" has no place in Spinozist didactics, and the Spinozist virtues are what is already healthy for us—they are what we already are and become in nature.

References

Bang, L. (2017). In the maw of the Ouroboros: An analysis of scientific literacy and democracy. *Cultural Studies in Science Education, 13*, 807–822. https://doi.org/10.1007/s11422-017-9808-2

Bang, L. (2018). The inquiry of Cyclops: Dewey's scientific inquiry revisited. In K. Otrel-Cass, M. K. Sillasen, & A. A. Orlander (Eds.), *Cultural, social and political perspectives in science education* (Vol. 15, pp. 49–69). Springer.

Bang, L., & Valero, P. (2014). Chasing the chimera's tails: An analysis of interest in science. In T. S. Popkewitz (Ed.), *The 'reason' of schooling: Historizing curriculum studies, pedagogy and teacher education*. Routledge.

Bazzul, J., Wallace, M. F., & Higgins, M. (2018). Dreaming and immanence: Rejecting the dogmatic image of thought in science education. *Cultural Studies of Science Education, 13*(3), 823–835.

Colebrook, C. (2016). 'A grandiose time of coexistence': Stratigraphy of the Anthropocene. *Deleuze Studies, 10*(4), 440–454.

Crutzen, P. J., & Stoermer, E. F. (2000). The 'Anthropocene'. *Global Change NewsLetter, 41*.

Curley, E. (Ed.). (1985). *The collected works of Spinoza. Volume 1*. Princeton University Press. (Original works published 1632–1677).

Curley, E. (Ed. & Trans.). (2016). *The collected works of Spinoza. Volume II*. Princeton University Press.

DeLanda, M. (2013). *Intensive science and virtual philosophy*. Bloomsbury Academic.

Deleuze, G. (1994). *Difference and repetition* (P. Patton, Trans.). Continuum International Publishing Group.

Deleuze, G. (2004). The method of dramatization (M. Taormina, Trans.). In D. Lapoujade (Ed.), *Desert Island and other texts 1953–1974. Semiotext(e)*. (Original work published 1967).

Deleuze, G., & Guattari, F. (1983). *Anti-Oedipus: Capitalism and schizophrenia*. University of Minnesota Press.

Deleuze, G., & Guattari, F. (1987). *A thousand plateaus: Capitalism and schizophrenia*. University of Minnesota Press.

Deleuze, G., & Guattari, F. (1994). *What is philosophy?* (H. Tomlinson & G. Burcell, Trans.). Columbia University Press.

Della Rocca, M. (2003). A rationalist manifesto: Spinoza and the principle of sufficient reason. *Philosophical Topics, 31*(1/2), 75–93.

Feynman, R. P., Leighton, R. B., & Sands, M. (1963). *The Feynman lectures on physics* (Vol. 1). Addison-Wesley Publishing Company.

Foucault, M. (1984). What is enlightenment? In P. Rabinow (Ed.), *Foucault Reader*. Pantheon Books.

Grosz, E. (1994). *Volatile bodies: Toward a corporeal feminism*. Indiana University Press.

Laplace, P. S. (1995). *Philosophical essay on probabilities* (A. I. Dale, Trans.; 1825 ed.). Springer-Verlag.

Lawlor, L. (2008). Following the rats: Becoming-animal in Deleuze and Guattari. *SubStance, 37*(3), 169–187.

Lord, B. (2010). *Spinoza's ethics*. Edinburgh University Press.

Lovecraft, H. P. (2002). *The call of Cthulu and other weird stories*. Penguin Books Ltd..

O'Donnell, A. (2018). Spinoza, experimentation and education: How things teach us. *Educational Philosophy and Theory, 50*(9), 819–829.

Rouse Ball, W. W. (1960). *A short account of the history of mathematics*. Dover Publications. (Original work published 1908).

Spinoza, B. D. (1996). *Ethics* (E. Curley, Trans.). Penguin Group. (Original work published 1677).

Van Strien, M. (2014). On the origins and foundations of Laplacian determinism. *Studies in History and Philosophy of Science Part A, 45*, 24–31.

PART II

Anti-colonial Anthropocene(s)

CHAPTER 6

Envisioning Non-elite and More-Than-Colonial Environmentalisms

Aline Carrara and Ritodhi Chakraborty

INTRODUCTION

Over the past few years, with a brutal global pandemic, worsening impacts of climate change, and dysfunctional governance across various scales, visions of an impending apocalypse have been proliferating. These coming end times are described by some to be the outcome of destructive human–nature relationships; others see it as the Earth, a superorganism, resetting itself, bringing back equilibrium, beginning a new age, and for yet others, the apocalypse can be averted, if only our institutions listened to the opinions of those attempting to foreground the needs of nature and the Earth in our politics (Awuh et al., 2021; Bosworth, 2021; Fine & Love-Nichols, 2021). These views shared and re-shared on social media platforms, headlined in media cycles, and spoken in fear and frustration remain oblivious

A. Carrara
University of California-Davis, Davis, CA, USA

R. Chakraborty (✉)
University of Canterbury, Christchurch, New Zealand
e-mail: ritodhi.chakraborty@canterbury.ac.nz

© The Author(s) 2024
S. Tolbert et al. (eds.), *Reimagining Science Education in the Anthropocene, Volume 2*, Palgrave Studies in Education and the Environment, https://doi.org/10.1007/978-3-031-35430-4_6

to both the beginnings of the apocalypse and how human–nature relationships underscore many kinds of humans, many kinds of natures. Such a plurality is starkly absent from mainstream environmentalism and climate action, which remain tethered to Malthusian overpopulation scenarios, usually in the majority world, authoritarian protectionism through conservation policies and climate adaptation/mitigation projects predicated on visions of "pristine" nature, and ecological stewardship rules which nominate the individual as the critical and thus fail to hold accountable the powerful machinery of the market and state alliance (D'Souza, 2019; Kashwan, 2013; Tindall et al., 2022). Echoing similar sentiments, Zoe Todd asks:

> What does it mean to have a reciprocal discourse on catastrophic end times and apocalyptic environmental change in a place where, over the last 500 years, Indigenous peoples faced (and face) the end of the world with the violent incursion of colonial ideologies and actions? What does it mean to hold, in simultaneous tension, stories of the Anthropocene in the past, present, and future? (2016, ¶5)

In the following chapter, inspired by the work of indigenous, feminist, anti-racist, anti-casteist, anti/de/post-colonial thinkers and doers, we consider Todd's critical question and interrogate the problematic roots of modern, mainstream environmentalism and its role in supporting certain visions of the Anthropocene. In doing so we propose a reorienting of our epistemic and political frames. Our intention is to highlight the myriad ways in which humans are entangled with the more-than-human beings that challenge environmentalism's reductionist human–nature binaries, which act as a tool of enclosure, exclusion, and displacement. Additionally, such a reframing questions the value of the planetary scale within environmentalism as a discursive tool and an organizing device, while also acknowledging the material cleavages of a deeply unequal world. In doing so, we hope to highlight the plurality of relationships that need to be nurtured, to move from the ideological and material prison of hegemonic mainstream environmentalism (HME) and the just futures it promises.

Mainstream Environmentalism's Problematic Past and Present

Environmentalism, or a series of practices and ideas to care for the more-than-human world of nature, is often seen as a modern movement, emerging as a response to the impacts of exploitative and extractive industrial development and natural resource management. However, environmentalism can also be defined as an ongoing exploration of viable human–nature relationships, which predates the beginning of colonial industrial state-building. But, to return to the former definition, an overt sense of "protectionism" runs like a throughline within most of the mobilizations surrounding environmental policy, planning, and ideologies within much of the minority world during the nineteenth and twentieth centuries. This vision of a world on the verge of collapse due to humanity's unregulated growth arguably rests on the establishment of the "scientific" ideals of the Enlightenment in Europe, its support of the vision of colonial human–nature management since the fifteenth century, and ultimately, the creation of an industrial global economy, trading in the labour and bodies of humans and more-than-human beings. Thus, environmentalism, as a modern global set of ideals and politics, solidifies post-World War II in the minority world. Along with the support of protected areas for conservation, as imagined and executed in the United States in the nineteenth century and inspired by colonial land management, environmentalism sought to address the exacerbating death and loss of biodiversity due to unregulated industry and the supposedly exploding human population which was nearing the limits of ecosystemic and planetary carrying capacities (Davies, 2020; Johnson, 2020).

However, given the plurality in ideological, cultural, and political positions, even within the minority world, environmentalism has stratified into several different avatars, coalescing around significantly different objectives and pathways to those objectives. These range from ecocentric ideas such as deep ecology to corporate environmentalism, which sees the capitalist free market as the best tool for planetary sustainability. They also include more authoritarian and violent ideas such as ecofascism, which rests on social Darwinist ideas of racial and ethnic superiority and the protection of such populations and their "nature," violently, if necessary, against less suitable people. Finally, authoritarian environmentalism is an exclusionary process through which powerful institutions (often the state) can create and apply top-down policies to manage "nature," irrespective of, and at times even against, the aspirations of impacted community members. Our framing of hegemonic mainstream environmentalism (HME) rests on the foundations of such ideals. First, it biases certain scales of governance, driven by the tools of colonial land management, instituting evaluation metrics that assess human–nature relationships, not for their ability to address the holistic wellbeing needs of a variety of communities and ecologies but for their ability to produce material and political assets for powerful elite institutions (e.g., the state, transnational corporations). Second, it attempts to explore and present human–nature relationships through a Cartesian, positivist lens, cleaving humans and nature into discrete autonomous units—artificially cleaving relational, entangled systems to often create oppositional binaries. Finally, third, it essentializes and romanticizes certain historical moments, advocating a return to such times as critical in restoring the equilibrium that is lacking in our current human–nature relationships. In this framing, the agency of both humans and nature is subsumed to support a narrative that can only exist when populated with caricatures, whose validity is tethered to the acquisition of certain elite political objectives (e.g., the trope of the noble savage, the presentation of pre-colonial human–nature relationships as the equilibrium state) (Bosworth, 2021; Lo, 2021; Mansfield et al., 2015; Smith, 2021). So, HME in many ways fails to address its colonial, authoritarian, essentializing overtures, which continue to insidiously motivate much of environmentalism and environmental policy. Here we want to make something clear: our use of HME in this text is only as a heuristic. We are not proposing yet another explanatory framing to distil and examine environmentalism. There is much ongoing scholarship and activism which addresses this issue. Our objective is to merely create a cohesive entity, which, we argue, encapsulates much of environmental ideology, policy, and activism in the world today.

The discourse and practice of HME in the majority world are quite variegated. In India, despite some recognition of communal land rights and decentralized institution building, the spectre of colonial forestry is still resonant, with displacement and evictions of forest-dependent communities, often along with caste/religious/ethnic differences, both for conservation and industrial development, a commonplace affair. In China, authoritarian environmentalism with state ownership of all land, and through it all of nature, often brings issues of social justice into collision with top-down environmental governance. In Kenya and Tanzania, fortress conservation to protect African wildlife from local Africans has led to the creation of heavily surveilled and militarized human–nature relationships for the benefit of the elite. In Brazil, the territorial aspirations of the settler state are juxtaposed to the international environmental organization "industry," which consistently engages in power tussles with other land-use stakeholders, undermining the complicated and diverse political mobilization of indigenous communities. Despite such differences, some key common threads emerge in the manifestations of HME in the majority world. These include the discursive and material domination by elites within national environmental ideologies, policies, and management; patriarchal foundations of institutionally codified human–nature relationships leading to a flourishing of technocratic management; an importation of both environmental activism and environmental management techniques from the minority world and marginalization of place-based "environmentalisms"; and finally, a significant lack of focus on root causes of exploitative and extractive human–nature relationships, leading to a focus on crisis resolution through top-down tools, instead of understanding the historical structural inequalities connected to ownership, access, and management.

However, there are also ongoing challenges to this powerful edifice of HME. Many have emerged from indigenous resistance to a variety of political and managerial tools, attempting to further alienate communities from nurturing and responding to the many changes in their human–nature relationships. Others have coalesced around the marked exclusion of women from decision-making pathways, highlighting their unique relationships with more-than-human beings. Yet others have critiqued the omnipresence of shallow technical solutions, proposing that the care of the natural world and the wellbeing of communities cannot be treated as separate projects. Ultimately, questions of justice and equality are being foregrounded by many who believe that they serve as critical yardsticks by which to evaluate the health and future viability of our human–nature relationships (Agyeman et al., 2016; Nightingale et al., 2020; Schmidt, 2022; Tschakert et al., 2021; Zanotti, 2014).

Complicating such non-elite mobilizations is the discursive and material reality of climate change. The significant changes in global climatic patterns, especially related to temperature and precipitation, brought about by fossil fuel-driven industrial growth are impacting key human–nature relationships and social-ecological systems. Nevertheless, the unjust precarity experienced by marginalized communities and ecologies, both in the majority and minority world, making them more susceptible to changing climate and society trends, is a result of historic structural processes such as colonization, predatory capitalism, patriarchy, casteism, and techno-managerial state-building. Therefore, while the urgency enshrined in global climate politics, building on visions of a coming apocalypse, is seemingly responding to a common condition, it ends up homogenizing complicated climate–society relationships. Ultimately, critiques of HME identify similar concerns in planetary and state-scale mitigation and adaptation knowledge and policies and propose instead a radical reimagining of the Anthropocene. They push back its genesis to the advent of colonial conquest and reframe the ultimate objective as a reconciliation of historic injustices within the life of modern nation-states and the nurturing of social-ecological wellbeing at sub-national scales, through more than mere carbon management (Dalby, 2017; Jackson, 2020; Larsen & Harrington, 2020; Lorimer, 2012; Mathews, 2020; Simpson, 2020).

Given such a plural and at times contentious existence, challenges to HME exist in many forms across the majority world. In the next section, we explore a temporary conceptual frame, the goal of which is to provide a habitat for our rendition of a variety of environmentalisms emerging parallel to HME.

Non-elite and More-Than-Colonial Environmentalisms (NEMCE): A Temporary Frame

Non-elite and more-than-colonial environmentalisms (NEMCE) is a frame we are using in this chapter to capture a whole host of ideologies and actions that challenge the validity of HME. Similar to HME, it is a heuristic device and, in this case, analytically employed to present two very different case studies, emerging out of very different intersections of colonization, state-building, and communal agency. Our choice of words to describe these environmentalisms is deliberate and is predicated on two considerations. First, in recent years there has been a burgeoning of "decolonial" scholarship, and such "attempts" at decolonization have appeared across multiple disciplines, institutions, political mobilizations, and collaborations. However, indigenous-led decolonization advocating for an end to the settler-colonial project and the return of material and political control to communities historically battling colonial violence and erasure is very different from the emergence of settler and white scholars wielding the ideological and intellectual premise of decolonization in the ongoing culture of appropriation, focusing on rhetoric rather than material relations. This trend is problematic when decolonial thought and action seem to be led by minority world institutions, racially privileged scholars and activists, and ethnically/culturally dominant elites from the majority world. Adequately examining the tussle at the heart of decolonial praxis is beyond the scope of this chapter; however, the problematic connotations of the word and also its existence adjacent to other terms like post-colonial and anti-colonial have inspired our choice of using the term "more-than-colonial." For us, more-than-colonial is a term rooted in hope and in the possibility of relationalities, ideologies, and materialities that extend beyond the colonial imaginary. In doing so, we also support the claim that it is impossible to "extract," "sever," or "eliminate" the many vestiges of colonization, since our communities and ecologies have emerged from the colonial encounter. Instead, by invoking more-than-colonial our understanding is that human–nature relations and environmentalism, even though formed of certain colonial elements, can and do become more than the sum of their parts (Curley et al., 2022; Halvorsen, 2019; Hope, 2020; Mollett, 2020; Tuck & Yang, 2012; Zanotti et al., 2020).

Second, given the very different experiences of socio-cultural and political hierarchies in the majority world, terms like racism, sexism, or ethnic and religious othering don't often translate across space and time. Therefore, to signpost existing power structures without essentializing the experiences of different communities, we are using the term "non-elite," which contains multitudes, doesn't nominate a certain kind of discriminatory process, and allows us to navigate across a variety of socio-ecological relationalities (Campbell, 2012; Gergan & Curley, 2021).

In this next section, we explore the three foundational "habitus" of NEMCE. These are emerging from multigenerational and multispatial encounters between marginalized non-elite communities and the hegemonic processes of the state-science-market triad. While not all these encounters challenge the edifices of HME, their complicated presence reveals the messy, unfinished-yet-generative attempts by a variety of agents to support certain place-based human–nature relationships. In doing so they repoliticize and pluralize environmentalism (Accetti, 2021) and highlight the inability of elite discursive and material tools to understand and ultimately extend allyship to the spectrum of human–nature relationships (Thomas, 2015; Whyte, 2020) (Fig. 6.1).

NEMCE as a culture of practice is almost impossible to categorize or articulate without getting discursively hijacked by the essentialist tropes

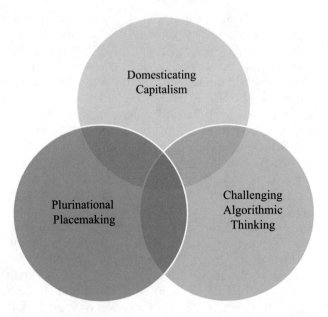

Fig. 6.1 NEMCE at work

that HME itself is mired in. However, witnessing a variety of mobilizations across the majority world (both through our engagement with communities and a review of ongoing projects), we notice three emerging processes (see Fig. 6.1).

DOMESTICATING CAPITALISM

Supporters of market-based solutions to address "environmental issues" remain fiercely (and even violently) opposed to more anti-capitalist discourses and mobilizations, and vice versa (Borras et al., 2021). However, many non-elite mobilizations seem to be pursuing tactics which attempt to "tame" or "domesticate" the extractive overtures of capitalism, while also wielding it to achieve their own political goals. We see such mobilizations in the Pacific, where indigenous economic development, pursued through a diversity of socio-culturally embedded practices, is defying extractive surplus distribution pathways. Instead, through building upon existing fluid relationships between various human and non-human agents, economic enterprise, practised by certain indigenous communities, is enabling a possible move from appropriation to circulation of surplus value and wealth (Amoamo et al., 2018; Vunibola et al., 2022). Another example is the Xavante indigenous group from the lower Amazon basin and Cerrado (savannah) regions of Brazil. Despite suffering both cultural/material genocide and territorial loss of their homelands, their strategy is one of "taming the waradzu (white man)." Currently, Xavante reclamation of frontier urban spaces, as a form of territorial control, through a variety of strategies complicates indigenous stereotypes. Their use of colonially constructed cultural stereotypes to take back the control of local- and regional-scale economic activity from the settlers is a result of the colonial state's historical failure at protecting indigenous sovereignty over territorial homelands, through protected area policies (Carrara, 2020; Welch & Coimbra, 2021).

Plurinational Placemaking

The limits of the nation-state, in the majority world, as both a representative of the diverse socio-ecological relationships it contains within its borders and the ultimate arbitrator and adjudicator of cross-scalar socio-ecological contentions have been well explored (Shawoo & McDermott, 2020). However, given the Euro-colonial world order, the state is here to stay. Nevertheless, through a variety of strategies, non-elite actors are challenging more monolithic state institutions, often pointing to the enduring elite control over such institutions across a variety of post-colonial spaces. Some of these challenges also incorporate a restructuring of intra-community governance, critiquing at once the vagaries of the post-colonial state and historical oppression by certain elite groups. We see such strategies underway in lower caste engagements with dictates of top-down land management policies in the Central Himalaya, in India. These strategies, while circumventing the increasing control of the state, often through the Forest Department, also defy upper-caste gatekeeping of communal land and institutions. A production and reproduction of human–nature relationships, while pursuing forms of extraction, similar to those of the elites, situates their actions within a historical throughline of material exclusion and severance from the land, both as a "resource" and as a "refuge" (Chakraborty & Sherpa, 2021; Sharma, 2022). We also observe such strategies in the "working-class environmentalism," being led by indigenous and peasant organizations within the oil and agricultural sectors in Ecuador. While some have pointed to the problematic existence of "indigeneity" as a symbolic resource and an essentialized political tool, the construction of Ecuador as a plurinational state is emerging from such a diverse, at times contradictory realm of identity positions. The emergence of labour organizations which go beyond the traditional labour unions, focused more on worker rights and working conditions, to also address "environmental" concerns, is an example of non-elite visions of human–nature relationships which ground themselves in questions of ownership and use (Uzzell, 2021; Vela-Almeida, 2018).

CHALLENGING ALGORITHMIC THINKING

There is an increasing presence of digital environmental governance within the auspices of environmental policy making. This emergence is predicated upon the rise of GIS-based Earth information systems, predictive models of socio-ecological change, and the use of artificial intelligence to explore possible current and future trends (Machen & Nost, 2021; Nost & Colven, 2022). This proliferation of "algorithmic thinking" has been touted as an answer to the burgeoning "science denialism." However, in the process, the hard-won battles of knowledge equality and justice, legitimizing knowledge production through indigenous, local, feminist methods, have been undone. Challenging such "science imperialism," a variety of non-elite actors are mobilizing, fighting for both the utility of their knowledges and the creation of a system of accountability to address the extraction and misuse of renditions of their human–nature relationships by members of the scientific establishment. We observe such challenges in the horizontal and plural knowledge production initiatives across the majority world, especially in regard to exploring transforming climate–society relationships. These include knowledge co-production, with the Waorani in Ecuador (Manuel-Navarrete et al., 2021), with the Sherpa in Nepal (Sherpa, 2014), with the Maasai in Tanzania (Goldman et al., 2018).

Taken together the three processes above present a powerful response to HME. They highlight its insidious reproduction of certain elite subjectivities, ideologies, and institutions, while claiming to support planetary visions of ecological wellbeing and through them social sustainability. However, we also acknowledge that the construction of NEMCE itself is an act of essentialism, an attempt to subsume a multitude of relationships under the auspices of a category, held together in most instances by its positioning as an alternative to HME. Additionally, we think, it is deeply problematic to claim that non-elite environmental politics is merely reactionary, and non-elite subjects are combative counterweights to the whims of powerful HME actors. This is why we refer to the NEMCE heuristic as a temporary one, its existence contingent on its conceptual utility, and we hope it is replaced by more inclusive and effective representations of non-elite aspirations.

BEYOND HEGEMONY: SOME ASPIRATIONAL CONCLUSIONS

As we consider NEMCE and its abilities to decentre HME, we begin with a provocation. A provocation that others have arrived upon as well and at this juncture allows us to bookend our argument. *We think the planetary focus, the scalar bias towards a global future, or for that matter, past, and a search for international harmony and dare we say, peace, inadvertently (or intentionally) undermines the many lives of NEMCE.* The Anthropocene is included in this mix as are more seemingly progressive mobilizations such as Earth jurisprudence and more ecomodernist ones such as Sustainable Development Goals. Over the past few centuries, the modern nation-state has emerged as the fundamental spatial, cultural, and ecological unit of our planet. It can be argued that our planet is inherently visualized as formed of international relations. In this hegemonic planetary mythology, and it is hegemonic, the plurality of human–nature relationships is consistently held hostage by the material and cultural aspirations of that national spatial organism. Is the solution then a spatial dissolution, a global "melt-down of borders" releasing both sovereignty and belonging from what currently exists as a powerful national imaginary? However, the imbalances of power within our nations predate their formal establishment. How would those relationships endure in a post-national planetary order? We don't really have the answer to this question and accept that the aspirational and the possible, while intertwined, can exist together, albeit in different forms. Embedded within a planetary body organized as such, what is the role of NEMCE?

First, it serves to rupture the various ideologies that are vying for political space to control the present and future of human–nature relationships. This advances through the dismantling of identity categories like "peasant," "indigenous," and "environmentalist," and also spatial categories such as "household," "village," "tribe," or "nation." In examples of domesticating capitalism which seem "unenvironmental," non-elite communities at once subvert a legal system which remains tethered to discrete notions of identity and spatial categories and serve to reward those who align most with such categories. This encountering of the essentialization of non-elite agency is necessary when working towards just futures. Additionally, indigenous inclusion in land management is increasingly observed as moments of plurinational democracy in action, through a successful weaving together of techno-managerial science and indigenous knowledge. The inclusion, we think, should catalyse around the human–nature relationship and its dynamic manifestations, having passed through a colonial rite of passage, and not be driven by the seeming directives of almost a rights-based approach to the institutions of the modern settler colony (Holst, 2016; Kashwan, 2013; Kvanneid, 2021; Laing, 2020).

Second, it calls out the artificial (and farcical) construction of strife between the quest for social justice and the needs of " nature." In recent scholarship highlighting the differences between the colonial co-optation of the political project of decolonization and the indigenous-led mobilization to decolonize, Curley and colleagues state that "native, settler, slave— these are categories that are posed in both powerfully effective and troubling ways—they may provide crucial starting points in understanding how white settler enslavers set the terms of the game against liberation and sovereignty. If we take them up too easily, however, might they reify and entrench settler-enslaver truths and bind us in relation to one another in ways that make it difficult to imagine and enact abundant futures?" (2022, p. 1056). A similar politics is at work (and play) in the ideological construction of a scarce planet on the brink of collapse and the actions of the non-elite, the most vulnerable, and a justifiable act of survival, which inadvertently enables the apocalypse (D'Souza, 2019; Whyte et al., 2019). Narratives of scarcity and humanity's reach past planetary boundaries, yet again, echo a scalar obfuscation. The life of the state is transposed on the life of the non-elite human and more-than-human agent and their many entanglements. How is this just? Furthermore, within the discourse of HME, the aspirations of non-elite communities are side-lined, replaced by a combination of an anxious response to the Anthropocene and the machinations of a political bloc, organized around the idea of "protectionism," vying for power over land, against the neo-extractive march of the settler state (Anthias & Radcliffe, 2015; Klenk, 2004).

Ultimately, NEMCE challenges a discrete global environmental ethic/science/policy and through it a master mythology of human–nature relationships. Instead, it proposes place-based mobilizations deeply rooted in spatio-temporally relevant injustices, whose goals don't have to be sacrificed for the promise of some imagined planetary future. The human–nature relationships emerging from very situated spatio-temporal encounters cannot be organized within a vision where scales are additive and higher order inferences can be drawn by extrapolating such situated encounters across space and time. NEMCE provides a key alternative to HME visions of planetary wellbeing—the human–nature relations being nurtured at one scale, spatio-temporal unit, place, whatever you want to call it, do not have to adhere to the aspirations of some grand narrative. Neither does the discomfort of encountering a pathway to equality and justice, which extends beyond what the state can offer, insinuate some internal collapse, and point to the victory of the apocalypse. Because, the truth is, for many non-elite communities the apocalypse has already happened, and they survived, and they are still resisting. Their politics resonate with a hopefulness which is in stark contrast to the anxiety now palpable in HME.

REFERENCES

Accetti, C. I. (2021). Repoliticizing environmentalism: Beyond technocracy and populism. *Critical Review, 33*(1), 47–73. https://doi.org/10.1080/0891381 1.2021.1908023

Agyeman, J., Schlosberg, D., Craven, L., & Matthews, C. (2016). Trends and directions in environmental justice: From inequity to everyday life, community, and just sustainabilities. *Annual Review of Environment and Resources, 41*, 321–340.

Amoamo, M., Ruckstuhl, K., & Ruwhiu, D. (2018). Balancing indigenous values through diverse economies: A case study of Māori ecotourism. *Tourism Planning and Development, 15*(5), 478–495. https://doi.org/10.1080/ 21568316.2018.1481452

Anthias, P., & Radcliffe, S. A. (2015). The ethno-environmental fix and its limits: Indigenous land titling and the production of not-quite-neoliberal natures in Bolivia. *Geoforum, 64*, 257–269. https://doi.org/10.1016/j.geoforum. 2013.06.007

Awuh, H. E., Elbeltagy, R., & Awuh, R. N. (2021). In the midst of every crisis, lies great opportunity? Analysing environmental attitudes in the face of the Covid-19 pandemic. *GeoJournal, 4*. https://doi.org/10.1007/s10708-021-10512-4

Borras, S. M., Scoones, I., Baviskar, A., Edelman, M., Peluso, L., & Wolford, W. (2021). Climate change and agrarian struggles: An invitation to contribute to a JPS Forum. https://doi.org/10.1080/03066150.2021.1956473

Bosworth, K. (2021). The bad environmentalism of 'nature is healing' memes. *Cultural Geographies.* https://doi.org/10.1177/14744740211012007

Campbell, J. M. (2012). Between the material and the figural road: The incompleteness of colonial geographies in Amazonia. *Mobilities, 7*(4), 481–500. https://doi.org/10.1080/17450101.2012.718429

Carrara, A. F. A. (2020). *The struggle for indigenous territories in the Brazilian Amazon* (Unpublished doctoral dissertation). University of Florida, Gainesville, Florida.

Chakraborty, R., & Sherpa, P. Y. (2021). From climate adaptation to climate justice: Critical reflections on the IPCC and Himalayan climate knowledges. *Climatic Change, 167*(3–4), 1–14. https://doi.org/10.1007/s10584-021-03158-1

Curley, A., Gupta, P., Lookabaugh, L., Neubert, C., & Smith, S. (2022). Decolonisation is a political project: Overcoming impasses between indigenous sovereignty and abolition. *Antipode, 54*(4), 1043–1062. https://doi.org/10.1111/anti.12830

D'Souza, R. (2019). Environmentalism and the politics of pre-emption: Reconsidering South Asia's environmental history in the epoch of the Anthropocene. *Geoforum, 101*(January 2018), 242–249. https://doi.org/10.1016/j.geoforum.2018.09.033

Dalby, S. (2017). Anthropocene formations: Environmental security, geopolitics and disaster. *Theory, Culture & Society, 34*(2–3), 233–252.

Davies, A. R. (2020). Environmentalism. *International Encyclopedia of Human Geography,* 259–264. https://doi.org/10.1016/b978-0-08-102295-5.10791-7

Fine, J. C., & Love-Nichols, J. (2021). We are (not) the virus: Competing online discourses of human-environment interaction in the era of COVID-19. *Environmental Communication, 0*(0), 1–20. https://doi.org/10.1080/17524032.2021.1982744

Gergan, M. D., & Curley, A. (2021). Indigenous youth and decolonial futures: Energy and environmentalism among the Diné in the Navajo Nation and the Lepchas of Sikkim, India. *Antipode, 0*(0), 1–21. https://doi.org/10.1111/anti.12763

Goldman, M. J., Turner, M. D., & Daly, M. (2018). A critical political ecology of human dimensions of climate change: Epistemology, ontology, and ethics. *Wiley Interdisciplinary Reviews: Climate Change, 9*(4), 1–15. https://doi.org/10.1002/wcc.526

Halvorsen, S. (2019). Decolonising territory: Dialogues with Latin American knowledges and grassroots strategies. *Progress in Human Geography, 43*(5), 790–814. https://doi.org/10.1177/0309132518777623

Holst, J. (2016). Colonial histories and decolonial dreams in the Ecuadorean Amazon. *Latin American Perspectives, 43*(1), 200–220. https://doi.org/10.1177/0094582X15570837

Hope, J. (2020). Globalising sustainable development: Decolonial disruptions and environmental justice in Bolivia. *Area*, April, 1–9. https://doi.org/10.1111/area.12626

Jackson, M. (2020). On decolonizing the Anthropocene: Disobedience via plural constitutions. *Annals of the American Association of Geographers, 111*(3), 698–708.

Johnson, B. H. (2020). American environmentalism and the visage of a second gilded age. *The Journal of the Gilded Age and Progressive Era, 19*(2), 246–252.

Kashwan, P. (2013). Forest policy, institutions, and REDD+ in India, Tanzania, and Mexico. In M. G. B. Lima & J. Gupta (Eds.), *Global Environmental Politics, 13*(August), 46–64. https://doi.org/10.1162/GLEP

Klenk, R. M. (2004). "Who is the developed woman?": Women as a category of development discourse, Kumaon, India. *Development and Change, 35*(1), 57–78. https://doi.org/10.1111/j.1467-7660.2004.00342.x

Kvanneid, A. J. (2021). Climate change, gender and rural development: Making sense of coping strategies in the Shivalik Hills. *Contributions to Indian Sociology, 55*(3), 392–415. https://doi.org/10.1177/00699667211059723

Laing, A. F. (2020). Re-producing territory: Between resource nationalism and indigenous self-determination in Bolivia. *Geoforum, 108*(May 2018), 28–38. https://doi.org/10.1016/j.geoforum.2019.11.015

Larsen, T. B., & Harrington, J., Jr. (2020). Geographic thought and the Anthropocene: What geographers have said and have to say. *Annals of the American Association of Geographers, 111*(3), 729–741.

Lo, K. (2021). Authoritarian environmentalism, just transition, and the tension between environmental protection and social justice in China's forestry reform. *Forest Policy and Economics, 131*, 102574.

Lorimer, J. (2012). Multinatural geographies for the Anthropocene. *Progress in Human Geography, 36*(5), 593–612.

Machen, R., & Nost, E. (2021). Thinking algorithmically: The making of hegemonic knowledge in climate governance. *Transactions of the Institute of British Geographers*, April. https://doi.org/10.1111/tran.12441

Mansfield, B., Biermann, C., McSweeney, K., Law, J., Gallemore, C., Horner, L., & Munroe, D. K. (2015). Environmental politics after nature: Conflicting socioecological futures. *Annals of the Association of American Geographers, 105*(2), 284–293.

Manuel-Navarrete, D., Buzinde, C. N., & Swanson, T. (2021). Fostering horizontal knowledge co-production with indigenous people by leveraging researchers' transdisciplinary intentions. *Ecology and Society, 26*(2). https://doi.org/10.5751/ES-12265-260222

Mathews, A. S. (2020). Anthropology and the Anthropocene: Criticisms, experiments, and collaborations. *Annual Review of Anthropology, 49*, 67–82.

Mollett, S. (2020). Hemispheric, relational, and intersectional political ecologies of race: Centring land-body entanglements in the Americas. *Antipode, 53*(3), 810–830. https://doi.org/10.1111/anti.12696

Nightingale, A. J., Eriksen, S., Taylor, M., Forsyth, T., Pelling, M., Newsham, A., Boyd, E., Brown, K., Harvey, B., Jones, L., Kerr, R. B., Mehta, L., Naess, L. O., Ockwell, D., Scoones, I., Tanner, T., & Whitfield, S. (2020). Beyond technical fixes: Climate solutions and the great derangement. *Climate and Development, 12*(4), 343–352.

Nost, E., & Colven, E. (2022). Earth for AI: A political ecology of data-driven climate initiatives. *Geoforum, 130*(March 2021), 23–34. https://doi.org/10.1016/j.geoforum.2022.01.016

Schmidt, J. J. (2022). Of kin and system: Rights of nature and the UN search for Earth jurisprudence. *Transactions of the Institute of British Geographers, 47*(3), 820–834.

Sharma, M. (2022). Caste, environment justice, and intersectionality of Dalit–Black ecologies. Environment and Society, 13(1), 78–97. https://doi.org/10.3167/ares.2022.130106

Shawoo, Z., & McDermott, C. L. (2020). Justice through polycentricity? A critical examination of climate justice framings in Pakistani climate policymaking. *Climate Policy, 20*(2), 199–216. https://doi.org/10.1080/1469306 2.2019.1707640

Sherpa, P. (2014). Climate change, perceptions, and social heterogeneity in Pharak, Mount Everest region of Nepal. *Human Organization, 73*(2), 153–161. https://doi.org/10.17730/humo.73.2.94q43152111733t6

Simpson, M. (2020). The Anthropocene as colonial discourse. *Environment and Planning D: Society and Space, 38*(1), 53–71.

Smith, J. K. (2021). *The (re) emergence of eco-fascism: White-nationalism, sacrifice, and proto-fascism in the circulation of digital rhetoric in the ecological far-right* (Doctoral dissertation).

Thomas, A. (2015). Indigenous more-than-humanisms: Relational ethics with the Hurunui River in Aotearoa New Zealand. *Social & Cultural Geography, 16*(8), 974–990.

Tindall, D. B., Stoddart, M. C. J., & Dunlap, R. E. (2022). The contours of anti-environmentalism: An introduction to the *Handbook of anti-environmentalism*. In D. Tindall, M. C. J. Stoddart, & R. Dunlap (Eds.), *Handbook of anti-environmentalism* (pp. 2–22). Edward Elgar.

Todd, Z. (2016). Relationships. *Cultural Anthropology, 21*. https://culanth.org/fieldsights/relationships

Tschakert, P., Schlosberg, D., Celermajer, D., Rickards, L., Winter, C., Thaler, M., Stewart-Harawira, M., & Verlie, B. (2021). Multispecies justice: Climate-just futures with, for and beyond humans. *Wiley Interdisciplinary Reviews: Climate Change, 12*(2), e699.

Tuck, E., & Yang, K. W. (2012). Decolonization is not a metaphor: York at New Paltz. *Decolonization: Indigeneity, Education & Society, 1*(1), 1–40.

Uzzell, D. (2021). The Palgrave handbook of environmental labour studies. In *The Palgrave handbook of environmental labour studies*. https://doi.org/10.1007/978-3-030-71909-8

Vela-Almeida, D. (2018). Territorial partitions, the production of mining territory and the building of a post-neoliberal and plurinational state in Ecuador. *Political Geography, 62*, 126–136. https://doi.org/10.1016/j.polgeo.2017.10.011

Vunibola, S., Steven, H., & Scobie, M. (2022). Indigenous enterprise on customary lands: Diverse economies of surplus. *Asia Pacific Viewpoint, 63*(1), 40–52. https://doi.org/10.1111/apv.12326

Welch, J. R., & Coimbra, C. E. A. (2021). Indigenous fire ecologies, restoration, and territorial sovereignty in the Brazilian Cerrado: The case of two Xavante reserves. *Land Use Policy, 104*, 104055. https://doi.org/10.1016/j.landusepol.2019.104055

Whyte, K. L., Talley, J. D., & Gibson, J. (2019). Indigenous mobility traditions, colonialism, and the Anthropocene. *Mobilities, 14*(3), 319–335. https://doi.org/10.1080/17450101.2019.1611015

Whyte, K. (2020). Indigenous environmental justice: Anti-colonial action through kinship in Environmental justice (pp. 266–278). Routledge.

Zanotti, L. (2014). Hybrid natures?: Community conservation partnerships in the Kayapó lands. *Anthropological Quarterly, 87*(3), 665–693.

Zanotti, L., Carothers, C., Apok, C., Huang, S., Coleman, J., & Ambrozek, C. (2020). Political ecology and decolonial research: Co-production with the Inupiat in Utqiagvik. *Journal of Political Ecology, 27*(1), 43–66. https://doi.org/10.2458/v27i1.23335

Indigenous Spiritual Geographies: Rosalie Little Thunder and "What Does It Mean to Be a Good Relative?"

Amanda Holmes

she walks

she walks
with Buffalo
under the ground
they are coming
across the sky
they are coming
through the needles of stone and pine
they are coming
she is waiting
not knowing
that this seed of bone
she will carry
has already found her

A. Holmes (✉)
Independent Scholar, Tucson, AZ, USA
e-mail: skennen7@gmail.com

© The Author(s) 2024
S. Tolbert et al. (eds.), *Reimagining Science Education in the Anthropocene, Volume 2*, Palgrave Studies in Education and the Environment, https://doi.org/10.1007/978-3-031-35430-4_7

there in the circle
where they sing
old songs
laid down long ago
they knew she would be coming
feet bloodied hands rubbed raw
their plastic cuffs
digging into skin
piercing
dragged across the rusting grates
made for cattle and trucks
to the pens
where the newcomers fabricate disease and lies
to justify the ways they slaughter
Buffalo and the People
walk over ground
they have known
since Creator shaped them
long ago
placing them there
together
blood and grasses
encircle the Hills
but today
they will be slaughtered
here
in their homes
she is coming
and you will stand here together
this long winter
of blood and snow
standing in the wind
turning to each other
to remember
it is time
to return
it is time
to go
home

— for Rosalie
  Beauty is an act, not just a painting. — Manulani Meyer

INTRODUCTION

Rosalie Little Thunder was a Lakota Elder and a protector of the Buffalo. She walked more than 500 miles during the winter, year after year, in solidarity with the Buffalo, the Elder brother to the Lakota, to stand with them, to draw attention to their suffering at the hands of the National Park Service, Montana Department of Livestock, and other federal and state agencies who were slaughtering the last of the wild Buffalo herds in North America around Yellowstone National Park. She called others to join her, to literally walk their ancestral ways of knowing who they are as Lakota People. Her walks for and with the Buffalo clarified for her this question she would ask, continually, "What does it mean to be a good relative?"[1] (Rosalie Little Thunder, personal communications). And as she asked it, she never stopped walking into this question—and into its answer—with her life.

In her walking of it, she asked all of us to embody this question in our *own* lives. "We need to become good relatives again," she would say (Rosalie Little Thunder, personal communications). Rosalie Little Thunder was a philosopher, whose questioning, perception, and expansive thought came from being deeply placed, rooted, and grounded within her Lakota worldview. This was a "quintessentially Rosalie question" and one that she always asked people to reflect on. It is a question that emerges from deep within her Lakota philosophy, understanding, and ways of seeing the world, as it emerges from within Lakota epistemology, philosophy, and ethics. Her question reflects this particular Lakota orientation to the Universe.

She asked this question not only to invigorate thinking and memory, but because within it there is an urgency, this call to action—and just like everything she did, it was about putting moccasins to the ground and walking it, living it. That's how Indigenous philosophies are—they are enacted, embodied. She called this *spiritual activism*—spiritual action and engagement—lived and reflected upon, practiced, and committed to daily, as Lakota ethical praxis. "Remind yourself, every morning, every morning, every morning. I'm going to do something. I've made a commitment.

---

[1] *"What does it mean to be a good relative?"*—this question is a question attributed to Rosalie Little Thunder in oral practices and communications, through formal and informal talks, interviews, keynote addresses, and so on, that she gave around the world, from approximately 1990 to 2014. Should you come across this question in other academic writing, please know that it should be attributed to Rosalie Little Thunder.

Not for yourself, but beyond yourself. You belong to the collective. Don't go wandering off, or you will perish" (Little Thunder, from Buffalo Field Campaign).

This question holds particular significance—cultural touchstones, reference points—of meaning and memory to her Lakota People, ways of knowing and being that only they will know, remember, comprehend, from within the longevity and profound depths of relationship and *being relation* in their places, being Lakota in those places, and being in their Universe as Lakota. And she also asked this question of others—non-Lakota, non-Indigenous people—to wake up the human beings in this time of massive human destruction of all life and the life systems of the Earth. She asked because she understood the urgency of the question and the necessity, the urgency, for non-Indigenous human beings to engage with this concept, though these worldviews may not be familiar to them, may not be immediately understood, not commonsense, not embedded within their knowledge systems and cultural practices. As she asked it, she was pushing the human beings to see ourselves—to imagine ourselves—as relatives, as a starting point. Her question reveals to us that it is already known in Lakota philosophy that human beings are relatives, that we are related to all-that-is, but it is in the *quality* of the relationship, the *being* of the relationship, the *embodiment* of the relationship that makes the difference.

Rosalie's question is a wakeup call that frames human survival from within a Lakota perspective of the world—Lakota epistemologies, ontologies, and ethical principles are asking the world's human beings—"what does it mean to be a good relative?" (Rosalie Little Thunder, personal communications). It is a Lakota re-imagination of the world. Being a Lakota Elder embedded within Lakota worlds, ethical principles, and oral-relational ways of knowing-being, her perception holds the possibility for radical re-imagination of different possibilities, possibilities for transformative shifts in consciousness, and thus the possibilities for different futures. Centering Indigenous worldviews allows for different questions and ways of questioning, different understandings and ways of understanding that hold the potential to wake us up to the urgent need to see, understand, perceive differently.

Being connected to all of life, as relative, is real within Indigenous worldviews, a collectivity that reaches far beyond the self and the human world. What would it mean to see yourself as *relation*, inextricably connected not only to the entire human world but to the rest of the universe? What does it mean to be a good relative? When an Elder repeats something, it is time to pay close attention. This question emerged from within Rosalie's consciousness as a Lakota thinker, where *All My Relations*, a guiding Lakota ethical principle, calls to the people to *live as relation*, in relationship to the rest of Life—*as a good relative.*

Her question asks us to perceive differently, to self-reflect, to look more deeply at what we think we know, and the ways we think we know it—a critical questioning that recognizes the urgency of engagement and calls for a re-thinking of the ways we are in the world. For Indigenous Peoples, these ways of knowing and being are ancient, known as "the way it is." Lakota knowledges, cultural sensibilities, and ways of knowing-being are articulated in the question itself. Rosalie offers this way of knowing in order to remind, renew, and restore the ways that Indigenous Peoples have always known, always done, always been, long before the onslaught of the multiple violences and brutality of Western ways of knowing-being that targeted these ancient Indigenous cultural knowledge systems and practices for extraction and disappearance.

In her articulation of the question, Rosalie lights a spark, calling for a deepening reflection on the meaning of these critical core belief systems and ways of knowing the world. Her question shifts our perception toward generosity, diversity, humility, connectedness—relatedness—as it *requires* human beings to shift not only our vantage point, the place from which we act in the world, but also our *way* of knowing the world and our assumptions about the nature of reality. Her question illuminates the idea of living life as a conversation not focused on the human world but heard and recognized among the living world, interacting with each other since the beginning of time. Rosalie's question reminds us too that it is *they* who ask us, who call us, even now, to be relatives, to remember our relationality. Indigenous Peoples recognize that they are calling out to the human beings in ways that they previously have not done—an invitation to wake up, to listen, to pay attention, to enter into relationship, with increasing urgency now. Is it any wonder? Are their lives not hanging in the balance, dependent on what the humans will do, on the choices the human world will make?

RELATIONALITY AND A TRANSFORMATION OF CONSCIOUSNESS

Indigenous Elders and Beyond-Human Elders are calling out the dire need for attention to this severance from relationality and thinking deeply about how to address the alienation from and violence toward the natural order. Rosalie's question is a call to action that emerged out of her lived Lakota praxis—to speak to what it is that needs to be most urgently communicated. As Seneca Elder John Mohawk (1990) articulated, in order to pull back from this edge of devastation to the natural world and the planet, we need a transformation of consciousness.

What, then, is relationality? What does it mean to see ourselves as related? Another layer of Rosalie's question calls for and invites accountability, reciprocity, respect, care, and responsibility to the relationship. To live as a good relative, how are you called to conduct yourself? How does perceiving yourself as a relative to every element of the universe alter your perception of yourself? How does it alter your perception of time and experience—of your generational past, your present, and of the generations yet to come? How might living a reality in which you are known as a relative to all of life shift your perception? How might you *relate* differently?

What is Rosalie asking? She is asking for us to consider a different way of being, an older, wiser ethic of understanding self, which is to say the self in connection, and the living world in connection to each other, practices that Indigenous Peoples have been living as embodied for millennia. Rosalie is talking about a spiritual reality that is known to Indigenous Peoples intimately, an everyday knowing that informs and shapes every aspect of Indigenous realities and engagement. Indigenous ethical principles, practices, and protocols of relationship—and the responsibility to that relatedness—emerge from particular Places, those lands, beings, elements, energies who are present together, carrying their original instructions of relatedness as a living way of perceiving and participating. How to live as a good relative gives meaning, relevance, and context to those instructions of how to live, of how to conduct yourself in a way that *you will make sense to* the rest of that particular world that is watching you and waiting.

For Rosalie, this quality of *being a relative*, a Lakota relational orientation to the world, is recognizable and coherent as each element finds life in each other. Rosalie's centering orientation was to that Lakota relational universe, and her question reflects that central concern of being recognizable, accountable, and coherent to that universe—being known by that universe as a relative. Her question calls us to reflect more deeply on how to live in places, within the coherence of that particular place and all that is held there—its energies, sacredness, spirits, knowledges, languages, beings—the knowing that is held in that place as living coherence, living systems. Thinking about what it means to be a good relative calls us to reflect more deeply from the standpoint of the beyond-human world, worlds that are waiting for the human beings to come to understand themselves as connected.

Rosalie's call to action emerged out of her lived Lakota wisdom praxis of relationality, a *radical relationality*, which is to say, a relationality *at the roots*, embedded within the lands, beings, spirits, and languages who belong to a particular place, the universe that *is* that particular place.[2] This braided thread of Rosalie's questioning emerged from this coherent, radical relationality, these ways of behaving and conducting oneself, orienting oneself as individual rooted within and inextricable from the collective. These are the relationships that are reflected in the ethical principles, living ancestral memory, ways of knowing-being and perceiving each other, and the deep cultural practices of being present to each other.

Rosalie was a visionary, an Elder who knew and understood her Lakota ways of knowing the universe in the ways of her grandparents and their grandparents, reaching through time and their collective intergenerational experience of memory and knowing of their particular place on Earth, the Lands where their Creator placed them among the spiritual beings, formations, and elements who are meant to be there and who interact with each other *there* in those places, as nowhere else on Earth—their own *spiritual geographies*. Indigenous Peoples know their Lands and the beings, energies, and relationships with whom they share their universe and their Original Instructions, as related, from long experience with each other. They exist within a relational, storied matrix, spoken and heard, felt and lived in words and through silences sung, prayed, and envisioned within their own cycles and remembered in the spiritual geographies of their Lands. These are ways of knowing-being that reflect a different way of perceiving the universe—as an interrelated universe of interaction and exchange—and that humans have a place within it that is no more

---

[2] By radical relationality, I want to be sure to make clear that it is not at all "radical"—in the Western sense of that word—for Indigenous Peoples who have lived and continue living these ways since time began.

significant or insignificant than any other element. A teaching in *radical humility*. How to understand this was one thread of Rosalie's lifework.

INTERGENERATIONALITY AND *FINDING LIFE* IN EACH OTHER

The perception of *being relation* calls us to hone our capacities to *perceive, imagine, and envision intergenerationally*, within the universe that is embedded within the very ground of our places, our ancestral homelands. Re-envisioning intergenerational connection is about living within cycles where Ancestors are remembered, and the ways they responded to their world are actively considered, as is what has been handed on to you to carry. Honing that perception, that capacity to envision what their work might be for us to continue, we engage in a quality of relationality that is intergenerational perception and envisioning, which begins with our Ancestors and extends to perceive intergenerational relationships with the rest of Creation and the beyond-human world.

From within cyclical time and an orientation to relationship, the living presence of Ancestors and their experience, knowledge, and wisdom become alive and embodied in an intergenerational ethic of reciprocity—what are our Ancestors asking of us and what is our unfinished work that we will be leaving for future generations. The quality of being a good relative and the self- and intergenerational-reflexivity embedded within Rosalie's question call us to perceive the intergenerationality of reciprocity and relationship—the intergenerationality of relationality. Re-imagining ourselves as part of intergenerational cycles of relatedness supports us to find life in each other.

WISDOM AS VERB: LIVING WISDOM, *LIVING* WISDOM

Wisdom as verb—living wisdom, *living* wisdom, wisdom as praxis—what does this look like, feel like, sound like, *mean*? Are we going to continue to center the same Western-dominant epistemological framework that is poisoning us and our world from our hearts and minds outward, to provide us with the "solutions" that we now all so urgently need? Indigenous praxes of embodied wisdom are so often only paid attention to when they are added on to Western knowledge to validate it. There is little recognition of what is being silenced—and the profound, desperate need for living, relational wisdom. Putting our lives in the service of open, clear-heartedness, as *related* to all-that-is, we re-align ourselves with the natural world and ways of being of the Universe—and our lives begin to take on that shape. We need to find spaces for this older wisdom to emerge, which will spark shifts in perception, so that the human beings are able to transform the ways they relate to the world and to themselves. Rosalie invited, called for, a return to ways of being and ways of being present that are rooted deeply within Lakota ethical principles. She asked people to engage in a more fully developed, deeper, wiser, more humble, more attentive sense of self, one that is grounded within the collective consciousness of relationality.

WHAT THEN, ROSALIE ASKED. WHAT NOW?

Rosalie never gave up trying to awaken something deeper, an orientation to relationship, to connectedness. She thought long and deep about how to do this, how to make these ways of knowing that were so implicit to her make sense to people in a society where relationships are so broken as to barely matter, a society built upon its own refusal to allow people to count, to exist, to be present as relative, much less to envision, to imagine, life as relational, or to live as if a relational world *mattered*.

Rosalie had the belief that Whitestream (Denis, 1997) people and their society needed to hear other ways of knowing, so that they would know that other ways of knowing-being exist. She was committed to giving voice to Lakota ways of knowing the world, in the radical hope that it might awaken the human beings, that they might hear something that would spark differently, engage a different part of their heart, mind, and spirit—and begin to listen differently, pay attention differently. I hope that writing this chapter and sharing it with you, the reader, honors and continues to bring her ways of knowing, her concerns, and her voice to new audiences, to those who would hear her message, her understanding, her perception, and her vision that she so hoped would make an impact and cause a shift in perception. I know that she hoped to engage with others in an older way, that they might begin to realize that their society's ways of understanding the world might, in fact, be missing something significant. And that the multiple, widespread violences at every level of a society that we see careening wildly out of control are the direct result of histories and choices based in an epistemology that is *anti-relational*. That has little respect, little care, little time, little need for *being* relative, for a world that is relational. She invited, in her soft-spoken way, people to listen. To come back to listening again. To listening that could be done differently, in ways that might begin to re-imagine what it means to live as connected, to re-establish caring, ethical relations with the rest of life, to live present to how we interact with the world, to live as related, a good relative indeed.

A DIFFERENT EPISTEMOLOGY

Different epistemologies. We are in desperate need of a different world-view, a different way of being in and seeing the world. Not a *new* one, lest the human beings become lulled yet again into believing they must chase a new shiny object, one more distraction. The Western way of knowing is forever looking for "the discovery," "the exploration," "the frontier," "the uncharted territory," as if no one else has ever been there, seen this before (regardless of the fact that peoples, and multitudes of beings, have lived there, experienced this, since time immemorial). The violence of such a worldview. When life is viewed as an endless resource to be "discovered," extracted, isolated, engineered, developed, and altered, the inevitable result is violation of Being-ness. It is a worldview that does not allow for—because it does not comprehend—the inherent Being-ness of every part of the natural world.

Relational wisdom emerges from knowledges that have been born out of longevity of experience within places, holding relational instructions that reach back to the beginning of time, in constant, cyclical interaction with each other and all the elements contained within their worlds. In this way, over countless generations, we come to understand the ways we *require each other to become fully who we are,* to be able to fulfill our promise to our Original Instructions of who we are and who we were meant to be.

BEING A GOOD RELATIVE ... AND RESISTANCE

I think for Rosalie, asking this question out of the heart of her Lakota ethical, oral, and embodied wisdom that emerged from within her own Elders' intergenerational wisdom praxis formed part of the core of her resistance and her resilience. Her resistance took many forms, including that quiet assertion of continuing Lakota presence, perception, pedagogies, and practices through Lakota language and philosophy. In asking this question of what it means to be a good relative, from within her Lakota language and knowledge system, she was at the same time resisting settler-colonizing conceptions and definitions of the world, resisting settler-colonizing claims of the right to define and cage the world in their image.

Centering her Lakota epistemologies and ethical principles provided a bedrock of resistance to colonizer conceptions of the world and her existence in it. She drew on this knowledge/knowing as a deep well that nourished her ability to resist ways of knowing that she saw as a brief, violent aberration in relation to the longevity of her Lakota People's presence. She talked about the Lakota ethical system, where generosity is a core cultural value and practice—and how this became so problematic for her People when in encounter with a Western system that is oriented around greed, accumulation, competition, and individualism. "To be a good relative" for her, her parents, her grandparents, and her great-grandparents, as she would say, meant a generations-long continuity of Lakota resistance to dominator paradigms of living and models of life, to predatory Western capitalist culture that views life as its commons to exploit at will, greed at the expense of any and all; a culture, a disease, that reduces the living world to commodity, that cannot envision beyond its linear, single-use view of the world as disposable. A worldview that cannot—and refuses to—envision a coherent, balanced, integrated, interconnected world that needs each other healthy, whole, connected to each other, and communicating with each other in their own unique ways, in order to exist at all, and certainly to thrive. This central, orienting way of knowing the Universe—oneself in the Universe and one's People in the Universe—*as relation* formed the resilient, unshakeable core of her resistance to the ongoing brutality of colonization.

Rosalie's question on Lakota relationality emerges from within the core of Lakota epistemology and views of the world, ways of knowing-being, and orientations to and relationship within a Lakota Universe, their lands and sacred sites, their language, cultural memory, philosophies and teachings, sacred histories, and ceremonial cycles (Holm et al., 2003). It is an invitation to her People, a welcoming them to return to (re)connect to who they are within their cultural memory and knowing, a particular framing centered within their own Lakota worlds of knowledge and knowing that extends far beyond, and will always elude the reach of, settler-colonizing violences.

Somehow—in ways that she navigated as an Elder—her particularly Lakota question was offered simultaneously as an invitation to others, to non-Lakota peoples, a call from deep within Lakota cultural knowing, to wake up—to listen to an older, wiser, more attuned, culturally skilled, skillful listening, whose orientations toward perception are profound and ancient. She shared this question and call to spiritual activism with the world because she loved the world that much.

Rosalie always circled back to the Lakota foundational philosophy and orientation of relationality through her embodied practice of *being in relation*, of *being* a good relative. The teachings she had received from the wisdom praxis of her own parents and grandparents and Elders made her deeply aware of how her People's traditional knowledge systems, practices, and language formed the core of their resistance and resilience—and would also form the heart center of their resurgence, as it had formed the core of the People's strength, courage, and resilience since the beginning of time. Part of the large and abundant work of her life was this reminder, this call, to the beauty and the rich abundance of their Lakota language and cultural ways and ethical principles, and the ways these carried the promise of health, healing, wholeness, and resilience held within them.

RELATIONALITY AND THE FUTURE

Elders like Rosalie have been saying for a long time now that the time has come to wake up and accept that the knowledges the Western world has attempted so long to subjugate, to prove as sub-human, pagan, savage, heathen are those that must now be recognized to hold the key to the continued existence of life on this planet. To be a good relative, it is imperative that human beings learn how to listen differently, from different places. It means practicing radical humility and radical listening, becoming better able to perceive what it means to be a good relative. Radical listening with an open heart means *listening as a good relative*.

What does it mean to embody relationality? I think it means ways of living, walking, breathing, singing, dreaming, comprehending, and perceiving the world that engage with and call upon this quality of Being-ness, this quality of interconnectedness, as one of being in continual interaction and exchange with the universe. This call to relationality suggests an ethical framework that springs from within different, and wholly divergent, worldviews. Indigenous Peoples not only hold on to and carry these ethical frameworks from deep within their epistemologies, they also *live* them as an intrinsic part of their ways of life, languages, knowledge systems, and ways of knowing-being, spirituality, and a deep richness of cultural practices. These imbue every aspect of life, as these continuities are present in cycles of regeneration.

CLOSINGS

Rosalie, her Elders and Ancestors, and generation upon generation of Lakota have been leaning into this question, this central, centering cultural philosophy of what it means to *Be a Good Relative* since their Creation Story. Within their knowledge of the quality of *being in relationship* are embedded protocols of relationality—the relational, collective narrative experience of being in ongoing, intergenerational conversation that is being held in the Lakota Universe and the collective remembering of embodying relationship together. Rosalie talked about how Lakota ways of knowing the world are guided by ancestral relationship with the lands and beings of those places, and by the sacred, by ways of knowing that comprehend and hold the capacity to perceive the natural world in its places as ongoing, continually regenerating conversation, a constantly re-emerging invitation to be in relationship. These conversations and relations are held in Places, on Lands, by Original Instructions, ways of being that orient, invite, carry, and call to being in real ways the necessity, the imperative of relationality—whose core, critically, is love.

Indigenous Elders carry the ancestral collective knowledge systems that are the seed of renewal and regeneration for Indigenous communities. They hold a critical place at the nexus between worlds—Ancestors in one world and younger generations in another—carrying the longevity of knowledge ways, resilience in the face of constant flux, and the ongoing reality of transformation. This, I think, is what Rosalie was guiding us toward. A different worldview that understands us as relatives—something fundamental and inherent within not only Lakota language, but within every aspect of a Lakota orientation to the world and the practice of living—relationality as the critical necessity of Being-ness, without which the world does not, cannot, exist.

This is the worldview and lifeworld from which Rosalie spoke and lived her spiritual activism—her spiritual responsibility to be a good relative to all-that-is. Her walk here among those of us who loved her left us way too early—but I am certain that she continues to help us here to continue to learn what it means to be a good relative. I know she is walking over there now with all her Relations, and occupying a special place among them, I am sure, are those she loved so much—her grandfathers, teachers and leaders, the Buffalo Nation. Maybe they will all continue to help us if we gather our hearts and minds and spirits together in a good way, for the benefit of all our relatives and our coming generations.

All Our Relations.
----
we lay you down
deep in earth
who has always been
your home
we lay you down
in the soft bed
of your mother
you so loved
we lay you down
singing
we lay you down
there
among your relatives
your family
where your people
have lived and walked on
since emergence
into prairie grass
and wind
hills of endless song and sky
song
magpie picks up
tossing it to prairie dog
runs it over
to buffalo
passes her
to lightning
this song they have carried
passing gently in the night
a bundle
wrapped up tight
protected
from prying hands and eyes
of those who refuse
and
intrude on our knowing
of all this
      of all this
            of all this
we sing now

to lay you down
we sing now
to lift you up
we sing now
to send you on
we sing now
to wrap you up
gently
we lay you down
singing
as Buffalo
they are coming
now
hooves pounding earth
for they know
their relative
friend
warrior
who stood up
and walked
    and walked
        and walked
carrying your voice
among the humans
so they would know
who you are
so they would know
who you were
so they would know
you as their home

they are gathering now
there
beyond the hill
of grass and wind
earth shakes
with their steps
they are coming
as we sing
their song
as we sing
you are becoming

their song
now
as we sing
you gently
into earth
to see you
this one
last
time

— for Rosalie

References

Denis, C. (1997). *We are not you: First nations and Canadian modernity.* Broadview Press.

Holm, T., Pearson, J. D., & Chavis, B. (2003). Peoplehood: A model for the extension of sovereignty in American Indian studies. *Wicazo Sa Review, 18*(1), 7–24.

Mohawk, J. (1990, April 18). *Perspective of Mother Earth.* From WinterCamp Chronicles. GENERATIONS Native American Radio Archives. http://twoelk2.tripod.com/Generations/WCCJohnMohawk.html.

CHAPTER 8

The Social Focus Framework: Antiracist and Anticolonial Conscientization, Consequence, and Presencing in Science Education

Anastasia Sanchez

PROPHECIES OF TOMORROW GUIDE US TOWARD TRANSFORMATIVE TEACHING

Thomas Banyacya was a Hopi elder chosen to share Hopi prophecy around the world. In one of his talks in 1995, he shares the image of Prophecy Rock. "This rock drawing shows part of the Hopi prophecy of two paths. The first path is towards life determined by technology that is separate from natural and spiritual law which leads to chaos. The lower path is one that remains in harmony with natural law. Here we see a line that represents a choice like a bridge joining the paths. If we return to spiritual harmony and live from our hearts, we can experience a paradise in this

A. Sanchez (✉)
University of Washington, Seattle, WA, USA
e-mail: sapient@uw.edu

© The Author(s) 2024
S. Tolbert et al. (eds.), *Reimagining Science Education in the Anthropocene, Volume 2*, Palgrave Studies in Education and the Environment, https://doi.org/10.1007/978-3-031-35430-4_8

world. If we continue only on this upper path, we will come to destruction. It's up to all of us, as children of Mother Earth, to clean up this mess before it's too late" (Param, 2020). "The Hopi tell us that this story has repeated itself many times for as far back as human memory reaches, and they have predicted where we will end up if we don't change course immediately" (Thomas Banyacya as cited by McLeod, 2020).

Thomas Banyacya and Prophecy Rock photo credit-Christoper (Toby) Mcleod (2020)

Fast forward to 2022, while in an eighth-grade science classroom during a Social Focus lesson that positioned students to grapple with the question, *how should science and technology serve all communities and the future?*, a student shared, "Ya know, what I really think it's about is right now, I mean, it's like two roads, one where adults care and do something and one they just don't, that will determine my future the most I think."

Each of these profound prophecies locates educators, science educators in particular, as guides and narrators of learning and doing, with the power to steer students toward an Anthropocene of catastrophe or a hopeful otherwise. Within these shared prophecies is acknowledgment of hegemonic structures that funnel the masses onto a path of destruction and chaos via coloniality and white supremacy blocking healthful pathways of kincentric living. These prophecies gift us clarity on the urgency and potential outcomes of our actions/inactions, learning/unlearning, and (un)met obligations to youth, multispecies kinfolx, and planet.

Holding these prophecies as truth, I invite others to consider the trajectory of their science pedagogies, as they materialize consequential realities. It is at this point of conformity and departure that I endeavor to disrupt the shepherding of youth and educators toward settled tracks of destruction and injustice. In this chapter I provide an antiracist and anticolonial framework for science education to act as compass and map to forge new science pedagogical pathways toward collective thriving and liberation in alignment with natural laws.

REMEMBERING FORWARD

Before setting a pathway ahead of us for engaging in dreaming and designing for liberatory pathways toward science education transformation, we must first face the nightmares that colonial schooling has brought to life. This requires *remembering forward* to be critically cognizant of the white supremacy of colonial schooling that has been designed to continuously erase and silence Black, Brown, and Indigenous livingness, rightful knowledges, and ways of knowing—to speculate, dream, and design forward, toward a just otherwise. This means deeply knowing that as replicas of colonial society, schools function as sites that reproduce systems and structures that maintain the education racial contract (Leonardo, 2013). Schools mirror society's hyper-validation of whiteness and coloniality through rhetorics of capitalistic success with a model of learning that is transactional, inflating the value of Eurocentric epistemologies and ontologies. Our educational systems have busied themselves up with achievement gap nonsense and racist disciplinary rate data, failing to see that these outcomes are inevitable when the infestation of white supremacy and coloniality is simultaneously ignored and defended. In short, dominant education has bamboozled the masses from pre-kindergarten to higher education about the purpose and promise of learning and set us on this course of socio-ecological discordance and turmoil.

The field of science education uniquely propels us toward and fuels discordance with its over saturation of utilitarian narratives and progress agendas rooted in settler eco-logics of invasion, extraction, and expansion (Dietrich, 2016). Through positivist messages of innovation within reductive and *settled science learning* (Bang & Marin, 2015), students are conditioned to displace their ontological intuitions and values of relationality as well as reciprocity with white supremacy values of individualism and human supremacy. And while this may secure white interests and comfort for the time being, by failing to heed the cultural and generational science knowledge of BIPGM (Black, Indigenous, and People of the Global Majority) communities who have been long-standing designers of socio-ecologically caring innovation and have endured multitudes of Anthropocenes (Yusoff, 2018), as science educators are we not guilty of shepherding students toward their own peril, both socio-culturally and ecologically? In short, yes; therefore, if we intend to be the elders that our youth and our multispecies kin need for collective continuance (Whyte, 2013), we must identify and overturn epistemic and pedagogical stones that maintain white supremacy and coloniality.

To begin this upheaval and (re)navigation, we start with the critique of practices and advances in science and engineering education that have been heralded as promising yet merely give shallow, and often false, approaches of equity and justice. For instance, the Next Generation Science Standards (NGSS) have been influential in the vertical alignment of science learning for all K-12 instruction; however, it must be asked what knowledge is being taught and who is absent in the authoring of these standards, and toward what definition of equity are they aligned (Rodriguez, 2015)? Furthermore, the marginalization of socio-ecological human impacts within NGSS maintains human dominance and settler innocence, by not providing clarity as to how standards and pedagogical methods are linked to colonialism, coloniality, racism, and white supremacy culture and traits—nor does NGSS identify how these ideological and epistemological orientations, that require ecological exploitation and extraction, are rooted in science and engineering standard-based learning, thereby contributing to the threatening and foreclosing of students' well-being and futurities. New standards, perspectives, and ways of doing, learning, and knowing science and engineering should be authored and implemented that address matters of consequentiality over traditional methods of reductive teaching and learning. Therefore, to reimagine science education in the Anthropocene, guidance is needed that provides principles, dimensions, and enactments of science learning in which content can be set on a trajectory of antiracism and anticoloniality towards a new education Other*wise*.

Remembering forward toward a new trajectory of science learning also calls for critiquing claims of equitable learning and instruction through the implementation of science core instructional practices, also known as high-leverage practices. While core practices such as conceptual models, summary charts, discourse moves, and phenomena-based units may increase repertoires of student participation and inclusion, by being enclosed within dominant culture curriculum and pedagogies, they merely veil Westernized science's epistemic goals and allow instruction to remain race-neutral (Shah, 2021). As we consider science instruction practices and learning, we must ask ourselves toward what end? Toward what purpose are we utilizing instructional practices? For far too long, science instruction and practices have been promoted for their ability to support students' participation in white, settled, transactional learning pathways—learn the fact(s), pass the assessment, get the grade, get access to the white institutional spaces of learning, then join the capitalist society, get the stuff, be successful. This rutted, circular path of false equity teaching practices sets generations of youth on a path toward destruction, destruction that has not only been prophesied but is here.

We are well past the point of accepting the rebranding or rearranging of science education with allegedly neutral standards and instructional practices that support and maintain goals of epistemic and human domination. These approaches within the colonial machinery of schooling busy us up, feeling like action is being taken, but merely mimic the rearranging of chairs on a sinking ship. For science educators committed to radical change and care, the classroom can and should be a place of movement toward (un)learning, unraveling, and undermining logics of white supremacy and coloniality. However, this requires new ways of seeing, knowing, and sensing. To this point, in the first edition of *Reimagining Science Education in the Anthropocene*, the editors cite Cash Ahenakew (2017) who states that "'the work of decolonization is not about what we do not imagine, but what we cannot imagine from our Western ways of knowing' (p. 88)." "We need new ways to (re)open what we can even imagine within science education as we respond to the Anthropocene(s)" (Wallace et al., 2022, pp. 6–7). We need new ways to critically see, sense, and make seen what is missing from science teaching and learning, to move toward "regenerative present futures" (p. 3). Given this understanding and call for transformation, I pose the question: *What would it mean and what would it take for science education to honor BIPGM critical historicities and be beholden to present and future youth, BIPGM communities, more-than-human kin, and Land Air Water Stars[1] (Sanchez, 2023)?*

By applying this question to science education together we can aim to critically notice what is missing and necessary to (re)orient science education toward a pathway aligned to natural laws and toward an acute awareness that our youth are on the front lines in our classrooms, well aware that the tapestry of our global community is unraveling and on fire, but not lost. In hopes of supporting science educators in this endeavor and new trajectory, I offer up the Social Focus framework, which provides three (inter)relational and multidimensional liberatory principles to be leveraged and elevated as guiding standards for desettling and transforming science and engineering learning, prekindergarten and beyond. In addition to the theoretical grounding of the Social Focus principles, I share classroom and curricular enactments, as evidence of the actualized power of utilizing the framework as lens for seeing what is missing, needed, and possible, and as compass to guide the development and implementation of antiracist and anticolonial science education.

The Social Focus Framework: A Principled Approach for the Cultivation of Antiracist and Anticolonial Science and Engineering Education

The Social Focus framework has been iteratively developed, researched, and implemented for over six years alongside science educators and students, in K-12 public school science classrooms in the Pacific Northwest, with the explicit intent to serve as a tool for serving Black, Brown, and Indigenous youth through a commitment to radically care for their onto-epistemic

[1] Land Air Water Stars is a term used to attend to the problematic ways that dominant, Westernized culture collapses Lands, Air, Waterways, into terms such as nature and environment, disconnected from Stars, in effort to honor each entity and connection I capitalize each while placing them together.

security (Bang & Marin, 2015). The Social Focus framework aims to call out, desettle, and counter the ways dominant science instruction and materials habitually decontextualize science and science learning from the complex socio-political-ecological entanglements, temporal tapestries, and global sinew in which they pulsate. By purposefully embracing radical transdisciplinary boundary crossing, the Social Focus animates and stories science to weave critical and relational knowledge, with multiple ways of knowing so students focus their learning on worlding, dreaming, and creating worlds that sustain collective thriving. The Social Focus seeks to foster critical consciousness by unapologetically elevating socio-ecological matters of consequential concern and antiracist, anticolonial counternarratives through the critical and liberatory presencing of knowledges, values, and brilliant beingness of those who have been made absent.

PRINCIPLE 1: CRITICAL CONSCIOUSNESS

Critical consciousness has been a long-standing endeavor and theoretical lens in the analysis of education in multi-scalar ways. Channeling Paolo Freire (1985), critical consciousness, or *conscientization*, is the development of critical awareness of social realities that determine the conditions and possibilities of living and defining one's own reality. Critical consciousness also encompasses awareness of others' realities and requires critical reflection and reflexivity of social inequities forced upon marginalized peoples by white-dominant culture and ongoing coloniality.

All too often, students are limited by teachers' settler consciousness (Kulago, 2019), animated by adult supremacy logics, fakequity, and standardized learning access agendas, which together maintain a course of ontological and epistemic colonization and racialization. In the classroom, a pedagogical goal of critical consciousness is to design learning to dissolve boundaries of empire which decontextualize content from the known and unknown socio-political and ecological realities of students' lived worlds during wicked times. To do so would require the transformation of dominant science education by unmasking the multi-scalar ways its inhabitation within society and the classroom perpetuate harm by concretizing socio-racial hierarchies and erasure—the invisibilizing of diverse knowledges and divergent ways of thinking, being, and doing science. To be critically conscious as an educator means to tear oneself away from the narcotic haze of white normativity that engages educators and students in relationships that further necropolitical practices which maintain within our academic system and science learning BIPGM students, multispecies kin, and Land Air Water Stars (Sanchez, 2023) are and have been set up to die, to fail, to be sacrificed. This haze dulls the ability to see beyond fake-equity rhetorics of participation and access to a false meritocracy as salvation, as if participation and access are anything but harmful and leading toward pathways of prophesied destruction. Critical *conscientization*, as it is taken up in the Social Focus framework, dimensions and pedagogical approaches, requires an eyes and heart wide-opening of productive clarity to care for ushering youth toward healing and critical hope in our classrooms. This would mean that learning is (re)purposed to center the loving care of students' "intellectual health" by providing students "healthful ways to deliberate about the world and to think about the world" and to think about the consequences and "challenges of Eurowestern systems, things like decolonization, things like understanding colonial systems, but also things like, how is it that we generate anew, always?" (Bang, 2020, 20:06–20:36).

By setting critical consciousness engagement and nurturing as our purpose, science and engineering education can hold greater possibility to disrupt settler colonial logics through critical dialogic conversations. Elevating critical consciousness as a classroom purpose sets a learning trajectory of mutuality that promotes radical caring (Hobart & Kneese, 2020) and healing because "rarely, if ever, are any of us healed in isolation. Healing is an act of communion" (hooks, 2000, p. 215). The Social Focus framework asserts that science learning can and should aim to nurture students' critical consciousness not merely to increase understanding about social and ecological injustices but rather in the hopes of "forging critical consciousness as part of creating a world we can get behind rather than only describing the one we reject in front of us" (Said, 1983, p. 234). Educators committed to this critical hope engage students in this just worlding project by designing science learning that attends to the critical consciousness dimensions of self, others, and society. Within these dimensions, students—and teachers—are able to see themselves as vital to the perpetuation of harm or healing by promoting a sense of reciprocity and valuing relationality with human and more-than-human kin. Critical consciousness enactments are actualized by directly connecting science content to complex socio-ecological contexts. The unfolding of contexts and the content are then investigated in the classroom as a radical learning community. Learning enactments explicitly decenter whiteness and coloniality through critical considerations about how and why things are the way they are and speculation about how science can and should be considered and carried out for the purpose of relationality, reciprocity, and responsibility.

The following are the dimensions and enactments of the Social Focus Framework's first principle, critical consciousness:

- **Dimension: Awareness of Self**

  - Science learning provides opportunities for engaging in critical self-reflection, critical noticing and developing self-awareness about one's identity in society to promote personal obligation, agency, and activism by positioning learners as essential contributors to science for complex problem solving as co-creators of change.

- **Dimension: Awareness of Others**

  - A sense of interconnectedness is developed by providing learning opportunities that position learners to value the experiences, stories, truths, and historicity of others in relation to their own identities, and holistic development through critical reflection, radical caring speculation, and anti-racist and anti-colonial science education.

- **Dimension: Awareness of Society**

  - Science learning fosters the development of critical and complex views of society through crriculum and instruction that explicitly promote awareness of, and caring for racial justice and multispecies/ecological well-being as a necessary justice priority for global well-being.

To elucidate the power of designing science and engineering learning that cultivates critical consciousness, I offer the following vignette. This vignette occurred in a seventh-grade classroom during a geology unit that was developed using the Social Focus framework. The Social Focus unit centered the question, "How and who should determine the value and use of Land?" This question situated students to critically consider the issue of having Bears Ears National Monument opened for uranium mining. Students moved through various stations, which provided a range of stakeholder perspectives, including those of the Ute tribe, archeologists, geologists, and government officials, using various authentic sources, including first-person videos, podcasts, and articles about the youth Land back movement. The following vignette was captured during a stakeholders' conversation between three white males about the push for uranium mining by the U.S. government:

S1:   "White dudes, we just do whatever we want."

S2:   "Seriously, oh, here's some Native land, yup I'll take that, just because I can…"

S3:   "Yeah, and you know what I am going to do with that land is make bombs!" (student mocks an evil laugh)

S1:   "Yay, I'm a white guy, a total d**k, great, but why?"

S2:   "We gotta be better, the Ute tribe, should have the land, I mean, it is their land, you know what I mean."

Student dialogues such as this vignette rippled through the classroom, with students deeply critiquing settler logics of expansion and progress and considering ways their identities are mirrored in our racially stratified society. While this lesson was placed at the beginning of the unit, this knowledge and further learning that nurtured critical consciousness were carried through the entirety of the unit. Ultimately, the nurturing of students' critical consciousness is directly related to, and dependent upon, the ways in which science learning is situated in socio-ecological contexts of consequentiality, as these contexts are defined and considered by the BIPGM communities most impacted by white supremacy and settler colonialism. Thus, the next principle of the Social Focus framework is consequential concern.

Principle 2: Consequential Concern

Years ago, during a lesson on clouds, a Samoan student interjected, "No disrespect, but I just don't get why we should care about clouds, we got real sh*t to worry about." And as I looked at my classroom, with black garbage bags covering the windows, gang affiliation symbols carved into the desks, knowing the lived and looming turmoil outside the classroom, I knew my response could not be, "Well because it is science," or "So you could get a good grade," or even, "Because weather and climate affect all of us." I realized, despite my institutional learning, science knowledge, and teaching degree, it was my obligation, not students', to connect science learning to multigenerational and temporal matters of consequentiality—sociopolitical and socio-ecological concerns facing and impacting students' communities, pasts, presents, and futurities.

Efforts to appropriately and holistically (re)evolve science learning to attend to consequential concerns of socio-ecological, cultural, and generational gravity requires critical attunement to positivistic and utilitarian narratives of Westernized-settler science education which feign objectivity and neutrality. Attunement to these falsehoods brings opportunity to pinpoint how and when science education has been mechanized for transactional learning, which purposefully ignores the wicked entanglements of science, ongoing racial and colonial harm, and need for future-facing radical care and reality-based learning. White normative science and engineering severs learning from consequential contexts and relation(s) with others and Othered, producing sterilize(ed) and malnourish(ed/ing) learning and progress agendas, void of radical livingness, and lived affects and effects of felt science (Wallace et al., 2022) and felt knowledges (Harjo, 2019). To do so requires that science educators become productively mindful and critically response-able (Higgins, 2021; Higgins & Tolbert, 2018) of the consequential concerns, consequential realities, consequential historicities, and consequential being(s)(ness) that embodied science promise and of the precarity in BIPGM lived worlds, the Land Air Water Stars, and our multispecies kin.

The Social Focus pedagogical framework, with roots in consequential learning (Jurow & Shea, 2015; Bang & Vossoughi, 2016), calls for consequential science instruction and materialities through *critical-reality pedagogies* (Sims, 2018) and *critical response-ability* (Kayumova & Tippins, 2021). Critical-reality pedagogies "not only looks at the macro level injustices, à la critical pedagogies, but it also positions students to understand, identify and begin to deconstruct and subsequently redress individualized issues while also helping students realize that the individualized, localized injustices that inform their lived experiences are part and parcel of a larger, macro-level system of oppression that is disproportionately injurious to poor people of color" (Sims, 2018, p. 9). Critical-reality pedagogies are concerned with the actions that are needed for students to thrive and to "develop critical analytical thinking so that they can use that knowledge to shift the socio-political constraints that oppress them" (Sims, 2018, p. 9). Complementary to critical-reality pedagogies, critical response-ability entails creating locations of possibility for the design of socially, culturally, environmentally sustainable and just learning contexts. Drawing from Karen Barad's theorizing of response-ability, Kayumova and Tippins (2021) state that critical response-ability should reopen STEM education toward its responsibilities so that "Black, Brown, Latinx, and Indigenous young people are recognized as authors and owners of their existing and emerging knowledge that they co-construct in affective and embodied ways within the complex web of human and more-than-human relations" (pp. 825–826).

Given the horrific events that are ongoing, unfolding, and yet to be unearthed due to interlocking systems of domination, as educators, do we not have an obligation to grapple with consequential matters alongside students? By not taking up matters of consequence, are educators and systems of education guilty of onto-epistemic injustice/violence and inter-generational incompetence? Is it not true that it is students' epistemic right to receive an education that best prepares them to navigate and spec-ulate about current realities and yet-to-come realities, especially given the weight of the roles they will inherit? *Pero, ya basta, es la tiempo para liber-tad, para todo gente y relaciones.* For science and engineering to be conse-quential requires fierce epistemic and pedagogical shifts away from normative, Eurocentric education norms that decontextualize learning from relational livingness on a shared planet.

While it is necessary to provide a compelling argument for consequen-tial learning in science that is grounded in theories of critical-reality peda-gogies and critical response-ability, I also hold the words of Fred Hampton, "Theory's cool, but theory with no practice ain't shit. You got to have both of them—the two go together" (1969). Therefore, I contend and offer up that teaching practice must be a practice of countering settled enclosures of science education through instructional moves that make vulnerable the positivism of Western modernity narratives and decenter whiteness through expansive student-led critique and contestation. In this vein, the Social Focus framework asserts that all science and engineering learning/units should be foregrounded and centered in "should we ques-tions" (Bang, 2020). By positioning students to engage in "should we" questions, educators engage in countering deficit frames and developmen-tal theories of youth as incomplete humans—or not-yet adults (Nxumalo, 2015). By posing consequential "should we" questions in science class-rooms we recognize youth as agentic beings and as vital, critically aware stakeholders in a world torn asunder from colonial projects and white supremacy, yet not without profound love and possibility. Clarification is needed about what is meant by "should we" questions. This means that all science units have a consequential concern that positions students to grap-ple with an expansively written question, authored to attend to each of the dimensions and enactments of the Social Focus framework. The Social Focus framework provides science educators guidance with identifying the consequential concerns that avoid the centering of white supremacy culture, concerns, values and comfort. The enactments are rendered as offerings and invitations to embark upon a path of onto-epistemic, critical antiracist and anti-colonial hope and justice.

The dimensions of Principle 2, consequential concern, include:

- **Dimension: Matters of Justice and Cultural Significance**

  - Science learning is designed to investigate contextualized issues of consequence that threaten non-white communities' rights and cultural ways of knowing by addressing systemic oppression and white supremacy, which limits access to resources, power, and physical and ecological security and well-being.

- **Dimension: Matters of Relational and Collective Well-Being**

  - Instruction and learning disrupt Westernized dominant paradigms, narratives, and practices of science that depoliticize science learning and promote human supremacy. Content moves beyond reductive "science for science's sake" furthering white progress narratives and traits and instead calls for critical responsibility and reciprocity. Teaching centers antiracist and anticolonial counter-narratives and methodologies of science that are relational, interconnected, and for collective thriving.

- **Dimension: Matters of Futurity & Ecological Caring**

  - Topics have social gravity as they position students to make/ nurture caring connections between science content and socio-ecological consequential concerns facing society. Learning aims to be generationally relevant and future-leaning as classes investigate topics that have significant impacts on the future well-being of society, ecosystems, and marginalized communities and cultures.

All branches of science taught in schools are interconnected and inherently, temporally, and geographically enmeshed in matters and contexts of monumental, consequential concern. For example, middle school science standards require teachers to cover the phases of the moon and sun. Using the Social Focus approach to science learning, eighth-grade teachers in Seattle have been exercising critical responsibility by connecting the content with the investigation of global and local socio-ecological impacts and causes of tidal flooding. These teachers did this by engaging students to co-think about the expansive question: *How are communities in Venice, Italy, and Miami impacted by tidal flooding, and what should be done?*

Just as critical consciousness about one's positionality and the positionality of others within various socio-political scales requires caution to avoid the centering and valuing of whiteness, the elevation of consequential concerns also risks being narrated and perceived from a lens of whiteness. Therefore, moves toward antiracism and anticoloniality necessarily nurture the (re)forming of connective tissue with others and Othered— Othered knowing, beingness, and mattering across timelines and a multiplicity of BIPGM justice projects (Tuck & Yang, 2018). In an effort to avoid falling into defaults of centering dominant culture values, perspectives, and knowledge, educators must be diligent about the critical and liberatory presencing of BIPGM peoples, Land Air Water Stars, and multispecies kin.

PRINCIPLE 3: CRITICAL AND LIBERATORY PRESENCING

Ultimately, teaching is storytelling: teachers tell the stories of facts and knowledge they have accumulated, either by living life and/or through academic institutions. As Shirley writes, "Teachers are storytellers who decide what to include in their curriculum" (p. 11). Understanding this to be true, coupled with the fact that most science and engineering educators are white, it is fair to say that even with the best intentions, by default, narratives of BIPGM, multispecies and Landir Water Stars kin will be negatively skewed due to white supremacy and colonial social programming of domination and exploitation of all, affirming their identities and protecting white innocence. Therefore, without explicitly storying learning with liberatory counternarratives of BIPGM communities, Land Air Water Stars, Earth kin and natural forces, their subordination will continue, impacting us all. Truly attending to justice in the classroom requires the desettling of false settler narratives that reify white identities as the originators, makers, and doers of science and engineering. To do so means designing and developing learning based on the principle of critical and liberatory presencing, which centers the rightful representation of the language; multiple ways of knowing (Warren et al., 2020); and historical, ongoing and global science contributions and perspectives of BIPGM peoples and cultures, contextually throughout learning.

Critical and liberatory presencing necessitates the discontinuing and demystifying of beliefs that racism will or can decline simply with anti-bias training, and/or access models and methods of diversity, equity, and inclusion within white-dominant culture spaces. These harmful approaches risk the perpetuation of tokenization, cultural and racial gaslighting, as well as extraction and exploitation of BIPGM as a human workforce resource for nuanced modes of settler comfort and saviorism. In schools this is often done with social justice accessorizing of dominant curriculum, or relegated to monthly "celebrations" of non-white cultures and/or by having students' or teachers' report on racial and *cultural unicorns*—individuals touted as having made it despite the odds. Of course, the odds are never identified as white supremacy culture and systemic oppression, sending a vile subversive message that it is not the structures in which we live in that have so many BIPGM struggling but rather their character, and thus sowing seeds of internalized racism and Othering. Within the materiality of the classroom, false equity practices are often dressed up as white gaze "cultural" approaches (Paris, 2019) and/or language accommodation approaches that are mere addendums, optional extensions, and cultural accessorizing of dominant curriculum—pretty paintings on the walls of empire. Inclusion and knowledge making-building without structural change that makes present that which has violently been made absent, merely reifies internal and external racism and anti-Indigeneity which are the foundation of deficit frames of BIPGM students and peoples.

Additionally, critical and liberatory presencing is also not about the inclusion of justice issues impacting Black, Brown, and Indigenous communities in ways that (re)victimize or essentialize them through damage-centered narratives (Tuck, 2009) which fail to disrupt racial injustice, much less center their liberation. Instead, the principle of critical and liberatory presencing asks educators to engage in the refusal of representation and storying of BIPGM as communities inherently bound to suffering, by replacing such narratives with, dignity-conferring and rights-generative (Espinoza & Vossoughi, 2014), honest representation of their legacies of and lived-living brilliance, joy, and relationality, through a process of *refiguring presence* (Nxumalo, 2016). Refiguring presencing as an orientation in "curriculum-making that does not shy away from the oppressive realities of settler colonialism and anti-blackness, yet simultaneously includes speculative curriculum-making practices that seek an otherwise decolonial future" (Nxumalo, Vintimilla & Nelson, 2018, p. 448). Refiguring presences, as a theory of change, invites educators and researchers to engage in the unsettling of what is seen as belonging in curriculum and critically consider making "what is invisible noticeably absent so that it can be remembered and missed" (Ahenakew, 2016, p. 337). This requires radical (re)searching and seeking out erased stories of Black, Brown, and Indigenous people and their relationality with more-than-human thriving. To be clear, the storying of science through critical and liberatory presencing means seriously countering white settler temporalities which make impossible the temporal co-presencing of Black, Brown and Indigenous peoples. This countering makes BIPGM peoples boldly present, not as myths or legends but as truths of the past, with embodied, self-determined and active nowness and futures of corporeal, spatial, spiritual and onto-epistemic sovereignty secured. Critical and liberatory presencing provides BIPGM students opportunities to safely, identify and (re)connect themselves and their cultures to scientific brilliance led by Black, Brown, and Indigenous communities currently and since time immemorial as acts of reclamation of birthright, as legacy. It must also be said that by decentering whiteness in the classroom, critical liberatory presencing disrupts the hyper-macro-affirmation of white students, which fails to nurture their humanity and ability to see BIPGM people as holistic beings-impeding their/the possibility for authentic and meaningful relationality and collectivity.

Therefore, if we are to imagine science and engineering learning as a methodology for moving toward pathways of being and knowing in alignment with natural laws and collective relationality and reciprocity, the Social Focus framework guides educators to design and develop learning based on the following dimensions of critical liberatory presencing.

- **Dimension: Restorative Justice-Oriented Representation**

    - Instruction and learning names the multi-dimensional, intersectional, injustices faced by racially, linguistically & ability diverse communities. It situates their experiences, solutions, healing and thriving throughout the context of science learning as restorative justice priorities. Designing, development and instruction prioritizes cultural resurgence (Bang et.al., 2012) and the desettling of STEM by centering justice goals that care for the futurity of BIPGM communities. Woven throughout learning, across units, the ways communities (human & more than human) have been, and continue to be extracted from, exploited and erased/marginalized due to embodies and enactments of colonization & white supremacy, thusly foreclosing lush potentialities for collective sustainability throughout time and in all spaces.

- **Dimension: Rightful Representation**

  – Science instruction and materials respectfully and rightfully ele-
  vate BIPGM scientists/engineers/leaders/cultural knowledge
  keepers as essential sources of expertise for complex and consequential learning and speculating and designing of solutions and
  innovations. Dignity-conferring and rights generative representations of BIPGM change makers are contextualized throughout the
  learning as best practice, as norm for generationally relevant, and
  responsible science instruction. Critical analysis is given to representation to ensure that problems of practice such as tokenization,
  essentializing, and (re)victimizing of marginalized communities
  are not perpetuated.

- **Dimension: Self-determined Representation**

  – Counters harmful narratives that pre-determine the lives of
  BIPGM, ecologies, and ecological kin by highlighting materials,
  research, innovation(s), knowledge, and stories from authentic
  sources/documentation generated by BIPGM and more-than-
  human communities as essential to understanding and advancing
  science. BIPGM students engage in self-determined representation and presencing of themselves within the learning content and
  context and as embodying epistemic sovereignty in STEM.

For context and clarity, I provide the following example of critical liberatory presencing that has been taken up in several local second-grade
classrooms. The unit required teachers to cover the concept that Plants
need Water, Sun, and Animals for Seed dispersal and growth. District-
mandated curriculum was identified as being absent of the cultural ways of
knowing and caring for Seeds and real-time consequential concerns, nor
did it identify humans in relation to plants, animals, or seeds, negatively or
otherwise. By designing learning in alignment to the Social Focus principle of critical liberatory presencing, students' learning was grounded in
the antiracist and anticolonial learning about African, and Central
American, Seed Guardians. Students learned about the radically loving
action of seed guardians from Africa, stolen as agricultural experts, weaving seeds into their hair, who knew/know that to love current and future
generations is to care for and love seeds (Penniman, 2018). Due to this
learning, students were able to ask about why people in Africa were stolen
and be told that they were stolen because of racism and colonialism, taken
for their knowledge of the Land—how to care for and be cared for by

Land. This provides students a counter story to slavery in white–Black history, which simplifies the reasons for the enslavement of Black peoples while collapsing those stolen as a mass of peoples, without a multitude of gifts, stories, brilliance, and promise. By centering the science and activism of seed guardians, students were positioned to challenge settler eco-logics of expansion and Land exploitation by researching why Seeds are endangered. This research led to classroom experiments on how overdevelopment "on stolen Land" (second-grade student) and the use of harmful agricultural practices impact seeds, rather than the mandated curriculum which had students learn that Plants need Sun and Water, facts they already knew. And with stakeholder lessons woven throughout the unit, students were positioned to also learn about seed guardians from Central America, as leaders and scientists addressing growing climate change realities, Students were then able to deeply and agentically consider the inter-dependent-relational, potentially reciprocal, bonds between LandAirWate rStars, human and more-than-human.

Teaching for the critical liberatory presence of Black, Brown, and Indigenous knowledges and ways of being, doing, and thriving fills the void of colonialism separatism with kinship and radical care for all Earthly beings measured by the securing of BIPGM futurities. And while the three Social Focus principles are presented individually, as is evident in each of the examples, each principle, dimension, and enactment are fortified by and built upon the others. Therefore, the framework should be taken up in its totality, interconnectedly. To do so would shift the way that science is felt and how it is either leveraged as a mechanism to maintain a course toward destruction or as a compass toward relationality, reciprocity, and alignment with natural laws.

Conclusion

Currently, we are neck deep in the muck of a convergence of many battles: battles to maintain the classroom as white property rather than infuse truth into the learning through critical race theory-based pedagogies, and battles between wielders of settler colonial extractivism causing ecological devastation and brave, Indigenous Land and Water Protectors and allies. And at the center of these ongoing battles that have seemed to reach a relational and ecological tipping point (Whyte, 2020) are the youth, future generations and our multispecies and natural relations. Given these urgent and ongoing realities, science education that severs the visceral connections between content and the socio-political-livingness and ecological-livingness in which science is situated is an insidious act of onto-epistemic harm and irresponsibility, further leading students toward a pathway of prophesied destruction.

Therefore, the Social Focus pedagogical framework takes seriously the stance, "If we do not do this work, if we do not collaboratively call into question a system of knowledge that delights in accumulation by dispossession and profits from ecocidal and genocidal practices, if we do not produce and share stories that honor modes of humanness that cannot and will not replicate this system, we are doomed" (McKittrick, 2020, p. 73). The Social Focus Framework, principles and dimensions, and embedded learning enactments shared contend and illuminate how science and science educators have the profound obiligation and capacity to transform our shared world and move us toward a path of liberation. Liberation toward worlding an Otherwise that embraces the un-dooming of futurities and youth through the reclamation of alignment with natural laws as the path towards antiracist and anticolonial collective well-being.

References

Ahenakew, C. (2016). Grafting indigenous ways of knowing onto non-indigenous ways of being: The (underestimated) challenges of a decolonial imagination. *International Review of Qualitative Research, 9*(3), 323–340.

Ahenakew, C. R. (2017). Mapping and complicating conversations about indigenous education. *Diaspora, Indigenous, and Minority Education, 11*(2), 80–91.

Arada, K., Sanchez, A., & Bell, P. (2023). Youth as pattern makers for racial justice: How speculative design pedagogy in science can promote restorative futures through radical care practices. *Journal of the Learning Sciences, 32*(1), 76–109.

Bang, M. (2020). *Central challenge of the 21st century: Cultivating just, collectively adaptable, sustainable, and culturally thriving communities.* Two Feathers—Native American Family Services.

Bang, M., & Marin, A. (2015). Nature–culture constructs in science learning: Human/non-human agency and intentionality. *Journal of Research in Science Teaching, 52*(4), 530–544.

Bang, M., & Vossoughi, S. (2016). Participatory design research and educational justice: Studying learning and relations within social change making. *Cognition and Instruction, 34*(3), 173–193.

Dietrich, R. (2016). Made to move, made of this place: "Into America", mobility, and the eco-logics of settler colonialism. *Amerikastudien/American Studies, 61*(4), 507–525.

Espinoza, M. L., & Vossoughi, S. (2014). Perceiving learning anew: Social interaction, dignity, and educational rights. *Harvard Educational Review,* *84*(3), 285–313.

Freire, P. (1985). *The politics of education: Culture, power and liberation.* Bergin and Garvey Publishers.

Hampton, F. (1969). You can murder a liberator, but you can't murder liberation. [Transcript] Retrieved from https://www.marxists.org/archive/hampton/1969/04/27.htm

Harjo, L. (2019). *Spiral to the stars: Mvskoke tools of futurity.* University of Arizona Press.

Higgins, M. (2021). The Homework of Response-Ability in Science Education. In: Unsettling Responsibility in Science Education. Palgrave Studies in Educational Futures. Palgrave Macmillan, Cham. https://doi.org/10.1007/978-3-030-61299-3_2

Higgins, M., & Tolbert, S. (2018). A syllabus for response-able inheritance in science education. *Parallax, 24*(3), 273–294. https://doi.org/10.1080/13534645.2018.1496579

Hobart, H. I. J. K., & Kneese, T. (2020). Radical care: Survival strategies for uncertain times. *Social Text, 38*(1), 1–16.

hooks, b. (2000). *All about love: New visions.* New York: William Morrow.

Jurow, A. S., & Shea, M. (2015). Learning in equity-oriented scale-making projects. *Journal of the Learning Sciences, 24*(2), 286–307.

Kayumova, S., & Tippins, D. J. (2021). The quest for sustainable futures: Designing transformative learning spaces with multilingual Black, Brown, and Latinx young people through critical response-ability. *Cultural Studies of Science Education, 16*(3), 821–839.

Kulago, H. A. (2019). In the business of futurity: Indigenous teacher education & settler colonialism. *Equity & Excellence in Education, 52*(2–3), 239–254.

Leonardo, Z. (2013). The story of schooling: Critical race theory and the educational racial contract. *Discourse: Studies in the Cultural Politics of Education, 34*(4), 599–610.

McKittrick, K. (2020). *Dear science and other stories.* Duke University Press.

McLeod, T. (2020, April 4). Hopi Prophecy-A Timeless Warning. Sacred Land Film Project https://sacredland.org/hopi-prophecy/

Nxumalo, F. (2015). Forest stories: Restorying encounters with "natural" places in early childhood education. In V. Pacini-Ketchabaw & A. Taylor (Eds.), *Unsettling the Colonial Places and Spaces of Early Childhood Education* (pp. 21–42). New York: Routledge.

Nxumalo, F. (2016). Towards 'refiguring presences' as an anti-colonial orientation to research in early childhood studies. *International Journal of Qualitative Studies in Education, 29*(5), 640–654.

Nxumalo, F., Vintimilla, C. D., & Nelson, N. (2018). Pedagogical gatherings in early childhood education: Mapping interferences in emergent curriculum. *Curriculum Inquiry, 48*(4), 433–453.

Param, A. (2020). Hopi prophecy for our times: Could we follow the other path—To live from our hearts? *Medium.* https://medium.com/middle-pause/hopi-prophecy-for-our-times-ef6563a6531

Paris, D. (2019). Naming beyond the white settler colonial gaze in educational research. *International Journal of Qualitative Studies in Education, 32*(3), 217–224.

Penniman, L. (2018, December 6). *Farming while Black: A legacy of innovation and resistance.* 38th Annual E. F. Schumacher Lectures [Video]. YouTube. https://www.youtube.com/watch?v=gw6eG1rQapo

Rodriguez, A. J. (2015). What about a dimension of engagement, equity, and diversity practices? A critique of the next generation science standards. *Journal of Research in Science Teaching, 52*(7), 1031–1051.

Said, E. (1983). *The world, the text, and the critic.* Harvard University Press.

Sanchez, A. (2023) Just worlding design principles: childrens' multispecies and radical care priorities in science and engineering education. Cult Stud of Sci Educ. https://doi.org/10.1007/s11422-023-10197-w

Shah, N. (2021). *Racial equity and justice in teaching and teacher education: Progress, tensions, and open questions.* Spencer Foundation.

Sims, J. J. (2018). *Revolutionary STEM education: Critical-reality pedagogy and social justice in STEM for black males.* Peter Lang Incorporated, International Academic Publishers.

Tuck, E. (2009). Suspending damage: A letter to communities. *Harvard Educational Review, 79*(3), 409–428.

Tuck, E., & Yang, K. W. (2018). *Toward what justice. Describing diverse dreams of justice in education.* Routledge.

Wallace, M. F. G., Bazzul, J., Higgins, M., & Tolbert, S. (2022). *Reimagining science education in the Anthropocene.* Palgrave Studies in Education and the Environment. Palgrave Macmillan. https://doi.org/10.1007/978-3-030-79622-8_1

Warren, B., Vossoughi, S., Rosebery, A. S., Bang, M., & Taylor, E. V. (2020). Multiple ways of knowing*: Re-imagining disciplinary learning. In *Handbook of the cultural foundations of learning* (pp. 277–294). Routledge.

Whyte, K. P. (2013). Justice forward: Tribes, climate adaptation and responsibility. In *Climate change and indigenous peoples in the United States: Impacts, experiences and actions* (pp. 9–22). Springer International Publishing.

Whyte, K. (2020). Too late for indigenous climate justice: Ecological and relational tipping points. *Wiley Interdisciplinary Reviews: Climate Change, 11*(1), e603.

Yusoff, K. (2018). *A billion black Anthropocenes or none.* University of Minnesota Press.

CHAPTER 9

Breaking the Paradigm: Storying Climate Change

Vandana Singh

Around mid-August in 2016, a week of incessant rain caused seven trillion gallons of water to fall on Louisiana. The resulting flood killed thirteen people, including a woman called Stacy Ruffin, who had been driving with her mother and a neighbor to see her dying brother in hospital. On their way back, the floodwaters swept the truck away. Her neighbor and mother survived, but Stacy was lost, leaving behind two children and a grieving family.

Stacy was forty-four years old; a loving mother and daughter, she lived with her mother and children in a mobile home and worked at the deli counter in a nearby Walmart. The story of her death, as recounted by CNN reporter John Sutter in "What Killed Stacy Ruffin?" (Sutter, 2017), is not only an account of an African American family dealing with a devastating double loss. What could once be understood as a natural disaster turned out to be a storm made 40% more likely because of climate change.

V. Singh (✉)
Department of Environment, Society and Sustainability,
Framingham State University, Framingham, MA, USA
e-mail: vsingh@framingham.edu

© The Author(s) 2024
S. Tolbert et al. (eds.), *Reimagining Science Education in the Anthropocene, Volume 2*, Palgrave Studies in Education and the Environment, https://doi.org/10.1007/978-3-031-35430-4_9

Sutter's article includes interviews with climate scientists in the area of weather attribution—that is, finding the climate footprint in extreme weather events—as well as interviews with the bereaved family. At the end it asks the question—if an extreme weather event is partly the result of anthropogenic climate change, and we can no longer consider it a chance misfortune or an "act of God," then what killed Stacy Ruffin? Who, or what, is responsible? Can we blame fossil fuel companies or the millions of us who drive gasoline-powered cars?

Such questions also arise in other parts of the world. One of the most climate-vulnerable states in India is Jharkhand, where climate projections include rainfall variability, always disastrous for agriculture, and heat waves. According to the Jharkhand State Climate Adaptation Plan, the heat waves are already happening: in one year alone (2010) there were a hundred of them. The region is vulnerable also because of its extreme poverty—it tends to have the lowest Human Development Index scores among Indian states.

In 2018 I had the chance to speak by phone to a woman called Parvati, who lives in a village in Jharkhand. She had to travel five kilometers before she could speak with me on a borrowed phone. She told me this remarkable story.

About twenty years ago, the water table in the area began to drop, following large-scale deforestation in the region. As a result, the two sources of survival for the villagers, forest produce and crops from subsistence agriculture, became unreliable. The thick, contiguous forests of the region, populated by a vast and diverse range of species including bears and tigers, were being cleared for mining, roads, and other projects of development. Summer temperatures began to rise, and malaria became widespread. In desperation, Parvati and other village women decided to do something about the problem.

They began by patrolling their dwindling local forest every morning in groups of three, confronting and driving away would-be loggers, a task that was sometimes dangerous. Understanding that a healthy forest is a biodiverse forest, they dug ponds for the animals and birds. They built mud check dams for the streams, ensuring that the water didn't escape to the desertified landscape outside and evaporate. Twenty years later, as confirmed by a friend who visited the area and made my conversation with Parvati possible, the forest is thriving. Tree trunks have thickened, streams are flowing, many animals and birds are back (but not the tigers). The two hundred hectares of regenerated forest have, Parvati says, restored water security, cooled local summer temperatures, and significantly reduced the malaria menace. This is despite the fact that deforestation and desertification have increased sharply in Jharkhand in the last few years. The women continue their work today, waking up early in the morning to patrol their forest. When asked what motivates her, Parvati says she is doing it for "our people and the animals and birds." A woman with no formal education, living in an impoverished village in a climate-vulnerable region, unlikely to be lauded on glossy magazine covers as an environmental champion, she speaks with passion, authority, and an infectious ebullience.

The climate crisis is a problem that confounds our usual frameworks for making sense of the world. This is, perhaps, one reason why education has not realized its potential as a climate mitigation tool; on the contrary, mainstream education is far more likely to reinforce the status quo. Among five roadblocks (Kwauk, 2020) that hobble education at the macro level is the lack of radical visions for climate pedagogy. This is an onto-epistemological problem at its root.

As an educator trained in theoretical physics, I began teaching climate science in my general physics college classes about twelve years ago. My motivation was primarily ethical, as young people are among those disproportionately affected by a climate-changed world. I wanted my students—mostly first generation and racially diverse, in an Eastern US public campus—to be informed change-makers in the face of an uncertain future. When I found that several of my students were, instead, becoming frustrated, angry, despairing, and apathetic, I realized that I could not *just* teach the science. Clearly, I had many more lessons to learn from the climate problem. Thinking of the *climate problem as teacher* helped me re-orient myself so that I could learn not only from climate scientists but also from the experiences of people at the forefront of climatic and other crises, and through them, from the non-human elements of the Earth system as well.

Stories like the ones I have related above led me to the formulation of a transdisciplinary, always-developing, radical pedagogy of climate change, about which I have written elsewhere (Singh, 2021). This pedagogy is built upon four dimensions of the climate problem: the scientific-technological, the transdisciplinary (which includes the socio-economic-ecological context), the onto-epistemological, and the psychosocial action dimension. Justice is the connective tissue of this approach. I make no claims to universality of application; instead, my hope is that this approach will add to and stimulate multiple transdisciplinary pedagogies in different geographies and contexts. In this chapter I focus on the onto-epistemological dimension, foregrounding the role that stories of different kinds can play when considering how justice and power are entangled with the climate issue.

If we consider the two stories above, some lessons become evident. The climate problem is a planetary phenomenon, yet it is felt and experienced locally, in very different ways. It also spans vast temporal scales—from a slow rise in temperature since the Industrial Revolution to the sudden fury of a storm made more likely by climate change. It is inherently transdisciplinary—the relevance of the socio-economic context and the history of colonialism and racism become apparent through the detailed examination of both these stories. The stories bring out the fact that the Earth's climate is a complex, dynamic system, consisting of parts—atmosphere, hydrosphere, biosphere, cryosphere, lithosphere, and modern industrial civilization (the latter term used as an imperfect but more accurate substitute for "anthroposphere")—that are linked through causal connections beyond simple linear causality (Grotzer, 2012). This inherent relationality of the human and the biophysical systems defies conventional modes of thinking about and conceptualizing the problem. Parvati's story, in particular, illustrates that climate change is not the only problem that communities face; in fact, all our social-environmental problems are inter-related, from species extinction to inequality. The planetary boundaries framework (Steffen et al., 2015), an evolving concept, identifies nine biogeophysical processes that define a space within which human societies can thrive. These processes, which include the carbon and nitrogen cycles, imply thresholds that cannot be transgressed without endangering the viability of the biosphere. According to this framework, five of these boundaries have already been crossed: climate, land-system change, biosphere integrity, novel entities, and the nitrogen and phosphorus cycles. Even a single boundary violation is dangerous—in part because these processes are not independent of each other. For example, climate change can exacerbate species extinction, and the imbalance in the nitrogen cycle also makes global heating worse. This behooves us to start to examine the root cause of these crises and thereby uncover the role of the current destructive globalized socio-economic system, including, but not limited to, capitalism.

Here lie considerations of power and inequality, ethics and justice. Consider the fact that the suffering of the villagers in Jharkhand in Parvati's story was not brought about directly, or even largely, by climate change, at least so far. It was due to deforestation in the service of "development." This destructive model of development, a colonial legacy from the West, enriches the middle and upper classes at the expense of "disposable" people and species. The villagers in Jharkhand do not benefit from such "development"; they are at the receiving end of the economic exploitation, violence, and habitat destruction that development makes necessary. In addition, they have not contributed significantly to the climate problem and yet are affected disproportionately by it. Although they have agency, creativity, and useful ideas, these attributes do not hold political power; they do not get villagers a seat at negotiating tables, nor do their experiences inform policy. Stacy Ruffin lost her life due to a storm directly connected with climate change; yet, as a Black woman living in a trailer park, she did not create or exacerbate the problem in any significant way. Thus, these stories demonstrate how issues of justice and ethics are inextricably entangled with our social-environmental problems.

Any effective pedagogy of climate change must, therefore, encompass these crucial features of the climate problem: that it spans vast scales of space and time; that it is inherently transdisciplinary, involving the (macroscopic) entanglement of the human and biophysical systems; that it is rife with nonlinear interconnections that propagate through complex causal webs; that climate is intimately connected with other violations of planetary boundaries, with whom it shares common roots; and that justice is central to any serious consideration of the problem.

The Newtonian or Mechanistic Paradigm, and Paradigm Blindness

As we confront the end of the biosphere as we know it, the question comes to mind: how did we end up at this fraught moment in planetary history? As historian of science Steve Shapin (Shapin, 2018) says, we can always attempt to trace threads from the complex tangle of the past in order to understand the present.

Extending Thomas Kuhn's (1962) original notion of a scientific paradigm to a *socio-scientific* framework of interconnected concepts through which societies construct their reality, the Newtonian paradigm might be considered an essential feature of modern industrial civilization. In science, the Newtonian paradigm encompasses the mechanical philosophies of Boyle, Kepler, Mersenne, Descartes, and Newton, among others (Boas,

1952), as well as the atomic materialism of Boyle and Descartes (Shapin, 2018). It considers the universe as reductionist, mechanistic, atomistic, impersonal, and deterministic. Newton's laws of motion and gravity helped reinforce and establish this mechanistic view of the world, with the clock as its metaphor. These laws appeared to reveal an orderly, clockwork-like universe, in which phenomena could be predicted and controlled, allowing humans to manipulate Nature for their purposes. This separation of human and Nature, which relegates a de-animated Nature to a store-house of "resources" for human exploitation, is a defining feature of modern industrial civilization and speaks to the influence of the Newtonian paradigm beyond science. We see this broad influence also in the formulation of mainstream economics, which is based in part on false analogies between physical and human systems, such as the idea of modeling human societies based on atoms of an ideal gas (Ackerman, 2018). The Newtonian paradigm arose from *and* helped engender large shifts in the Western conceptualization of the world, changes that included the Industrial Revolution, colonialist explorations, and—crucial to the climate crisis—the shift from water power to fossil fuels. According to scholar Cara N. Daggett (2019, p. 18), the role of science changed from describing the harmonies of Nature to harnessing the "energies of change" for human beings. "By the 19th century, not only progress, but unlimited progress had become an almost universal faith in the modern West ... the preference for constant motion, action, dynamism, growth."

A clock consists of parts—springs and gears that have a clearly defined function. Their relationships are straightforward, and neither their form nor function is subject to change. In my own field of physics, the Newtonian mechanistic view has been displaced by more accurate descriptions of Nature such as quantum physics and relativity. Yet, while physicists have provincialized Newton's laws, in a manner of speaking, the mechanistic paradigm is still pervasive as a kind of mental baggage—and, more to the point, its influence beyond the boundaries of physics continues to be strong. We see its manifestations not only in the Nature/culture divide and in the fantasies of capitalist economics but also in education, with its power hierarchies and its separation of the body of knowledge into water-tight compartments. Thus, rarely do physicists converse with poets or economists with biologists or sociologists with engineers.

The Newtonian paradigm, though somewhat simplistically described here, is a useful lens with which to become aware of the assumptions that underlie modern industrial civilization—to make visible the invisible scaffoldings of a framework is to recognize it as a construct and thereby allow us to question it and change it. The disjointed, atomistic view of the universe that is a key feature of the Newtonian paradigm manifests as the chopping up or fragmentation of space, time, and relationships within and beyond the human. We who live in modern industrial civilization are constrained by the Newtonian paradigm to think short term and local, to limit our empathetic reach, to separate our actions from their consequences, to acknowledge only simple, linear causality, and to deny our connections to multispecies others through the food we eat and the air we breathe. It is unsurprising that our economic and educational systems, as well as social arrangements in general, should reflect this reductionist, atomistic perspective. Compare these aspects of the Newtonian paradigm to the key features of the climate problem: spanning of vast spatial and temporal scales, inherent transdisciplinarity, complex causal connections within and between human and biophysical systems, and centrality of justice and power. When we recognize this mismatch between the Newtonian paradigm and the essential characteristics of the climate crisis, a troubling *paradigm blindness* becomes apparent: the application of Newtonian approaches to an inherently non-Newtonian problem.

Returning to the stories I told at the beginning of this chapter: what do they, and storytelling in general, have to do with paradigms? In their groundbreaking book, *Storylistening: Narrative Evidence and Public Reasoning*, Sarah Dillon and Claire Craig (2022, p. 59) make the point that stories—described generally as causal accounts of something happening that includes entities with agency—have, among other roles, an ontological function; they lock in certain framings, and they help build collective identities. Specifically, *dominant narratives* play this role: "Individuals act based on assumptions, recollections, and anticipations acquired from the dominant cultural, social, or other public narratives available to them." Stories, they aver, are "sites of power," whether hegemonic or resistive, as reflected in whose stories get told and which are suppressed or ignored.

Thus, certain dominant narratives can shore up a paradigm, while other stories can challenge it. I would add that crucial to the examination of stories as ontological entities is the consideration of audience. The listeners and receivers of stories also bring their pre-conceived notions, stereotypes, and cultural and personal alignments to the process of making meaning from stories. Sometimes the same story can mean entirely different things to different listeners. Therefore, providing a *context* for storytelling, which also involves understanding the audience—in my case, a diverse group of students in a public university in the Eastern US— becomes crucial.

I briefly note one classroom example of how the dominant narrative is reinforced through story, via the movie *Interstellar*.[1] In this movie, a white male hero on a dying Earth rekindles the hope of space exploration and the "destiny" of humankind to be spacefarers, that is, explorers rather than mere caretakers. "The Earth has turned against us," and similar statements denying responsibility are reminiscent of the abuser in a relationship of domestic violence blaming his victim before he moves on to better pastures. Student reaction to this movie, which was screened on our campus, was initially positive. Subsequent discussions made apparent the problematic onto-epistemological basis of the film.

WHAT MIGHT THE BEST STORIES DO?

In their book, Dillon and Craig relate how stories' most important roles may well be to present multiple points of view; contrary to popular belief, stories' key function may not be the generation of empathy in the reader, and, in fact, according to the authors, there seems to be little evidence that an empathetic response results in prosocial behavior (pp. 23–30).

Because stories are onto-epistemological tools, I believe that the presentation of multiple points of view can potentially shake us loose from dominant narratives and thereby free us from the hold that the Newtonian paradigm maintains over the imagination.

In this section I focus on certain kinds of stories I have found useful (when appropriately contextualized) in the classroom. In particular, I consider the role of these stories as boundary objects, a term conceptualized by Star and Griesemer (1989) in a different context, and applied more recently by a climate scientist in whose work I first encountered the term (Shepherd & Lloyd, 2021). I categorize the kinds of stories that I've used in my general physics classroom and explore their functions as boundary objects and paradigm-shifters based on response from students.

The term "boundary object" originally arose in the context of communication across social worlds comprising practitioners in different fields of science. It is specific to situations where different social worlds need to interact (for scientific projects requiring multiple disciplines, for example)

[1] https://en.wikipedia.org/wiki/Interstellar_(film)

but does not demand consensus or interdisciplinarity. I quote from Star and Griesemer (italics mine):

> *Boundary objects are objects which are both plastic enough to adapt to local needs and the constraints of the several parties employing them, yet robust enough to maintain a common identity across sites.* (1989, p. 393)

Boundary objects have interpretive flexibility that arise from their organizational and material structure, allowing for cooperation without consensus. Can we broaden the meaning of the term to help us navigate the apparent dichotomies of local/global, human/Nature, science/humanities? Climate scientist Ted Shepherd proposes what he calls storylining as a boundary object, to which I will return later. I propose certain categories of *stories*, more generally, as boundary objects that can help us travel across multiple boundaries, including those between contesting paradigms.

First, real-world stories like those I have mentioned above, when appropriately contextualized, can take us from the local to the global and back again. Climate change and its related ills manifest differently in different places around the globe; while climate is relatively abstract, the idea of *clime* as the relationship between weather, place, and culture is finding traction in recent work (Carey et al., 2014; Fleming, 2014). Multiple real-world climate stories set in distinct climes marry localness with a sense of the planetary. Further, these stories take us across various disciplinary boundaries—it becomes clear that considerations of economics, history, injustice, and politics are entangled with the biophysical changes wrought by global heating. Parvati's story allows us to interrogate a model of economics and development globalized throughout the world through old and new forms of colonialism. It also allows us to explore the ecological and climatic role of tropical forests. The meaning that forests hold for forest-proximate and Indigenous communities invites insights from anthropology and sociology. Thus, such stories provide an opportunity for short, disciplinary deep dives into economics, history, development, ecology, anthropology, and climate science.

Perhaps most crucially, these stories, along with the discussions that follow, present perspectives that challenge the dominant narrative. In Stacy Ruffin's story we get to know a family and its resilience in the face of tragedy that upends stereotypes about African Americans. We discuss the role of racist policies with regard to where Black people have been compelled to live. The story also opens space for considering the troubled

Earth, the disturbed global climate system, whose manifestation as the extreme rainfall event is like a larger-than-life character coming on-stage. Thus, stories of marginalized peoples and elements of climate that are not generally included in the dominant narrative take center stage in the classroom. Parvati's story further challenges the colonialist narrative of development-as-a-common-good and the rural poor as helpless victims who need to be "uplifted" to our standards and ways of life. Instead, it becomes apparent that the voices of people who experience and creatively manage their social-environmental challenges, despite social marginalization and oppression, need to be central in climate discourse. Further, the idea that the villagers are restoring the forest not just for people but also for animals and birds stands in sharp contrast to the human-centric orientation of the Newtonian paradigm, as does the ability to think long term, and the holistic-integrative (as opposed to reductionist) approach to engaging with a problem.

Parvati's story opens the way to a deeper consideration of Indigenous and local knowledge systems. In a freshman seminar on Arctic climate change, my students and I examine these through stories from and about Indigenous experiences. One such story ("The Moose Hunters and the Bear") is told by a Native American moose hunter in the Yukon region of Canada (Clark & Slocombe, 2009). He tells of an event when a moose had been shot, and he and his companions were cutting up the meat, planning to take it back to their homes in two trips. They then noticed a large grizzly watching them. Knowing that the grizzly would probably take the rest of the meat when they left with the first load, they covered the remaining meat with a tarp, leaving the moose head on top for the bear. They asked him not to take all the meat. When they returned for the second load, they found that the bear had, indeed, only taken the head. In gratitude they left some of the remaining meat for the bear. The hunter related a similar case of a bear watching moose hunters carve up a moose, but in that case the hunters were unaware of the Native tradition and did not leave a gift for the bear, nor did they ask the bear to leave some meat for them. When the hunters returned for the second load, the bear had dragged it away.

Stories like these illustrate what Chie Sakakibara (2016) calls "collaborative reciprocity," the idea that humans are not superior to other species but are connected to them through webs of relationship and reciprocity, and that the human and non-human spiritually co-constitute each other. This concept is absent in the Newtonian paradigm, which places humans at the pinnacle of all life and ignores the agency, intelligence, and emotional lives of other beings. Collaborative reciprocity also makes clear another, fatal gap in the Newtonian paradigm: the inability to acknowledge deep interconnections and relationality. In Fig. 9.1, we see that the Newtonian paradigm has another boundary along the z-axis—it is limited to describing simple systems rather than complex ones. *Complex systems* can be understood as those in which relationships between the parts are so important that they can change the Nature and function of the parts, often in sudden and surprising ways. Examples include ecosystems, human social networks, the endocrine system, and the global climate system. Consider the fact that in a clock—the archetypal Newtonian system— gears and springs have well-defined, unchangeable functions and relationships to each other. The climate system can also be considered to be made up of parts—oceans, icy regions, atmosphere, biosphere, etc. However, the ocean, which is currently a net carbon absorber, can, under different circumstances, become a net carbon emitter, thus drastically changing its function as an element of the climate system. Complex systems therefore align with the Aristotelian dictum that *the whole is greater than the sum of its parts.* A purely reductionist, Newtonian approach does not help us understand complex systems as *systems*. Unfortunately, while our world is

A non-Newtonian Universe

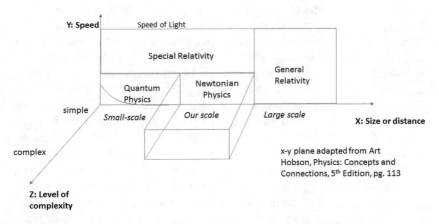

Fig. 9.1   A non-Newtonian Universe

filled with complex systems, our thinking—and our approach to education—remains largely Newtonian.

This is where the best stories, placed in the appropriate context, can
serve the role of complexifying our understanding of the world. Through
these stories the gaps and holes in the Newtonian edifice become apparent. Indigenous knowledge systems can then be seen as onto-epistemologies
that view the world as a priori complex. Carefully curated stories can serve
as boundary objects that allow travel between paradigms and onto-
epistemologies, enabling a decentering from the Newtonian paradigm.

Often, a single story can mean different things to different students,
depending on their backgrounds and contexts, and the extent to which
they are influenced by dominant narratives. Therefore, at the beginning of
our exploration of the story, I give time to listen respectfully to various
student interpretations. What emerges includes recognizable commonalities—the framework, characters, and happenings of the story—as well as
impressions arising unconsciously from dominant narratives, along with
varied individual responses. The discussion that follows helps students
make crucial connections and onto-epistemological distinctions. Thus, the
story-as-boundary-object becomes a place of multiple meanings, but
ambiguity and multiplicity do not prevent a kind of broad consensus from
emerging, a direction away from the Newtonian paradigm toward a complex, holistic understanding of the world.

There are other useful kinds of stories. As a theoretical physicist, it has
always been apparent to me that "inanimate" matter is active, not passive,
in the universe. I think of physics as one way of eavesdropping on some of
the conversations that matter holds with matter, tuned and filtered through
the methods of science. As a writer of speculative fiction, it is also apparent
to me that there is a sense in which inanimate matter can be thought of as
possessing agency, as is apparent in so many cultural stories where matter
becomes animated through character. In modern industrial civilization,
Nature is for exploitation (hence "natural resources"), other species are
(in a Cartesian sense) only machines, and matter is raw material for humans
(in power) to bend (with knowledge of physical law) to their own aims
and devices. Thus, matter and non-human life are merely a backdrop to
the human drama. Bringing the non-human on to the stage then becomes
a radical act for multiple reasons. For example, I have experimented with
embodied learning in the physics classroom, where students enact physical
processes through "physics theater," thus crossing the subject-object separation   so   central   to   science-as-we-know-it   toward   more   of   a

participant-observer role. It is one thing to learn about the oscillation modes of the carbon dioxide molecule, which is relevant to understanding its role as a greenhouse gas, and quite another to add on to that by performing its "dance" as part of enacting the greenhouse effect. Not only does this enhance cognitive understanding, but by "storifying" natural processes we are able to acknowledge that matter *matters* in our not-just-human world.

One kind of scientific "storification" is climate scientist Ted Shepherd's concept of storylining: "A physically self-consistent unfolding of past events or of plausible future events or pathways" (Shepherd & Lloyd, 2021). An example cited is the destruction of the Mackenzie River Delta freshwater ecosystem in Canada, following a severe storm; presented as a storyline, we can see causal connections between global climatic change, such as sea ice loss and sea level rise, and local conditions. Storylining is an example of an unfolding epistemological broadening in climate science: the use of narrative to make the science meaningful and usable for decision-makers and communities.

So far I have talked about the roles that real-life stories and science-as-story can play in the classroom. Another category of useful stories is speculative fiction and cultural stories in which matter and non-human life can walk on to the stage of the story as characters in their own right. Cultural stories are often teaching stories that transmit values as well as cultural-ecological knowledge across generations. For example, a story from the Iñupiat of the North Shore of Alaska ("The Mouse on the Tundra," as retold by Iñupiaq Elder and educator Dr. Edna MacLean [Chance, 2002, p. 13]) tells of a mouse who lives underground in the tundra and decides to dig a hole up to the surface so he can know the world. When he emerges, he finds that he can touch the ceiling and the sides of the world and concludes that he must be the biggest thing in the universe. The punchline of the story is that the poor mouse has come up into an upside-down Iñupiaq boot, the soles and sides of which constitute the boundaries of what he thinks of as the universe. This story illuminates not only the dangers of hubris but also the limitations of preconceptions and paradigms that can blind us to crucial aspects of the world. With regard to science fiction, while it is true that popular science fiction often reproduces a colonialist, frontier mentality, at its best it can broaden our horizons beyond anthropocentrism to speculate on our relationship with the non-human—whether a planet, a landscape, or another lifeform—through an interplay of the literal and metaphoric. It does so both by extrapolation of a trend or aspect of the present day in order to creatively imagine otherwise unforeseen consequences and by asking what-if questions, not only about technology and the physical universe but also about the human condition. "What if things were different?" can be a radical question (Singh, 2008), as exemplified by authors like Ursula K. Le Guin, who imagined different social arrangements on alternate worlds.[2] Examples of science fiction stories I have used in the classroom include Isaac Asimov's "Nightfall"[3] and Carrie Vaughn's "Amaryllis."[4] While Asimov is a science fiction writer from the "classic" age of white male-dominated techno-fetishist science fiction, this particular story—set on a planet that has six suns and knows only the light of day—dramatically illustrates how a challenge to a dominant paradigm brought upon by a natural phenomenon outside the onto-epistemological framework of that society—nightfall every two thousand years—can wreak havoc. Carrie Vaughn's story about the meaning of family and plenitude in an ecologically devastated world of scarcity forces the reader to reconsider what taken-for-granted concepts like family and happiness might mean under such extreme circumstances. Another good example is Nnedi Okorafor's "Spider the Artist,"[5] which reflects on the history of oil companies' devastation of the Nigerian landscape and peoples through a woman's phantasmagoric encounters with a robot created by the oil company. In these stories the non-human environment is not a backdrop but a character in its own right.

There is another crucial role that speculative stories can play in the classroom. Shaking us loose from the dominant paradigm is only the first step. By immersing us in alternate worlds informed by different concepts and paradigms, these stories can help us out of the trap of the imagination that makes it "easier to imagine the end of the world than the end of capitalism."[6] Emerging from paradigm blindness allows us to see other possibilities for social-ecological futures. Subgenres of speculative fiction such as solarpunk and hopepunk, along with the liberating ethos of movements such as Afrofuturism and Indigenous Futurism, are helping build new paradigms that foreground justice and are grounded in the more-than-human world we share with other beings.

[2] See, for example, the novels *The Left Hand of Darkness* (1969) and *The Dispossessed: An Ambiguous Utopia* (1974).
[3] https://en.wikipedia.org/wiki/Nightfall_(Asimov_novelette_and_novel)
[4] https://www.lightspeedmagazine.com/fiction/amaryllis/
[5] https://www.lightspeedmagazine.com/fiction/spider-the-artist/
[6] A quotation attributed to Mark Fisher.

Toward the end of the semester I engage my students in an exercise in speculative futurism. Drawing on their learnings about climate change, paradigms, and alternative epistemologies, students work in small groups to collectively create stories based on simple prompts. Instead of reductively dividing up a future desirable world into sectors like transport, power generation, agriculture, education, etc., for students to reimagine, I provide prompts that, at first sight, appear stupendously ordinary. One example is "A man goes to work." I ask students to reimagine every noun and verb in the sentence and then construct a future world in which they would want to live. The idea is not to think (as yet) about what is practical; rather, the purpose of the exercise is to free the imagination from the paradigm trap and, within the limits of physical reality (no flying pigs), to come up with the wildest possible scenarios. So, students get to interrogate default definitions of "man," "work," commuting, city, family, etc., and to come up with alternatives. The mini-stories that result from these are always instructive. Sometimes they serve to remind me of the depth of the imagination trap, but more often than not, students come up with ideas that surprise and delight them and me. Through the "ordinary" sentences of the prompts, they deconstruct taken-for-granted concepts and create something new and refreshing—a cityscape powered by mushrooms, a living forest-city, a world without cars where you cannot make homes taller than the local trees, cis-men who wear skirts and flowers in their hair, and much more. While these might be implausible or impractical based on current reality, the exercise serves to nurture and free the imagination. Students also report feeling positive emotions such as hope, anticipation, and enjoyment while immersed in constructing such futures. Further, such an exercise serves to undermine existing power structures in two senses. One, students' feeling of agency helps them feel that they need not surrender their intelligence and creativity to those in power—that they also have something to give. In addition, questioning the defaults of the dominant narrative and creating stories that provide alternatives help students take back, at least for the duration of the exercise, some of their power.

CONCLUSION: RE-STORYING AND RESTORING OURSELVES

In this chapter I have foregrounded the role of stories in teaching and learning about climate change in a college physics classroom. Real-life stories from and about marginalized communities dealing with climatic and related ills, stories from science in which matter and non-human life are protagonists, science fiction, and cultural stories all have important roles to play in the science classroom. With a multiplicity of such stories, the key features of the crises of our troubled world can emerge, and we can begin to see the climate problem as teacher. As boundary objects, these stories serve as portals and pathways that allow travel between disciplines, geographies, histories, and paradigms. They can talk back to power by positing alternative paradigms and ways of knowing. They can help make sense of problems that seem too abstract and too overwhelming at global scales. Student responses indicate that stories as carriers of complex information are more memorable, more meaningful, and speak to the emotional as well as the intellectual selves of students. Embodied learning and creating stories of alternatives collectively in small groups can help students toward a greater sense of agency and participation as well as a deeper understanding, disrupting the pyramidal power hierarchies of our society. Crucially, for these approaches to be successful, the classroom space must also be reimagined. Students should feel comfortable challenging the "authority" in the classroom, the teacher. To do this, I employ a number of techniques (Singh, 2021) to make the classroom a place for collaborative learning, where standards remain high and mistakes can be made safely as part of learning. A key aspect of this is to build genuine relationships with students and to consistently show them that their thoughts, ideas, feelings, and whole selves are valued. These inspirations from transformational learning theory and the works of other scholars help create an environment where students feel psychologically safe, allowing them to be intellectually audacious.

I end with a quotation by the great speculative fiction writer, Ursula K. Le Guin. From her National Book Award speech[7] in 2014: "We live in capitalism. Its power seems inescapable. So did the divine rights of kings. Any human power can be resisted and changed by human beings. Resistance and change often begin in art. Very often in our art, the art of words."

---

[7] https://www.youtube.com/watch?v=Et9Nf-rsALk

REFERENCES

Ackerman, F. (2018). *Worst-case economics: Extreme events in climate and finance.* Anthem Press.

Boas, M. (1952). The establishment of the mechanical philosophy. *Osiris, 412–541.* https://doi.org/10.1086/368562

Carey, M., Garone, P., Howkins, A., & Endfield, G. (2014). Forum: Climate change and environmental history. *Environmental History, 19*(2), 281–364. https://doi.org/10.1093/envhis/emu004

Chance, N. A. (2002). *The Iñupiat and Arctic Alaska: An ethnography of development.* Cengage Learning.

Clark, D. A., & Slocombe, D. (2009). Respect for grizzly bears: An Aboriginal approach for coexistence and resilience. *Ecology and Society, 14*(1). http://www.ecologyandsociety.org/vol14/iss1/art42/

Daggett, C. N. (2019). *The birth of energy: Fossil fuels, thermodynamics and the politics of work.* Duke University Press.

Dillon, S., & Craig, C. (2022). *Storylistening: Narrative evidence and public reasoning.* Routledge.

Fleming, J. (2014). Climate physicians and surgeons. *Environmental History, 19,* 338–345.

Grotzer, T. (2012). *Learning causality in a complex world: Understandings of consequence.* Rowman and Littlefield.

Kuhn, T. (1962). *The structure of scientific revolutions.* University of Chicago Press.

Kwauk, C. (2020). *Roadmaps to quality education in a time of climate change.* Brookings Institute. https://www.brookings.edu/wp-content/uploads/2020/02/Roadblocks-to-quality-education-in-a-time-of-climate-change-final.pdf

Sakakibara, C. (2016). People of the whales: Climate change and cultural resilience among Iñupiat of Arctic Alaska. *Geographical Review, 107,* 159–184.

Shapin, S. (2018). *The scientific revolution* (2nd ed.). University of Chicago Press.

Shepherd, T., & Lloyd, E. (2021). Meaningful climate science. *Climatic Change, 169,* 17. https://doi.org/10.1007/s10584-021-03246-2

Singh, V. (2008). A speculative manifesto. In V. Singh (Ed.), *The woman who thought she was a planet and other stories.* Zubaan. Antariksh Yatra: Journeys in Space, Time and the Imagination. https://vandanasingh.wordpress.com/2021/10/20/a-speculative-manifesto/

Singh, V. (2021). Toward a transdisciplinary, justice-centered pedagogy of climate change. In R. Iyengar & C. Kwauk (Eds.), *Curriculum and learning for climate action: Toward an SDG 4.7 pathway for systemic change.* UNESCO-IBE.

Star, S. L., & Griesemer, J. R. (1989). Institutional ecology, 'translations' and boundary objects: Amateurs and professionals in Berkeley's Museum of Vertebrate Zoology, 1907–39. *Social Studies of Science, 19*(3), 387–420. https://doi.org/10.1177/030631289019003001

Steffen, W., Richardson, K., Rockström, J., Cornell, S. E., Fetzer, I., Bennett, E. M., Biggs, R., Carpenter, S. R., de Vries, W., De Wit, C. A., Folke, C., Gerten, D., Heinke, J., Mace, G. M., Persson, L. M., Ramanathan, V., Reyers, B., & Sörlin, S. (2015). Planetary boundaries: Guiding human development on a changing planet. *Science, 347*(6223), 736–746.

Sutter, J. C. (2017, August 11). What killed Stacy Ruffin? *CNN.com.* https://www.cnn.com/2017/08/11/health/sutter-louisiana-flood-stacy-ruffin/index.html

PART III

Politics and Political Reverberations

From False Generosity to True Generosity: Theorizing a Critical Imaginary for Science Education

Sara Tolbert, Alejandra Frausto Aceves,
and Betzabé Torres Olave

In honoring Paulo Freire, whose 100th birthday we celebrated last year, we have taken some time to collectively reflect on the current status quo of science education. Consistent with the themes of this book, we have been asking ourselves, and each other in our science and education communities, why, what, and who are we educating for in science education?

S. Tolbert (✉)
University of Canterbury, Christchurch, New Zealand
e-mail: sara.tolbert@canterbury.ac.nz

A. F. Aceves
Northwestern University, Evanston, IL, USA
e-mail: alejandrafrausto2026@u.northwestern.edu

B. T. Olave
University of Leeds, Leeds, UK
e-mail: B.TorresOlave@leeds.ac.uk

S. Tolbert et al. (eds.), *Reimagining Science Education in the Anthropocene, Volume 2*, Palgrave Studies in Education and the Environment, https://doi.org/10.1007/978-3-031-35430-4_10

We recently wrote of how we came to see ourselves as working within a larger critical-liberatory paradigm for science education (Frausto Aceves et al., 2022) and as part of a broad and ever-expanding sociopolitical movement in science education (Tolbert & Bazzul, 2017). In this chapter, we aim to take up a provocation that emerged from that article (Frausto Aceves et al., 2022)—differentiating between *false* versus *true* generosities, as part of the project of what Freire referred to as building solidarity in and across differences.

Paulo Freire's ideology has been characterized as "equal parts Jesus Christ and Karl Marx" (Harvard Divinity School, 2022, n.p.). Freire's theories were certainly influenced by his Catholic upbringing, and the notion of generosity in Freire's writing has roots in Christianity, which later influenced the development of liberation theology (Gutierrez, 1988). In fact, Freire is known by some Latino-American theologians, such as Leonardo Boff, as "one of the founders of liberation theology" (2011, p. 241). Freire rejected two guiding pillars of Christianity, however: divine intervention (i.e., that salvation will be granted through prayer and moral living) and charity. He viewed both of these pillars as mechanisms for social reproduction; he believed that they do little to dismantle oppressive systems and, in many ways, make it possible for oppressive systems to stay intact. While Freire maintained a spiritual practice in his personal life, he rarely addressed in his writings and teachings—with some exceptions, that is, *Pedagogy of Hope*—the ways in which his spiritual practice influenced Freirean critical pedagogy. However, his work, his praxis, and his hopeful outlook and belief in the possibility of liberation can be characterized as a critical spirituality (c.f., Dantley, 2003). Dantley writes about critical spirituality in the context of relationships between African American spirituality and critical theory, while also drawing from Cornel West's (1988) concept of prophetic pragmatism. Boyd (2012) analyzes how Freire's work also brings together spiritual and critical pedagogical tenets—with similarities being a focus on conversion to the oppressed and a prophetic utopian, revolutionary vision. Along these lines, Freire himself constantly revised his notions of humanity and increasingly contested the false dichotomy between the material and the spiritual world. In a 1997 interview,[1] Freire argued, "the more I read Marx, the more reasons I encounter for my spirituality."

[1] https://www.youtube.com/watch?v=1ViM1oCPNoA

compromise is drowning out King's insistence that we cannot submit to the terms of white supremacy" (Barber & Wilson-Hartgrove, 2019). Appeals for compromise and polite discourse are often used to dismiss critical-liberatory projects, while radical or transformative ideas and activism are regularly dismissed as too far-fetched, or too impossible, or too agitating. Freire pointed out the impossibility of being a moderate, or maintaining neutrality, regarding matters of oppression; he remarked that this kind of neutrality always "works in favor of the dominant" (Horton & Freire, 1990, p. 104). And, as educator Kortney Hernandez has stated, "[P]edagogically deceptive notions of benevolent intentions coupled with well-meaning discourse serves only to camouflage so unapologetically the deep injustices that are constantly at work" (Darder, 2017, pp. 176–177).

In science education, this takes many forms. One of those forms we often observe are discourses which serve to position social justice work at the margins—or even outside—of science education. These discourses are ever present in curriculum and science education policy documents (e.g., justice-related issues in science are hardly addressed and only as Appendices in Next Generation Science Standards (NGSS)), or in manuscript reviews requiring citations of mainstream science education research, or in messages that get communicated to pre-tenure faculty about what is required to be "accepted," get jobs, or get tenure, in our field—all in the name of rigor and/or compromise. Alberto J. Rodriguez (2006) has referred to these discourses as part of a larger project of politics of domestication, that is, "a negative process of acculturation by which one's ideals and commitment to work for social justice are tamed and reduced to fit dominant discursive practices" (Rodriguez, 2006, p. 48).

False Generosity as Politics of "Inclusion"

False generosity often takes shape as liberal-progressivist initiatives, like inclusion and diversity which, as Sara Ahmed (2012) argues, "can be a way of maintaining rather than transforming existing organizational value" (p. 59). However, as Freire (1970) points out, "our insistence that the authentic solution of the oppressor-oppressed contradiction does not lie in a mere reversal of position, in moving from one pole to the other. Nor does it lie in the replacement of the former oppressors with new ones who continue to subjugate the oppressed—all in the name of their liberation" (p. 57).

Representational politics alone do not have the power to eliminate systems of oppression. Angela Davis reminds us that diversifying the workforce, such as the police, for example, does little to root out racism from a system that was built on white supremacy. [3] As Grande and Anderson (2017, p. 139) have also pointed out, multicultural or culturally responsive education that privileges "the cultivation of a respect for difference over critiques of power" is a form of false generosity that perpetuates the myth of meritocracy. Efforts at inclusive education are often driven by a commitment to bringing up members of marginalized groups into higher ranks of the current neoliberal, capitalist education system, and not on destabilizing the system that is designed to deprofessionalize teachers or commodify students for capital gain (Au, 2017; Lorde, 1984). (We point out here that even Raytheon has a vested interest in expanding the participation of girls in STEM and makes significant financial contributions to STEM education programs for girls.) Well-intentioned efforts at more inclusive science education can also have the (intended or unintended) effect of positioning minoritized students as lacking, in need of something, or, as Kirchgasler (2022) puts it, "not-yet-healthy citizens."

False Generosity as Dehumanizing

Freire wrote that while focusing on the preparation of scientists and technicians was critical for Brazil's development, an education for scientists and technicians must not lose sight of "the battle for humanization. ... It was essential to harmonize a truly humanist position with technology by an education which would not leave technicians naive and uncritical in dealing with problems other than those of their own specialty" (Freire, 1974, p. 34). False generosity creates dehumanizing conditions for teachers, who are positioned not as professionals-who-care but rather as technicians who must implement the (corporate-backed or state-backed or NGSS-backed) curriculum (Eaton & Day, 2020), teach to the tests, and/or teach standards with fidelity (Darder, 2017).[4] Deterministic views of

---

[3] "The technology, the regimes, the targets are still the same. I fear that if we don't take seriously the ways in which racism is embedded in structures of institutions, if we assume that there must be an identifiable racist ... who is the perpetrator, then we won't ever succeed in eradicating racism" (Davis, 2016, pp. 17–18).

[4] Furthermore, we often implicitly position "teachers" as the solution to all of society's inequalities (and the reverse of this is that teachers [or students] are positioned as the problem when inequalities persist)—yet rarely do we reflect upon or analyze our own practices as teacher educators and science education researchers, or our own role(s) in perpetuating those same inequalities.

science education such as teaching the canon or the practices of science in better or more inclusive ways so that students will learn better or acquire more science knowledge can be grounded in false generosities.

This denial of humanity is evident in science education, particularly in physics faculties, where disciplinary epistemic assumptions implicitly communicate that by merely knowing content knowledge you are qualified to teach it (Larsson et al., 2021) and, therefore, content is prioritized over pedagogy in the education of future physics teachers (Torres-Olave & Dillon, 2022). Such assumptions dehumanize students as well as teachers (teacher educators, in this case, who in physics education are often physicists). For example, physics teacher educators are positioned as holders of knowledge and wisdom, rather than in a constant process of learning, under the premise that physics is an objective type of knowledge that can be "deposited" into someone else's mind (see also Singh, 2023, this volume). This is a classic banking method approach; there is no dialogue, nor a process of transcendence, of *being more*. The banking model, therefore, also denies the possibility for scientists who teach to be more fully human, rather than "things" possessed by capitalist logics (Freire, 1970).

False Generosity as Apolitical Scientific Saviorism

In the case of science, false generosity can occur when, for example, we as science (or engineering) educators consider and promote only the value of technoscientific innovation but not how systemic oppression is constituted through innovation as part of larger socioscientific and sociopolitical entanglements (Gunckel & Tolbert, 2018). As an example, we turn to the reduction of greenhouse gas emissions as a resolution to the climate crisis. The transport sector generates approximately 23% of these gasses, which is why "green mobility" strategies such as replacing fossil fuels used by cars is one of the more common solutions proposed to greenhouse gas reduction (Jerez et al., 2021). However, this effort has simultaneously led to the exploitative extraction of lithium in the lithium triangle in South America. In South America, lithium is found in brines (i.e., underneath salt flats) contained in the salars, which are fragile ecosystems with high biodiversity and species endemism. In the case of Chile, particularly the Atacama desert, to obtain lithium, it is necessary to evaporate the water from the brines—in one of the driest deserts in the world. This process is not free of harm for local human and non-human communities and often results in overexploiting hydro-social territories (Gutiérrez et al., 2022; Jerez et al., 2021). Meanwhile, leftist governments organizing for state versus corporate control over lithium resources, and Indigenous activists organizing against lithium extraction in Latin America, are being framed as obstacles to carbon emissions reduction (Dube, 2022; see also Carrara & Chakraborty, 2023, this volume), while the super wealthy fly around in private jets (Milman, 2022). This is a clear case of eco-imperialism. We certainly depend on minerals and materials that are obtained from nature and need to be finding other sources of energy (and other ways to live). But at what costs and to whom? We ask, how much do we engage with these nuances and issues of climate justice and climate imperialism in science education? Whose "green" environment counts? In the Anthropocene, science and politics cannot be disentangled.

(Theorizing) True Generosities for Science Education

True generosity is about destabilizing and reimagining systems—not maintaining the status quo. Freire wrote that "True generosity consists precisely in fighting to destroy the causes which nourish false charity" (1970, p. 45) through the denunciation of oppressive conditions and proclamation of new liberatory ones. Drawing from the Freirean constructs of political clarity, *autonomía*, and solidarity, we propose a model for true generosity in science education. Differentiating between false and true generosity requires thoughtful reflection, particularly within a field that has for so long positioned itself as apolitical. We feel that a model for true generosity, therefore, must start with *political clarity*—what are we doing and who for, moving into a reclamation of our *autonomía*—the power to (re)define ourselves and our praxis, and, related to this, building *solidarities* across difference, in the name of justice.

Political Clarity

"Instead of reproducing the dominant ideology, an educator can denounce it, taking a risk of course" (Freire, in Horton & Freire, 1990, p. 118). Practicing true generosity requires what Freire referred to as political clarity. That is, "the educator must know in favor of whom and in favor of what he or she wants. That means to know against whom and against what we are working as educators" (Freire, in Horton & Freire, 1990, p. 100). Part of the challenge is that the field of science education has long established disciplinary boundaries in terms of what is or is not part of our charge as science educators, in ways that position politics or sociopolitical contexts as outside of our purview. However, while of course a science educator (a biology teacher, for example) must teach biology, Freire explained that it would be impossible to study "the phenomenon of life without discussing exploitation, domination, freedom, democracy, and so on" (p. 104). He further elaborated that

> I cannot put history and social conditions in parentheses and then teach biology exclusively. My question is how to make clear to the students that there is no such a thing named biology in itself. If the teacher of biology does that and the teacher of physics does that and so on, then the students end up by gaining the critical understanding that biology and all the disciplines are not isolated from social life. This is my demand. These two risks exist. The risk of putting in parenthesis the content and to emphasize exclusively the political problem and the risk of putting in parenthesis the political dimension of the content and to just teach the content. For me both attitudes are wrong. (Freire, in Horton & Freire, 1990, pp. 108–109)

Political clarity requires an understanding of the fact that education is always a political act. Science education is also inherently imbued with politics—and ethics—even (especially) when the ethico-political dimensions of science and education are not made explicit to students. On a basic level, even very young students can engage in problem-posing pedagogies, such as why do we only study science once a week (or less)? Or why does this citizen science investigation require that we kill the insects we are studying?[5] Or, perhaps for older students, why don't our state standards for science education address climate change? Or what is it about the political history of our community that led to the soil or water becoming contaminated in the first place (Morales-Doyle et al., 2019; Tolbert et al., 2016)? Or, as we referenced earlier, in the case of engineering education, how can we advance technologies that are not exploitative of nature (Sanchez, 2023)? On this point, Myles Horton also responded that

[5] https://www.sciencelearn.org.nz/resources/2619-ahi-pepe-and-tikanga

There's no science that can't be used for good or for evil. ... If you make people knowledgeable about the science and don't point out this fact, then you're saying, I withdraw from the battle, from the discussion of the ethics involved. I just stick to the facts. And that of course means that you've surrendered to the strongest forces. You say you're neutral in what you do, you aren't concerned with it. ... It's unavoidable that you have some responsibility ... regardless of what you teach or what your subject is or what your skill is. Whatever you have to contribute has a social dimension. (Horton, in Horton & Freire, 1990, p. 105)

Often liberal-progressive positionings, or scientific notions of neutrality and objectivity, lead educators to feel that they can't share personal viewpoints with students. But, since education is always political, educators, including science teachers and science education researchers, have a responsibility to share their ideas—and wonderings—about justice and to create pedagogical conditions and learning environments in which justice-oriented teaching and learning can happen. As Camille Rullán has argued, it is necessary for the public to understand

the ways in which capital and power influence and distort the production, use, as well as the nature of science and, more critically, of re-imagining the ways we practice science. There is no hero that can give us this. The only way forward is through collective action, for scientists to put their skills at the service of the people and against the oppressors, through a science for (and by, and of) the people. (Rullán, 2021, n.p.)

Freire points out that political clarity, however, is not the same as an authoritarian teacher imposing their ideas on students. Developing and practicing political clarity does not mean we will always have clarity about what is Right or what is Wrong, but rather that we make (our own and others') political wonderings and ethical complexities of "doing science" explicit to students, while we engage them in their own wonderings and ethical dilemmas (e.g., Krishnamoorthy & Tolbert, 2022; Moura, 2021). In other words, we can't shy away from political engagement in science classrooms, because politics always already are part of doing—and teaching and learning—science.

Autonomía

True generosity also entails reclaiming autonomy. We believe, as Freire states, that *autonomía* (autonomy), not *individual freedom*, should be the goal and condition of liberatory education. We cannot be free while others are not; therefore, we cannot be truly free in a neoliberal capitalist society in which "liberty" is understood as individual "freedom"—and the focus is often on individual choice over collective wellbeing. However, it is necessary to define and contrast what we mean by freedom (*autonomía*) and "freedom" in this context. The often-cited example contrasts neoliberal conceptions of freedom as the "freedom to," that is, the simple removal of legal impediments to a course of action. This contrasts with a more liberatory conception of "freedom from," that is, freedom from restrictions to the actor's own autonomy, which is often heavily restricted by market-based directives imposed from above with little to no democratic input, despite neoliberal lip service to the idea. In neoliberal policy contexts, the term autonomy has been co-opted, to frame what people can or cannot do, linked to the idea of individual *freedom to* and choice (Torres-Olave & Dillon, 2022). This version of "freedom" positions people as having individual responsibility for their own future and wellbeing. We hear more and more about the concept of "individual rights" or "individual freedom," for example, in response to the COVID-19 mandates. In a similar line, the growing cost of postsecondary education, the treatment of students as consumers of knowledge, and increasing academic precarity in the university system are largely a result of marketization (i.e., privatization of formerly public resources), cloaked in a discourse of individual autonomy, such as "equity and access" and "freedom to choose." Boaventura de Sousa Santos (2019) has written that, in this sense, individualistic autonomy is a cruel slogan.

The concept of autonomy needs to be reclaimed as a collective autonomy. As Freire states, being autonomous is fundamentally about our nature as interdependent beings, in communion with others: It "is the authority of the *not me, or you*, which makes me assume the radicality of myself" (Freire, 1998, p. 46). Such radicality, Freire argues, is put into practice when we embrace an ontology of relationships and differences that transcends the alienating character of individual freedom. Autonomy, in Freirean terms, is to embrace that we live in a plural world, with different dreams, but shared struggles. Ana Cecilia Dinerstein (2016) states that autonomy is a "doing," and she calls this practice *the art of organizing hope*, which entails working against individualism while imagining and practicing alternative horizons. As educators and researchers, we need to be, and encourage our students to be, autonomous and critical, but without losing sight of *the other*. Dinerstein argues that autonomy needs "cracks" which create opportunities for transcending self, for dialogue with others' dreams, in pursuit of other alternatives. That is what Freire refers to when he says, "not me *or* you," which can be linked back to his notions of *unfinishedness* and our ontological vocation to be *more fully human*. It is in that constant search that we may become aware of our limits, of what we can and cannot achieve in our individuality. It is in that search that we encounter a new radical possibility, of recognizing the "self" as interdependent. It is in that moment we recognize others (beyond our own familial sense of interconnectedness). Our existence is inextricably linked to others and, therefore, any type of "individual empowerment" at the expense of another goes against our "ontological vocation to be more fully human." A Freirean ontological perspective is one in which becoming fully human requires not only a freedom from one's own suffering but also a freedom from suffering for all of humanity (e.g., Chen, 2016). That is, real autonomy operates in the service of social responsibility. It moves away from vertical dependence and false generosity, toward horizontal relations in solidarity—that support autonomous collective flourishing over an individualistic "independent life."

Solidarity

As Antonia Darder has written, "Paulo Freire believed till his death that 'to change what we presently are, it is necessary to change the structures of power radically'" (Freire, 1997, cited in Darder, 2011, p. 155). Freire recognized that radical change in the structures of power required solidarity, preceded by a "conversion to the people," that is, "a spiritual transformation that brings one into identification, solidarity and common struggle with those that are oppressed" (Boyd, 2012, p. 772)—or in Freire's words, "communion with the people" (Freire, 1970, p. 61). Part of any critical education, in the critical pedagogical tradition, is coming to see oneself as part of a larger community and to understand the immense power of the collective. Freire (1970) wrote that

> Solidarity requires that one enter into the situation of those with whom one is solidary; it is a radical posture. ... [T]rue solidarity with the oppressed means fighting at their side to transform the objective reality which has made them these "beings for another." The oppressor is solidary with the oppressed only when he stops regarding the oppressed as an abstract category and sees them as persons who have been unjustly dealt with, deprived of their voice, cheated in the sale of their labor—when he stops making pious, sentimental and individualistic gestures and risks an act of love. ... To affirm that men and women are persons and as persons should be free, and yet to do nothing tangible to make this affirmation a reality, is a farce. (pp. 49–50)

Freire believed that forging alliances across differences was integral to the work of the critical educator. While solidarity does not require an abandonment of group-specific goals and struggles, it means finding common ground across groups through movement-building, through what we previously mentioned as *the cracks of autonomy*. Solidarity is about risk; solidarity is an act of love and hope:

> The "different" who accept unity cannot forgo unity in their fight; they must have objectives beyond those specific ones of each group. There has to be a greater dream, a utopia the different aspire to and for which they are able to make concessions. Unity within diversity is possible, for example, between anti racist groups to overcome the limits of their core racial group and fight for the radical transformation of the socioeconomic system that intensifies racism. (Freire, 1997, p. 85)

Other critical educators have also underscored the importance of solidarities that bring together marginalized communities in formation toward a broader sociopolitical movement. Sandy Grande (2018) has asked, for example, "What kinds of solidarities can be developed among marginalized groups with a shared commitment to working beyond the imperatives of capital and the settler state?" (p. 48). Angela Davis (2016) reminds us that finding connections between movements strengthens their collective power and impact. Insisting on the intersectionality of movements, she wrote, "Initially intersectionality was about bodies and experiences. But now, how do we talk about bringing various social justice struggles together, across national borders?" (p. 19).

Antonia Darder (2011) emphasized that teachers can play a key role in movement-building. She highlighted how a "unifying, albeit heterogeneous and multifaceted, anticapitalist[6]" movement can be built by teachers, in partnership with other organizations, and sustained through critical education. Both Freire and Darder have pointed out that alliances must be formed across class (including across class positions), race, gender, and other social categories, to "rescue the concept of power from its diffused and immeasurable position" ("of being everywhere and nowhere") (Naiman, cited in Darder, 2011, pp. 155–156) back to the immense power of collective action:

> History has repeatedly shown that significant institutional change can truly take place only as a result of collective work within social-movement organizations. True, legal and policy strategies have had some impact, but ultimately the collective pressure of the masses has had the greatest impact in quickly mobilizing these forces. (Darder, 2011, p. 156)

It is through the power of the collective that we can achieve environmental justice, living wages, fully funded schools, and healthcare, etc.— and build an alternative vision of/for science education and schooling (Darder, 2011). We (Sara, Alejandra, and Betzabé) have witnessed the power of teachers coming together to "channel the fears, guilt, rage, and despair into productive action" (Darder, 2011, p. 156) such as a science teacher collective in which curricular and pedagogical tools are the participatory design work of teachers, community organizers, scientists, researchers, and youth (Morales-Doyle & Frausto, 2021). We have witnessed how science teachers with whom we have worked moved from frustration and dismay about the low status of their professions, and the political constraints they faced as educators, to collective empowerment, for example, becoming part of recent "glocal" teacher movements (i.e., #REDforED) and multi-day strikes, demanding and securing better working conditions for teachers and support staff, and better learning conditions for students, (e.g., Torres-Olave et al., 2019; Williams & Tolbert, 2018, 2021). We, like Freire, Darder, Horton, and other critical pedagogues, believe that

> [t]hrough the building of ethical communities for struggle and change, we can develop the critical strength, reflective ability, political knowledge, social

---

[6] Antonia Darder (2011) underscores the importance of explicit attention to class struggle: "Such a process requires that we remain ever cognizant of the increasing significance of class and the specificity of capitalism as a system of social and political relations of power" (p. 155).

commitment, personal maturity, and solidarity across our differences necessary to reinvent our world. (Darder, 2011, p. 156)

We see powerful examples of ethical communities such as those Darder describes in the recent global student #SchoolStrike4Climate movements, scientists protesting against cuts to public funding for scientific research (as well as the elimination of some forms of funding such as climate change research under the Trump administration in the U.S.A.), the #NoDAPL movement (Dakota Access Pipeline protests), Black Lives Matter, youth-led Free Palestine movements,[7] scientists fighting and taking positions of power for the right for water in Chile (e.g., MODATIMA[8]), and countless others.

But How Do We Build Pluralistic Solidary Communities for Liberatory Science Education?

As we move, not only our gaze, our words, but also our worlds toward imaginaries that mobilize our pockets of resistance as transnational solidarities, we must recognize that will alone is not enough to move from a neoliberal and individualistic to a just and liberatory science education. Scholars like Arturo Escobar have given us powerful language with which to (re)imagine a science education whose corazón (heart) thrives for our whole Earth system. It will take practice and humility to build pluralistic solidarity communities that honor "a multiplicity of worlds and peoples coexisting … always flowing, constantly changing owing to interdependence of all aspects of living systems" (Escobar, 2018). There will be things and ideas we must leave or change and things we must practice and build as we consider moving from false toward true generosity. Of those things, two that have been weighing on us the most lately have been courage and imagination and the role these play in both our collective and individual praxis. In our work toward *justicia y libertad*, we can at times become lovers of problems (Ginwright, 2021) that cloud and limit our critical imaginations. False generosity is good at further convincing us of those limits, that we cannot imagine worlds that live outside of the systems that oppress us. We have struggled and learned to hold on, maybe even survive. Will we talk honestly and undeviatingly about what should exist?

---

[7] https://www.aljazeera.com/opinions/2021/5/28/generation-z-will-free-palestine
[8] https://www.frontlinedefenders.org/en/profile/rodrigo-mundaca

Our courage will inevitably be tested as we move forward into new realms and modes of being, inspired by our collective imaginations—but will we really let go? Will we decide to be truly generous? Will we be willing to make the sacrifices needed? Will we maintain our hope?

Reimaging the Heart of Science Education: A Collective of Critical Imaginaries

Hope is an indispensable seasoning in our human, historical experience. Without it, instead of history we would have pure determinism. History exists only where time is problematized and not simply a given. (Freire, 1998, p. 69)

Hope is a collective project enacted through communities, through praxis of the people, fomenting pockets of resistance, with the political clarity to witness and name past-present history and acting to (re)imagine and change present-futures. For Freire, hope is a verb, *esperanzar*, and *tener esperanza*. To have hope is a practice. It implies action, such as (re) claiming spaces for solidarity building and dialogue that nurture our hope and *autonomía*, together as (science) educators. In order to counter a "well-meaning" yet dismissive false generosity, we will need to be humble enough to slow down and listen, with an open mind and heart. True generosity will require us to have the courage to be uncomfortable, collaborative, and honest, asking and listening to what we need from each other. We will have to catch ourselves, pull our peers aside, or listen with humility as we are pulled aside, because we will not always get it right. In fact, we acknowledge the sheer impossibility of always getting it right. Yet, a commitment to true generosity is lifelong, walking (and making) the roads with others, sometimes stepping up and sometimes stepping back, because of our love, solidarity, and responsibility to our collective liberation.

What we are proposing in this chapter are ways to advance toward a critical imaginary of/for science education. An imaginary "encode[s] not only visions of what is attainable through science and technology but also of how life ought, or ought not, to be lived; in this respect, they express a society's shared understandings of good and evil" (Jasanoff, 2015, p. 6). For us, such a critical imaginary for science education (and beyond) is rooted in the three principles we have outlined: political clarity, *autonomía*, and solidarity. Our task now is to reflect upon our own practices and contemplate how we can enact such principles in our research, as/for a reimagination of our future selves, future scientists, future science teachers, and future engineers.

Just as a seed will not sprout without water and light, what is seeded together among us as science educators will not germinate without its own form of water and light. What practices will support us to reach and stretch in pluralistic ways that are emergent, responsive, and relevant? We have called attention to the importance of *para qué, para quién, con quién enseñamos* (for what, for whom, and with whom we teach). While advancing this critical imaginary, we must have the courage to name and reject false nourishments, like representation alone, that without power, only serve to manipulate and contain possibilities and imaginations. In *Pedagogy of the Oppressed* (1970), Freire explains this myth as

> the model of itself which the bourgeoisie present to the people as the possibility for their own ascent. In order for these myths to function, however, the people must accept the word of the bourgeoisie. ... Through manipulation, the dominant elites can lead the people into an unauthentic type of "organization", and can thus avoid the threatening alternative: the true organization of the emerged and emerging people. The latter has two possibilities as they enter the historical process: either they must organize authentically for their liberation, or they will be manipulated by the elites. (pp. 147–148)

Freire's words should give us pause and much to think about. The myth of "the model of itself" for ascension is part of the playbook of the oppressor. The myth limits our imagination and in doing so keeps us oppressed. Then there is the why: "avoiding the threatening alternative" of liberation. By holding onto the myth, we are blind to the possibilities of freedom, but if we just let go, if we decide that we cannot be a copy and instead take our place as "emerged and emerging people," we will articulate and dream worlds toward different ends. Thus, organizing so that we prioritize "for what, for and with whom" we dream our worlds will be vital to our nourishment. Reimagining the heart of science education is a collective project, sustained by and grounded in community and place, inclusive of all beings in those places, and their relationships and perspectives. As we look forward when we walk together, we must also look to our sides, at who is walking with us, as well as what we are walking by and on. Science education must be dynamic enough to emerge in and with the world, for the world, because of the world.

Toward the Ethical and Just Worlds We Deserve

Honoring and practicing our commitments will require that we take an honest look at what we bring with us to our work. How can anchoring ethics with courage and imagination help us guard against apolitical scientific saviorism that does not serve the collective? As Freire reminds us in *Pedagogy of Freedom* (or rather, Pedagogy of *Autonomía*), because we are not predetermined, we must read and write the ethical and just worlds we all deserve.

> If I am a pure product of genetic, cultural, or class determination, I have no responsibility for my action in the world and, therefore, it is not possible for me to speak of ethics. Of course, this assumption of responsibility does not mean that we are not conditioned genetically, culturally, and socially. It means that we know ourselves to be conditioned but not determined. (1998, p. 26)

As we practice true generosity and build solidarities toward a liberatory science education, we will need to consider our contribution and responsibilities to others. Reflection and dialogue will be important in helping us recognize what we carry and need from each other. We will need to name and hold not just the beautiful things that bring us joy and hope but also the limiting and problematic things that perpetuate injustice and oppression. We will need courage to have uncomfortable conversations about difference, our human centricity, and universalizing tendencies, and we will have to carry the tensions as we move toward beautiful elsewheres. We will appropriate science as a tool for social justice, yet we will also humbly acknowledge that dominant forms of science are ways, of many ways, to understand or describe our world(s). It will take courage to confront, renounce, and transform systems of oppression that erase differences and perpetuate exclusion, hate, imperialism, dehumanization, and environmental devastation—a courage that cannot be manifested alone, but one that grows from and is sustained by the power of our collective radical imaginaries.

REFERENCES

Ahmed, S. (2012). *On being included*. Duke University Press.

Au, W. (2017). When multicultural education is not enough. *Multicultural Perspectives, 19*(3), 147–150.

Barber, W., & Wilson-Hartgrove, J. (2019, January 18). MLK warned us of the well-intentioned liberal. *The Nation.* https://www.thenation.com/article/archive/martin-luther-king-trump-wall-jim-crow/

Boff, L. (2011). The influence of Freire on scholars: A select list. In J. Kirylo (Ed.), *Paulo Freire: The man from Recife* (pp. 235–269). Peter Lang.

Boyd, D. (2012). The critical spirituality of Paulo Freire. *International Journal of Lifelong Education, 31*(6), 759–778.

Carrara, A., & Chakraborty, R. (2023). Decolonizing climate justice and environmentalism. In S. Tolbert, M. Wallace, M. Higgins, & J. Bazzul (Eds.), *Reimagining science education in the Anthropocene* (Vol. 2). Palgrave Macmillan.

Chen, R. H. (2016). Freire and a pedagogy of suffering: A moral ontology. In *Encyclopedia of educational philosophy and theory.* Springer.

Dantley, M. E. (2003). Critical spirituality: Enhancing transformative leadership through critical theory and African American prophetic spirituality. *International Journal of Leadership in Education, 6*(1), 3–17.

Darder, A. (2011). *A dissident voice: Essays on cultural, pedagogy, and power.* Peter Lang.

Darder, A. (2017). *Reinventing Paulo Freire: A pedagogy of love* (2nd ed.). Routledge.

Darder, A., & Torres, R. D. (2004). *After race: Racism after multiculturalism.* New York University Press.

Davis, A. Y. (2016). *Freedom is a constant struggle: Ferguson, Palestine, and the foundations of a movement.* Kindle Edition. Haymarket Books.

Dinerstein, A. C. (2016). Organizing hope: Pluriversal concrete utopias against and beyond the value form. *Educação & Sociedade, 37*(135), 351–369.

Dube, R. (2022, August 10). The place with the most lithium is blowing the electric-car revolution. *Wall Street Journal.* https://www.wsj.com/articles/electric-cars-batteries-lithium-triangle-latin-america-11660141017

Eaton, E. M., & Day, N. A. (2020). Petro-pedagogy: Fossil fuel interests and the obstruction of climate justice in public education. *Environmental Education Research, 26*(4), 457–473.

Escobar, A. (2018, February). *Farewell to Development.* Interview by Allen White. Great Transition Initiative. http://greattransition.org/publication/farewell-to-development

Frausto Aceves, A., Torres-Olave, B., & Tolbert, S. (2022). On love, becomings, and true generosity for science education: Honoring Paulo Freire. *Cultural Studies of Science Education, 17,* 217–230.

Freire, P. (1970). *Pedagogy of the oppressed.* Bloomsbury.

Freire, P. (1974). *Education for critical consciousness.* Bloomsbury.

Freire, P. (1997). *Pedagogy of the heart.* Bloomsbury.

Freire, P. (1998). *Pedagogy of freedom: Ethics, democracy, and civic courage.* Rowman & Littlefield Publishers.

Ginwright, S. (2021). *The four pivots: Reimagining justice, reimagining ourselves.* North Atlantic Books.

Grande, S. (2018). Refusing the university. In E. Tuck & K. Wayne Yang (Eds.), *Toward what justice?* (pp. 47–65). Routledge.

Grande, S., & Anderson, L. (2017). Un-settling multicultural erasures. *Multicultural Perspectives, 19*(3), 139–142. https://doi.org/10.1080/15210960.2017.1331742

Gunckel, K. L., & Tolbert, S. (2018). The imperative to move toward a dimension of care in engineering education. *Journal of Research in Science Teaching, 55*(7), 938–961. https://doi.org/10.1002/tea.21458

Gutierrez, G. (1988). *Liberation praxis & Christian faith.* Orbis Books.

Gutiérrez, J. S., Moore, J. N., Donnelly, J. P., Dorador, C., Navedo, J. G., & Senner, N. R. (2022). Climate change and lithium mining influence flamingo abundance in the Lithium Triangle. *Proceedings of the Royal Society B, 289.* https://doi.org/10.1098/rspb.2021.2388

Harvard Divinity School. (2022). *Paulo Freire.* https://rpl.hds.harvard.edu/faq/paulo-freire

Hernandez, J. (2022). A California school district is asking families to rent rooms to teachers. *NPR.* https://www.npr.org/2022/09/07/1120849458/a-california-school-district-is-asking-families-to-rent-rooms-to-teachers

Horton, M., & Freire, P. (1990). *We make the road by walking: Conversations on education and social change.* Temple University Press.

Jasanoff, S. (2015). One future imperfect: Science, technology, and the imaginations of modernity. In S. Jasanoff & H. Kym (Eds.), *Dreamscapes of modernity: Sociotechnical imaginaries and the fabrication of power* (pp. 1–33). University of Chicago Press.

Jerez, B., Garcés, I., & Torres, R. (2021). Lithium extractivism and water injustices in the Salar de Atacama, Chile: The colonial shadow of green electromobility. *Political Geography, 87,* 1–11.

King, M. L., Jr. (1963). *Letter from Birmingham Jail.* Great Neck Publishing.

Kirchgasler, K. L. (2022). Science class as clinic: Why histories of segregated instruction matter for health equity reforms today. *Science Education.* https://doi.org/10.1002/sce.21756

Larsson, J., Airey, J., & Lundqvist, E. (2021). Swimming against the tide: Five assumptions about physics teacher education sustained by the culture of physics departments. *Journal of Science Teacher Education, 32*(8), 934–951.

Lorde, A. (1984). *Sister outsider.* Crossing Press.

Krishnamoorthy, R., & Tolbert, S. (2022). On the muckiness of science, ethics, and preservice teacher education: contemplating the (im) possibilities of a 'right'-eous stance. *Cultural Studies of Science Education, 17*(4), 1047–1061.

Milman, O. (2022, July 21). A 17-minute flight? The super-rich who have 'absolute disregard for the planet.' *The Guardian.* https://www.theguardian.com/

environment/2022/jul/21/kylie-jenner-short-private-jet-flights-super-rich-climate-crisis?CMP=Share_iOSApp_Other

Mohn, T. (2017, November 15). Howard Schultz, Starbucks, and a history of corporate social responsibility. *The New York Times.* https://www.nytimes.com/2017/11/15/business/dealbook/howard-schultz-starbucks-corporate-responsibility.html

Morales-Doyle, D., & Frausto, A. (2021). Youth participatory science: A grass-roots science curriculum framework. *Educational Action Research, 29*(1), 60–78. https://doi.org/10.1080/09650792.2019.1706598

Morales-Doyle, D., Frausto, A., Chappell, M. J., Childress-Price, T. L., Collins, D. A., Levingston, A., Aguilera, A., Canales, K., & Herrera, E. (2019, April). *Beyond PCK: Science teachers building critical historical knowledge for environmental justice.* Presentation at the annual international conference of NARST, Baltimore, MD.

Moura, C. (2021). Science education research practices and its boundaries: on methodological and epistemological challenges. *Cultural Studies of Science Education, 16*(1), 305–315. https://doi.org/10.1007/s11422-020-09984-6

Philanthropy News. (2014). Zuckerberg, Chan, donate $120 million to Bay Area schools. *Philanthropy News Digest.* https://philanthropynewsdigest.org/news/zuckerberg-chan-donate-120-million-to-bay-area-schools

Rodriguez, A. J. (2006). The politics of domestication and curriculum as pasture in the United States. *Teaching and Teacher Education, 22*(7), 804–811.

Rullán, C. (2021). Se reapproprier la science. *Science for the People.* https://magazine.scienceforthepeople.org/online/se-reapproprier-la-science/

Sanchez, A. (2023). Just worlding design principles: children's multispecies and radical care priorities in science and engineering education. *Cultural Studies of Science Education.* https://doi.org/10.1007/s11422-023-10197-w

Singh, V. (2023). Breaking the paradigm: Storying climate change. In S. Tolbert, M. Wallace, M. Higgins, & J. Bazzul (Eds.), *Reimagining science education in the Anthropocene* (Vol. 2). Palgrave Macmillan.

Sousa Santos, B. de (2019). *Educación para otro mundo posible.* Corporación para la Educación y el Desarollo de América Latina y el Caribe (CLACSO).

Spauster, P., & Campbell, M. (2022, August 30). The bitter fight for unions at Starbucks, one year later. *Bloomberg Asia Edition.* https://www.bloomberg.com/news/articles/2022-08-30/the-bitter-fight-for-unions-at-starbucks-one-year-later

Tolbert, S., & Bazzul, J. (2017). Toward the sociopolitical in science education. *Cultural Studies of Science Education, 12,* 321–330.

Tolbert, S., Snook, N., Knox, C., & Udoinwang, I. (2016). Promoting youth empowerment and social change in/through school science. *Journal for Activist Science and Technology Education, 7*(1), 52–62.

Torres-Olave, B., Aviles, N., & Leon, V. (2019). La búsqueda de buen vivir como manifestación de agencia colectiva e identidad profesional: Una exploración teórica del movimiento docente con foco en Chile y Latinoamérica [The search for wellbeing as an expression of collective agency and professional identity: A theoretical exploration of teachers' movement in Chile and Latin America]. *Paulo Freire Revista de Pedagogía Crítica, 17*(22), 49–73.

Torres-Olave, B., & Dillon, J. (2022). Chilean physics teacher educators' hybrid identities and border crossings as opportunities for agency within school and university. *Journal of Research in Science Teaching.* https://doi.org/10.1002/tea.21774

Williams, J., & Tolbert, S. (2018). Finding the freedom to resist: Connecting everyday and spectacular resistance. *Environment and Planning D: Society and Space.* Online forum on walking out: Teaching, working, and striking on the neoliberal campus. http://societyandspace.org/2018/06/27/finding-the-freedom-to-resist-connecting-everyday-and-spectacular-resistance/

Williams, J., & Tolbert, S. (2021). 'They have a lot more freedom than they know': Science education as a space for radical openness. *Cultural Studies of Science Education.* https://doi.org/10.1007/s11422-020-10016-6

CHAPTER 11

Anti-racist Praxis in (Science and) Education

Mahdis Azarmandi, Kelli Gray, Rasheda Likely,
Huitzilin Ortiz, and Sara Tolbert

The idea for this chapter emerged following our town hall discussion for the Science Educators for Equity, Diversity, and Social Justice (SEEDS) 2021 conference, which was broadly focused on anti-racist interventions in science and education. The town hall session was organized as a transnational conversation with panelists and participants who participated virtually from around the world. After the town hall, five of us from the panel came together to reflect further on the many topics and themes that

M. Azarmandi (✉) • S. Tolbert
University of Canterbury, Christchurch, New Zealand
e-mail: mahdis.azardmandi@canterbury.ac.nz; sara.tolbert@canterbury.ac.nz

K. Gray
Universidad de Playa Ancha, Valparaíso, Chile

R. Likely
Kennesaw State University, Kennesaw, GA, USA
e-mail: rlikely@kennesaw.edu

H. Ortiz
University of Minnesota, Minneapolis, MN, USA
e-mail: huitzilin@email.arizona.edu

© The Author(s) 2024
S. Tolbert et al. (eds.), *Reimagining Science Education in the Anthropocene, Volume 2*, Palgrave Studies in Education and the Environment, https://doi.org/10.1007/978-3-031-35430-4_11

emerged during the live session, meeting (virtually) several times over a series of months. Our conversations were open ended and focused on sharing experiences or discussing the difference between our different locations and subjectivities. Mahdis and Sara are currently based in Aotearoa New Zealand, Kelli in Valparaiso, Chile, and Huitzilin and Rasheda in the United States. We all brought our diverse geopolitical locations and positionalities to bear in our conversations about racism and science education.

Mahdis is a woman of color who has worked in Germany, Denmark, the United States, and Aotearoa New Zealand and tries to think about race and anti-racism across different contexts. As a person of color living on Indigenous land, she is interested in how to teach for justice that centers Indigenous sovereignty and how to best build solidarities across difference. Kelli is a bilingual (English-Spanish) Black woman who was born in DC and raised in Maryland. Prior to becoming an educator, Kelli worked as a freelance American Sign Language [ASL]-English interpreter in the DC metropolitan area. As a teacher educator in Chile, Kelli is interested in exploring with pre- and in-service teachers how to disrupt and counter the stock stories (Bell, 2010) that serve to maintain the status quo in education. Rasheda is a Black woman born and raised in Northwest Florida with a previous career as a medical testing scientist and research scientist in Northeast Florida. Her career path shifted after a challenging conversation with seventh-grade Black girls who questioned her identity as a scientist. She is dedicated to expanding and desettling normalized perceptions of science and scientists for K-12 students through incorporating activities such as sports and gaming, DIY hair care product making, and cooking into classroom learning. Huitzilin is a Xicana from metro Los Angeles with raíces in Michoacán and Chihuahua, Mexico. Her current research focuses on the creation and implementation of environmental justice project-based curriculum in the physical sciences. She is a doctoral student and science teacher in Wisconsin. As someone deconstructing her induction into cultural values of whiteness and Western modern science, Huitzilin strongly believes that no student should have to leave their identity at the classroom door. Sara is a white[1] woman from the southeastern

---

[1] As authors, we discussed capitalization of racial identifications and agreed that while we would capitalize Black, we would not capitalize white, for reasons outlined here: https://www.cjr.org/analysis/capital-b-black-styleguide.php. See also, Crenshaw, K. (1990). Mapping the margins: Intersectionality, identity politics, and violence against women of color. *Stan. L. Rev.*, *43*, 1241.

United States (metro Atlanta), with parental roots in Appalachia and southern Louisiana, and a former science, ESOL (English to speakers of other languages), and environmental educator, now teacher educator. Sara sees her role in anti-racist education and anti-racist teacher education as part of a larger justice-oriented and critical pedagogical project which emerges from her inheritances as a white professor with institutional power, as well as through respons-able relations with (and obligations to) others.

Three of us (Rasheda, Huitzilin, and Sara) work directly in social justice-oriented science education, while two of us (Mahdis and Kelli) have expertise in anti-racism and anti-racist education more broadly. We organized our contributions to the chapter around the three overarching questions that were posed to us during the SEEDS town hall, as well as around the topics that emerged from our conversations. While each of us answered questions individually, we also responded to each other's writing. Our transdisciplinary and transnational approach—bringing to bear a wide range of expertise among the geographically diverse group, from political science to science education, literacy education, teacher education, and social justice education—helped us engage with challenges and possibilities from a more robust and multidimensional perspective.

We encourage readers of this chapter not to engage with our conversation from the starting point of, "What does this have to do with science education?" Countless others—too many to name here—have made clear how whiteness and racism are embedded in the fabric of science and science education. And racism and white supremacy[2] are part of the ongoing sociopolitical ills that define what geologists refer to as the Anthropocene. And so we are not trying to make a case for anti-racism in science education, or why we should take on anti-racist projects as part of life and living together in the Anthropocene. (We know most of you probably don't need us to!) Rather, our goal through these conversations was to bring our own experiences to bear on how anti-racist and social justice-driven education can be enacted and sustained, in science and education and beyond. We also found that our transnational and transdisciplinary perspective enabled us to collectively think and imagine beyond the limitations of our own disciplinary (and nationalist) socializations. Furthermore,

---

[2] And, according to some geologists, European colonization marks the beginning of the Anthropocene (Lewis & Maslin, 2015).

transdisciplinary and transnational thinking are imperative, in our view, for transcending and acting upon the problems that were, in large part, created by nationalistic- and isolationist- (including within disciplines) informed action.

Question 1: What Vision Do You Have for Cultivating Anti-racist and Social Justice-Driven Education?

Rasheda: When considering anti-racist and social justice-driven education, I would encourage one to ponder first what might it mean for them to be anti-racist personally. Understanding that racism is endemic to [U.S.] American society (Crenshaw et al., 1996), how are these ideals of racism reified through my own actions and values? By asking oneself "How are my actions upholding racism?" the solutions become deeply personal. Also, racism is a part of a larger interlocking system of oppression such that multiple identities can be experiencing oppression in one person. Therefore, considering racism as part of a largely intersectional problem to address, I think, provides an opportunity for true transformative education. My vision for cultivating a more socially just educational space would be to consider "How am I supporting oppression in my educational practices? What choices can I make toward social justice?"

Through identifying the ways in which racism is interlocking and operationalized through institutional systems and structures, I continuously seek opportunities to oppose the status quo through my personal choices in instruction by providing space for marginalized and minoritized cultures to be more broadly considered. Although these choices may not move a quantitative needle of assessment or curriculum development immediately, these choices do cultivate anti-racist and social justice education. For example, Black girls have been over-disciplined and pushed out in K-12 spaces, leading to behavioral difficulties and mental health trauma due to expectations and norms rooted in patriarchy and whiteness (Collins, 2002; Morris, 2016). The ways minoritized students experience trauma and harm through education require seeking new visions and pedagogical practices for their survival and thriving. The daily choices toward transgressing the status quo, identifying oppressive structures, shifting pedagogy toward liberation, and expanding assessment strategies support educational opportunities that are inclusive and diverse. Cultivating anti-racist and social justice-driven educations requires co-conspirators and agitators from within the educational system. However, these allies and actors would need to reimagine social justice in education for themselves first. Spaces for educators and stakeholders to interrogate the structures that continue to other and erase communities and populations would lead to identifying ways education could be broadened for all. Specifically, guided and continued conversations focused on desettling hierarchies within education across institutions, learning centers, and schools nationwide would be transformative. Also, requiring products such as strategic plans and budget justifications from these conversations would support actionable steps toward social justice. Often these conversations and meetings are a means to an end rather than tangible steps toward liberatory praxis. My vision for cultivating anti-racist and social justice-driven education requires daily choices.

Huitzilin: When I think about anti-racist teacher education, I first think of my own experience in teacher training. I did this in an alternative certification program through a small state university, and while I'm glad I had the opportunity to obtain my certification this way, I often wonder what my education might have looked like had I had the privilege of taking two years off work to complete coursework through a daytime, in-person program at a major state university.

When I imagine a future of anti-racist and social justice-driven teacher education, I think of prospective teachers who have the potential to be amazing educators with deep knowledge about themselves and their communities, but do not have the privilege of attending high-quality training programs for financial or other reasons. The bar in a lot of places is simply too high without adequate supports for those teachers, especially as we know high-performing countries have these types of programs as a matter of course for prospective teachers (Wei et al., 2009).

Having done some research, it appears that some new programs are now in place for paid pre-licensure apprenticeships in small private colleges, and these are absolutely a step in the right direction. Creating fertile ground for "grow your own" programs, paid training, and loan-for-service programs for these prospective teachers should be the cornerstone of programs that train teachers primarily serving diverse demographics. However, the answer to this prompt or question is incomplete without a vision for student teachers from more privileged backgrounds, who still comprise a majority of prospective teachers in traditional pre-licensure programs. These teachers must learn—and quickly—that whatever their motivations for becoming a teacher are, the realities of the marginalized students they teach must be taken into account. They must make peace with the idea of, and go to war against, systemic racism starting on day one. This can be uncomfortable or even anxiety inducing, but must be done.

Lastly, a teacher's education should not end the day a student obtains their licensure. More structured teacher induction programs, possibly as part of a district–university partnership, could function as a buffer or a sort of vaccine against first-year burnout and rates of attrition (as documented in Ingersoll et al., 2018). In short, I believe that an anti-racist and social justice teacher education program must give prospective teachers *the time they need* to be able to cultivate a reflective and successful practice.

Kelli: From October 2019 to March 2020 Chile experienced a series of massive demonstrations and riots nationwide known as the Estallido Social. These demonstrations were triggered by an increase in metro fares against which high school students planned a coordinated fare evasion campaign. Although the hike in metro fares triggered the demonstrations, protests continued for more than four months due to inequalities in health care, education, and social security, among other social inequalities. In addition, in September 2021, in the north of Chile in Iquique, citizens once again took to the streets protesting the large number of undocumented immigrants crossing the border and holding signs saying "No más ilegales." Although both the Estallido Social and the protest against the wave of undocumented immigrants entering the country have brought to the forefront many topics of injustice that plague Chilean society, there must be a concerted effort within the educational system to teach about and for social justice. This effort must start with our teacher preparation programs.

In 2014 the University of Playa Ancha in Valparaíso, Chile, like many other universities in the country, implemented changes to their program

of studies. At Playa Ancha we incorporated two courses on critical pedagogy. Although this is a good start, it is far from enough. Fundamental to
cultivating an anti-racist and social justice-driven educational curriculum
for schools is first educating preservice teachers within an anti-racist pedagogical framework. Furthermore, teaching programs must be committed
to intentionally recruiting and maintaining Black, Indigenous, and People
of Color [BIPOC] educators and educators from other marginalized
groups. Thus, for me a vision for anti-racist and social justice-oriented
education must include educators from marginalized groups working in
empowered spaces within teacher preparation programs, a curriculum for
teacher preparation that is based on anti-racist pedagogical theories and
social justice teaching, and school-based curricula for both elementary and
high school students that are based on critical social justice and anti-racist
education.

Sara: After a teacher education meeting one afternoon in Tucson,
          Arizona, a group of us (predominantly white) teacher educators
          stayed after to discuss ways in which we could intervene in the
          racist and white supremacist climate of schooling in the state of
          Arizona, in the wake of the anti-Ethnic Studies ruling in the
          Tucson Unified School District. Kelli Gray, who was teaching as
          a graduate student instructor at UA at that time, said to the
          group, "I don't know if I can trust any of you as 'anti-oppressive
          educators' until you take a look at your own selves, and your
          own programs." She called on us to do the work in our own
          "backyards," so to speak, first, before trying to engage with the
          racist and white supremacist climate of K-12 schooling in the
          state of Arizona. And she was right. Under her leadership, a few
          of us (mostly precarious workers including pre-tenure assistant
          professors, graduate students, and adjunct faculty) formed a
          committee to bring to the forefront and disrupt the ways that
          students of color, and faculty of color, in our "social justice edu-
          cation" programs, were experiencing micro- and macro-aggres-
          sions in our department, even by some faculty members who
          claimed a social justice agenda (Tolbert et al., 2014).

Anti-racist practice in education and (predominantly white) teacher
education often turns its gaze toward teaching students how to be anti-
racist, or preservice teachers how to be anti-racist, when anti-racist praxis

is first and foremost about critically analyzing and dismantling the racist systems in which we are already complicit. I have seen this play out within public schools and universities, in the United States and now here in Aotearoa New Zealand, where cultural competence in predominantly white institutions is conflated for anti-racist practice, while the university remains a hostile place for many people of color, particularly for women of color (Azarmandi & Tolbert, 2022). My vision for anti-racist and social justice-driven education is about willingness "to participate in a killjoy movement" (Ahmed, 2017, p. 267), one in which we collectively and intentionally cause disturbance, one that "begins by recognizing inequalities as existing" (p. 252)—a killjoy movement that starts right where we are.

Mahdis: When I think of anti-racism I often think of being willing to take a risk. If racism, as Ruth Wilson Gilmore says, "is the state-sanctioned or extralegal production and exploitation of group-differentiated vulnerability to premature death" (2007, p. 28), then it means that being racialized and living under white supremacy means being at constant risk of violence. Anti-racism then, as Goldberg points out, has to take a risk. Anti-racialism constitutes merely a stand "against a concept, a name, a category, a categorizing [which] does not itself involve standing (up) against (a set of) conditions of being or living" with racism (Goldberg, 2008, p. 10). For Goldberg, anti-racialism is devoid of that risk.

When looking at the university I currently see a lot of colleagues talking or wanting to talk about anti-racism; many assure me that what they teach is committed to anti-racism, and I always ask myself, what is the risk that is being taken here? If anti-racism means standing up and challenging those conditions of racism, to the extent of risking one's own position, resources, etc., what is it that we are willing to put on the line? Especially for those who are racialized white.

For example, in the last two years a number of colleagues have approached me to talk about their interest in understanding racism and committing to anti-racism. Many turned to Robin D'Angelo to understand "white fragility" and are surprised that it is not a book I would recommend as a starting point for anti-racism. While I find the notion of white fragility useful at times, starting with D'Angelo and focusing on what I describe as making "nicer" or more culturally competent white folks don't necessarily change the structures that produce premature death. It is therefore crucial to assess anti-racism in its commitment to such risk; that is, do groups work toward reducing racist incidents, or do they also question the very structures that enable racist culture in the first place? For me, anti-racism is thus also about shifting whose work we start with and how we engage in action beyond "raising awareness." Rushing to understand our fragility in some ways is like trying to find the fastest way to redemption—"hey look, I'm not racist, I'm one of the good ones."

For me as an educator, understanding structural racism and how my own role as a teacher may play in reproducing and disrupting it is central to anti-racism. Especially, as Sleeter (2017) points out, many programs claim to adopt a justice-oriented stance, yet do so in "incremental or symbolic ways" such as adding a multicultural education course to the curriculum or by hiring a professor of color to the faculty, while otherwise preserving programs "defined by White interests" and curricula reflecting white sensibilities (p. 158). I really like Dei's definition of anti-racism education "as an action-oriented educational practice to address racism and the interstices of difference (such as gender, ethnicity, class, sexuality, ability, language, and religion) in the educational system" (2014, p. 240). So, whenever I am told some workshop, intervention, or framework is supposed to be "anti-racist" or is supposed to make us all more anti-racist, I look for the action-oriented practice. Are we just trying to redeem ourselves and make sure we are not the bad ones, or are we taking a risk to question structures that we might be entangled in or those we are upholding and benefitting from?

WHAT ARE SOME CHALLENGES YOU HAVE EXPERIENCED IN ENACTING THIS VISION?

Mahdis: Nobody loves a killjoy. The biggest challenge for me has been getting folks to see that being a killjoy is not about killing joy but about working toward structures that create less premature death. The other challenge is being able to talk about racism and its complexities beyond false binaries. For example, much of the effort in education here in Aotearoa New Zealand have focused on cultural competency and cultural difference rather than looking at racism as a political process. When challenging the limitation of cultural competence (which is important), I find I am heard as saying we should maintain the status quo. I also see a lack of intersectional analysis and solidarity among racialized groups; for example, we need to understand how white supremacy manifests among groups of color, why it is important to think of Indigenous dispossession when trying to address Islamophobia in a place like Aotearoa New Zealand, etc.

Sara: I chose a career in teacher education for explicitly political reasons. Though I had initially planned to return to classroom teaching after completing my master's degree, my own teacher education coursework was so bereft of sociopolitical perspectives that I decided to pursue doctoral studies in education. At my first position as an assistant professor and teacher educator in Arizona, I hoped to facilitate a critical pedagogical approach to preservice teacher education—to explore, for example, what could conscientization (Freire, 1970) look like as a pedagogy of/for the privileged in science education (e.g., Schindel et al., 2021)? Fostering deeper understandings of intersecting contexts of oppression (e.g., racism, heteronormativity, sexism, classism) became part of the science teacher education curriculum. As one example, (predominantly white) students would read astrophysicist Neil de Grasse Tyson's convocation speech,[3] in which he talked candidly about his experiences of racism in school science. Many preservice teachers in my courses would complain about why he seemed so depressed; shouldn't he use his platform to celebrate his success? As Mahdis points out, no one loves a killjoy (Ahmed, 2010).

Rasheda: Some challenges I have experienced enacting a vision of antiracism through personal reflection is in order to engage with science learning experiences beyond education is to enter spaces where sociocultural systems have not been interrogated. From my experience as a scientist, there was little

---

[3] *Sara*: Since then (in 2011 when I started teaching at University of Arizona), I learned of the sexual harassment allegations against Neil de Grasse Tyson. This blog post by Chanda Prescod-Weinstein underscores the importance of centering Black women's perspectives in this and related cases: https://blogs.scientificamerican.com/observations/sexual-misconduct-allegations-against-neil-degrasse-tyson-reveal-the-complexity-of-academic-inequality/

opportunity given to critical self-reflection or critical questioning of systems and structures in place broadly (Mensah & Jackson, 2018). Having several identities outside of what is represented in science education as who scientists are and what scientists do, assimilation became the way to reconcile these tensions. Challenges enacting anti-racist visions are deeply internal and personal due to the personal choices of continuing and participating in one's own marginalization. The choices toward more equitable learning experiences mean to consider one's own marginalized experience and ways to not replicate similar systems of oppression.

I had an experience as a graduate student in a biology master's program where anti-racism should have been made as a personal choice. I was working in a biology research lab late on a Friday evening. As I was packing up to leave, I left the secured area to fill a water bottle. When headed back into the secure area, I was met by a white man coming out of the door as I was inserting my key into the door. I was a little startled since no one is usually in the lab area late on a Friday night. As I stepped aside so that he could come out of the area and I walked in, he blocked the door and said, "This area is for secure personnel only." What would have been the response if this man was used to seeing Black women in the science lab space as students, lab supervisors, or even professors? Additionally, I had not seen him before—was he a professor assuming a Black woman was not supposed to be in that area? Was he a student who had not seen another Black female student? Had his personal engagements been challenged or at least had there been conversations around equity and diversity, or lack thereof, among the students, faculty, and staff in the science department? Although conversations challenging structural and systemic biases have been happening at large institutional levels, these local-, community-, department-level conversations should be encouraged and used to shift experiences for marginalized and minoritized students.

Huitzilin: I had trouble identifying the source of the challenge for me and had to think on this one for quite some time, but now that I've had a chance to consider it, I imagine the biggest personal challenge is something like burnout. There are days where the task of achieving a socially just education system seems too big for even one person to carve out a niche—the monstrous inertia of the status quo is unbeatable. Choosing which battles are worth fighting, on top of all the other day-to-day decisions that must be made in a career where one makes thousands of decisions a day, is sometimes just a bit too much. Then I think about the multiplicative effect of thousands of other educators experiencing the same feeling and I start to see just the edges of the problem at hand.

Kelli: The biggest challenge that I have faced here in Chile is English as a Foreign Language [EFL] teachers not being familiar with how to teach from a critical perspective. In 2014, the Universidad de Playa Ancha [UPLA], the university where I work, along with some other universities, re-vamped their teacher education curricula in several teaching areas, English being one of them. In the Teaching English to Speakers of Other Languages [TESOL] teacher education program at UPLA, they included two courses on critical pedagogy: one is a foundation course, and the other is a practical course on how to take theory and translate it into practice in the classroom. This is a wonderful start; however, it is not enough. The concepts and theory around teaching from a critical perspective should be a theme or strand that is woven into every course in some way. It is my belief that if we want classroom teachers to be more justice-oriented in their teaching, teacher educators need to be more justice-oriented in their work with preservice teachers. In fact, our programs of study need to be justice-oriented.

DESCRIBE CREATIVE MANEUVERS OR INTERVENTIONS YOU OR YOUR COLLEAGUES HAVE ENACTED TO RESPOND TO COUNTER-RESISTANCE STRATEGIES

Huitzilin: One of the very first papers I read that served as an inspiration was a chapter by Rochelle Gutiérrez on political *cono-cimiento* in the teaching of mathematics (Gutiérrez, 2017). After reading that chapter, I have long considered that part of my personal strategy to counter-resistance to anti-racist teaching practices should include development of that *cono-*

cimiento in my science teaching. For my students, what that looks like on a daily basis is helping students become aware of the ways in which the science I'm teaching doesn't occur in a vacuum—neither in terms of the content that I'm teaching them nor the context I'm teaching it in. Essentially, putting information out there that might not be seen—making the invisible, visible—has been an effective strategy for dispelling resistance in students. Of course, when you start to operate in spaces beyond individual classrooms, countering that resistance becomes more difficult. An excellent example of resistance to anti-racist schooling in these larger spaces like the state arena is Proposition 203 in Arizona, which repealed bilingual education laws and effectively made us an "English-only" instruction state. Some creative insubordination—such as making bilingual materials available to *parents*—has been an antidote to that for myself and my colleagues. Lastly, there's the in-between spaces—interactions with other educators and administrators at the building and district levels—that require *conocimiento* to navigate. You have to learn the limits of how hard you can push back before the cost of the effort outweighs the effectiveness of the strategy. It's a fine balance between pushing in the ways that need to be pushed and preserving your emotional and spiritual energy in a way that you can still give of yourself to your students—the people that really matter.

Rasheda: For me, creative interventions build on previously developed strategies toward anti-racist research and education. As an educator who seeks to actively teach to transgress (thank you, bell hooks) in the science learning space, I deliberately seek opportunities to push against individualized experiences and achievement but rather support collaboration and small group work. Specifically for engaging in science and engineering practices, I want students to see learning and discovery in community rather than by one individual. "Students learn science by actively engaging in the practices of science, including conducting investigations; sharing ideas with peers; specialized ways of talking and writing; mechanical, mathematical, and computer-based modeling; and development of representations of phenomena" (Houseal, 2016, p. 1). Since students learn science through collaboration, shifting who is seen as an expert or scientist in K-12 science learning spaces is one practical way to transgress norms in science learning.

Kelli: Bell (2010) in her book *Storytelling for Social Justice: Connecting Narrative and the Arts in Antiracist Teaching* defines four story types used in the Storytelling Project Model that describe how people "talk and think about race and racism in the United States" (p. 22). These four story types are stock stories, concealed stories, resistance stories, and emerging/transforming stories. Stock stories—stories that maintain and even perpetuate the status quo on topics such as race and racism, or in the case of Chilean education inclusion, gender equity, immigration, or LBGTQ+ topics— are commonplace. However, just as commonplace to these stock stories are the concealed and resistance stories that serve as counternarratives or counter stories.

One major way I have tried to counter the stock stories I experienced and heard as a parent and as an educator was by starting my own school, an example of an emerging or transforming story. Built on the principles of an inclusive, multicultural, and social justice education, our school, through its climate and curriculum, was deliberately built "to challenge the stock stories, [to] build on and amplify concealed and resistance stories, and [to] create new stories to interrupt the status quo and energize change" (Bell, 2010, p. 25).

Sara: Mahdis and I have recently written elsewhere (Tolbert et al., 2022) about how a pedagogy of alienation might serve as a means to disrupt the culture of complacency (including our own, our students', our colleagues') that we feel is often the biggest impediment to social change. People become too comfortable or too resigned in feeling like it's too big of a problem for me to solve, or things are mostly "fine," while we carry on drinking our lattes—similar to what Zukin (2010) refers to as "pacification by cappuccino" (in the context of gentrification). It's easy for white allies to be susceptible to this as we do not experience the embodied effects of racism— which is why I think our responsibilities as anti-racist allies have to include real vulnerabilities and risks, as others have described here in this chapter. It's easy to "say" the things that make us look like "good allies"—in fact, these are things that are so often rewarded by the institution—because they are essentially ways "to not do things with words" (Ahmed, 2016). They don't actually challenge the status quo and can even exacerbate it through the illusion of progress. A pedagogy of alienation embraces refusal, rage, and anger as an unsettling alternative to "diversity and inclusion"; in becoming (unapologetically) disaffected and alienated, and seeking systemic change, together we forge new solidarities (Tolbert et al., 2022). This looks different for everyone, but for me it starts with rejecting the politics of domestication in science education—and in the academy (Rodriguez, 2006).

Mahdis: More recently, the main way I have managed to navigate the "everyday political whiteness" (Ambikaipaker, 2019) of academia has been finding others who find joy in "killing joy"—that is, building relationships and strategically organizing with those committing to dismantling white supremacy. For example, in institutional discussions about curriculum or teaching, we have tried to support each other and make sure that the burden of speaking up against racism does not always sit with Indigenous colleagues or colleagues of color. Building solidarity across minority scholars but also with white colleagues who choose to be accomplices and risk takers, rather than just fair-weather allies, is how we can survive the structure that is designed to exclude us. The other way, as Sara has mentioned, is the pedagogy of alienation and the politics of refusal: refusing to be loyal to an institution that doesn't love us. This doesn't mean that we cannot love the connections and solidarities we build, love the classrooms we co-construct, or love the political project of moving beyond the Westernized and settler colonial university. Ambikaipaker writes,

Mitigating the power of the institutions and rule of law to act in alignment with racial minorities' rights or claims of justice is hegemonic work that people investing in whiteness must undertake on an everyday basis to stabilize hierarchical power relations among other bodies, identities, epistemologies, and cultural lifeways. Hence, racism cannot be understood as simply an aberration in everyday institutional life. It is in fact necessary and habitual political work.

Thus, in order to change the university, we need to make disrupting white supremacy our habitual political work.

CONCLUSION

Huitzilin: This morning I was having a Socratic seminar with my students on the subject of climate change and of "wicked problems" (Rittel & Webber, 1973) in general. Climate change is one of those big, interconnected issues that are too big for any one person to solve, where a correction at one node can influence other nodes in positive and negative ways, and touches so many different intersections on social, economic, and political levels. Because of this, the issue of climate change has depth and complexity outside the realm of understanding for the average high school student and can seem much too big to fix. Looking back to my response to question 2, I can see my students' feelings of disillusionment and cynicism during this seminar mirrored in my previous words about the challenges of enacting an anti-racist and social justice-oriented science education. I realized in that response that I wasn't listening to my own lesson.

Whenever I teach climate change, I want students to leave my classroom feeling like there's *something* within their agency that they can do. So I ask them to identify *one thing* within their power, something that they could start doing tomorrow. A lot of them come up with wonderful ideas. They find their lane. For me, a conclusion to this discussion would be that we must do the same—find our lane. It might change from time to time, but knowing that you're doing your *one thing*, or your several things, is the way to start.

Sara: One thing that our conversations, as well as events occurring at our own institution (which Mahdis and I have spoken about as well over the past few weeks with you all), have made me contemplate is the question of who benefits from doing (or being seen as doing) anti-racist work? And also, who is viewed/positioned as an expert and who is not? Who should/can ethically lead the work, particularly in institutions like the academy, where whiteness is so ever-present yet so invisible, even to those who identify as anti-racist? I often marvel at anti-racism in science and education panels that are dominated by white folks, and at how anti-racist work becomes a platform for career advancement for some (white educators) while presenting considerable risks—and/or not being valued as intellectual or scholarly contributions—to/ for others (BIPOC educators). This is not to say that there isn't a critical role for white allies/accomplices in anti-racist movement but that these are ongoing ethical tensions that need to be at the forefront of our thinking (particularly for allies/accomplices/co-conspirators)—and why the idea of "praxis" is so critical. The Showing up for Racial Justice (SURJ) movement is an example of how an organization of predominantly white anti-racist allies can take up the call from civil rights leaders to support a Black and People of Color [POC]-led struggle, by "being in relationship with and following the leadership of liberatory Black and POC-led movement organizations" (https://surj.org). There is no "rule book," per se, but it is about the specificity of acting (versus not acting) in solidarity, with care and attention to one's own positionality and complicity in whiteness. Of course, we will make mistakes along the way, so we need to be prepared to humbly own and apologize for—and if/when possible repair the harm from—those mistakes. And I love thinking about it in the context of Huitzilin's commentary above—what is the one thing we could do that is within our power, something that we could do tomorrow? And for each of us, this will (must) look and be different, and must be informed by an ethic of good, humble, and respons-able relations.

Mahdis: I want to connect to what Sara has mentioned about who is centered in conversations about anti-racism and what kind of anti-racism is advocated. For example, are we using culturally responsive pedagogy to mean anti-racism? And who are the so-called experts? Often the attempts to be anti-racist push for incremental change and focus on individual learning/ change, as if racism is some kind of moral failure rather than a system of white supremacy. To work within our own power also means understanding where we are implicated and complicit in this system, that is, to understand where and how we keep white supremacy in place. What helps me continue doing this work is that we don't do it in isolation; it is the relationships and solidarities that make change possible, and I'd like to think how we can do that in ways that are locally grounded but also globally oriented and recognize the histories of empire and race outside our own specific contexts.

Rasheda: I want to revisit a quote by Toni Morrison that culturally sustaining pedagogy (Paris & Alim, 2014) follows: "What would our pedagogies look like if this [white] gaze weren't the dominant one?" (p. 86). True anti-racist pedagogies require centering someone else—someone else's culture, someone else's expertise, someone else's knowing, someone else's normal. What might education look like if whiteness were not centered or considered as the basis for learning? Considering other possibilities that do not replicate or mirror the same systems and structures of oppression is so exciting for me. Dreaming of learning experiences and settings that are celebratory and uplifting for the diversity of learners and teachers brings me deep joy and renewed strength to continue in this work. Beyond dreaming, every day, personal, socially just choices are amplified in a community of co-conspirators and agitators like my co-authors. To echo Mahdis, "What helps me continue doing this work is that we don't do it in isolation, it is the relationships and solidarities that make change possible." I am more encouraged and better situated to dismantle harmful and violent learning spaces and build something much better when I know that I am not ever alone. Future generations deserve the just and equitable changes we are seeking now.

Kelli: For me writing this chapter came at the perfect time. I have felt alone at my university as I try to center something other than whiteness, maleness, neurotypicalness, and heterosexuality among other dominant structures. To use the words of Huitzilin, my "lane" has been reaffirmed through our collaboration in writing this chapter. I have been energized to look for ways to continue to move forward in speaking my truth. Furthermore, Rasheda's question "What might education look like if whiteness [add any other dominant structure] were not centered or considered as the basis for learning?" is pushing me to think more deeply about my own work in teacher education and how I am (or am not) deliberately designing my classes to challenge the stock stories in education in general and second language education in particular. It is refreshing, joyful, and empowering to envision ways my teaching and presence can counter oppressive educational structures in the hope of building new, transformative stories that truly celebrate the diversity of learners in my classroom and in classrooms across Chile.

References

Ahmed, S. (2010). *The promise of happiness*. Duke University Press.

Ahmed, S. (2016). How not to do things with words. *Wagadu: A Journal of Transnational Women's and Gender Studies, 16,* 1–10.

Ahmed, S. (2017). *Living a feminist life*. Duke University Press.

Ambikaipaker, M. (2019). Everyday political whiteness and diversity university. *Kalfou, 6*(2), 268–279.

Azarmandi, M., & Tolbert, S. (2022). On the non-performativity of 'being-well'—A critique of institutional wellbeing policy. In A. Kamp (Ed.), *Wellbeing: Global policies and perspectives*. NZCER Press.

Bell, L. A. (2010). *Storytelling for social justice: Connecting narrative and the arts in antiracist teaching*. Routledge.

Collins, P. H. (2002). *Black feminist thought: Knowledge, consciousness, and the politics of empowerment*. Routledge.

Crenshaw, K., Gotanda, N., Peller, G., & Thomas, K., Eds. (1996). Introduction. In D. Kennedy & W. Fisher (Eds.), *The canon of American legal thought* (pp. 887–891). The Princeton University Press. (Reprinted from *Critical race theory: The key writings that formed the movement* (pp. xiii–xxxii). The New Press.)

Dei, G. J. S. (2014). Personal reflections on anti-racism education for a global context. *Encounters in Theory and History of Education, 15,* 239–249.

Freire, P. (1970). *Pedagogy of the oppressed*. Bloomsbury.

Gilmore, R. W. (2007). *Golden gulag: Prisons, surplus, crisis, and opposition in globalizing California*. University of California Press.

Goldberg, D. T. (2008). *The threat of race: Reflections on racial neoliberalism*. Wiley-Blackwell.

Gutiérrez, R. (2017). Political conocimiento for teaching mathematics. In S. E. Kasteberg, A. M. Tyminski, A. E. Lischka, & W. B. Sanchez (Eds.),

Building support for scholarly practices in mathematics methods (pp. 11–37). Information Age Publishing.

Houseal, A. K. (2016). A visual representation of three dimensional learning: A model for understanding the power of the framework and the NGSS. *The Electronic Journal for Research in Science & Mathematics Education, 20*(9), 1–7.

Ingersoll, R. M., Merrill, E., Stuckey, D., & Collins, G. (2018). *Seven trends: The transformation of the teaching force—Updated October 2018.* CPRE Research Reports. https://repository.upenn.edu/cpre_researchreports/108

Lewis, S. L., & Maslin, M. A. (2015). Defining the Anthropocene. *Nature, 519*(7542), 171–180.

Mensah, F. M., & Jackson, I. (2018). Whiteness as property in science teacher education. *Teachers College Record, 120*(1), 1–38.

Morris, M. (2016). *Pushout: The criminalization of Black girls in schools.* The New Press.

Paris, D., & Alim, H. S. (2014). What are we seeking to sustain through culturally sustaining pedagogy? A loving critique forward. *Harvard Educational Review, 84*(1), 85–100.

Rittel, H. W. J., & Webber, M. M. (1973). Dilemmas in a general theory of planning. *Policy Sciences, 4*(2), 155–169.

Rodriguez, A. J. (2006). The politics of domestication and curriculum as pasture in the United States. *Teaching and Teacher Education, 22*(7), 804–811.

Schindel, A., Grossman, B., & Tolbert, S. (2021). Mobilizing privileged youth and teachers for justice-oriented work in science and education. In K. Swalwell & D. Spikes (Eds.), *Educating elites for social justice: Promising practices and lessons learned in K-12 schools* (pp. 147–158). Teachers College Press.

Sleeter, C. E. (2017). Critical race theory and the Whiteness of teacher education. *Urban Education, 52*(2), 155–169.

Tolbert, S., Gray, K., Stevens, V., Combs, M., Griego-Jones, T., & Dias, J. (2014, November). *"Becoming" and "being" a socially just department of teaching & learning: Encouraging a critical stance.* Annual meeting of the National Association of Multicultural Education (NAME), Tucson, AZ.

Tolbert, S., Azarmandi, M., & Brown, C. (2022). A modest proposal for a pedagogy of alienation. In P. Jandric & D. Ford (Eds.), *Postdigital ecopedagogies.* Springer.

Wei, R. C., Darling-Hammond, L., Andree, A., Richardson, N., & Orphanos, S. (2009). *Professional learning in the learning profession: A status report on teacher development in the United States and abroad.* National Staff Development Council.

Zukin, S. (2010). *Naked city: The death and life of authentic urban places.* Oxford University Press.

CHAPTER 12

Science Education: From an Ideology of Greed to an Ideology of Thriving

Kurt Love

We are still living in a time when greed is normalized and used as a motivation for politics, business, research, scholarship, and education. Greed is the root ideology that creates social injustices and unsustainability, but despite its destructive outcomes, greed continues to motivate us. Teachers, parents, and politicians tell students at an early age that their success in school is tied to their ability to earn a higher salary, which means that we are reinforcing greed as a motivator. Greed drives capitalism as it did feudalism. Greed begets exploitation for profit. Greed created slavery via the invention of race as assigned to groups of humans in order to generate and concentrate profit via exploitation. Greed is the evolving force behind globalization (slavery 2.0). Greed is the destructive force behind global climate change, which was created by scientists and engineers who fueled the exploitative Industrial Revolution with fossil fuel technology. Greed seeps into every aspect of mainstream education, and science education is influenced by a corporate-STEM movement to create employees for

K. Love (✉)
Central Connecticut State University, New Britain, CT, USA
e-mail: lovekua@ccsu.edu

© The Author(s) 2024
S. Tolbert et al. (eds.), *Reimagining Science Education in the Anthropocene, Volume 2*, Palgrave Studies in Education and the Environment, https://doi.org/10.1007/978-3-031-35430-4_12

companies that concentrate wealth, contribute to weaponizing governments into "global powers," and push profit-based technological solutions rather than more balance with nature and protection of the commons. Scholars across all content areas use a greed discourse for their own personal profit/status in their fields, and science education is no exception. Greed is normalized and moralized in the current corporate-STEM movement in science education, which is coupled with the willing omission of scientist and engineer contributions to climate change, widespread destruction via nuclear weapons, and labor exploitation.

This chapter uses a (semi-)fictional context (a device used by philosophers such as Plato, Friedrich Nietzsche, Voltaire, and James Baldwin) in order to more fully humanize thinking. This chapter is written in the form of a story in order to explore waves of greed ideology that flow into and through professional discourse from the perspective of a complete outsider looking in ... a perspective very much needed in order to better see ourselves and our entrenchments that blind us.

The unanswered aspects, the disequilibria, and the loose ends of this story are intentional—much like a movie that asks the audience to keep thinking about the storyline. It is my intent to leave spaces open, ambiguous, and awkward at times for the sake of the reader's pleasant discomfort. Consider it yoga for the mind.

Most importantly, let's implore each other to move toward an ideology of thriving, whatever that may be.

PART 1: SCIENCE EDUCATION TOWARD GREED AND *HOMO GREEDYUS*

"Interesting," said the alien, who was soun ding more like a counselor than ... well ... an alien (pejorative). The non-Earth being (NEB) (non-pejorative) sat on a single chair in the lobby of a major hotel chain in a major metropolitan city (in a country that thinks that it's "totally major") talking to a person with a majorly recognizable name around the "world" (a word with a rather confusing usage to the NEB) of science education researchers ... speaking only in the key of D minor because of its calming effect (while also being the most popular minor scale, which is to say, familiar and comforting, as the NEB hoped). The NEB, who was able to morph into various life forms after considerable training, sat in the hotel lobby with a laptop, an iced chai tea, and a universal translator in its fake human ear disguised as a very popular stylized white cordless earbud. The NEB assumed a pseudonym (as another layer of human social camouflage) based on the name of a very real science education researcher who typically avoided professional conferences unless absolutely necessary. "So, learning science in public schools is ultimately tied to broader socioeconomic outcomes like money, salaries, and businesses? Doesn't that just mean that science education is ultimately just a costly externality for societies that benefits STEM-based corporations?"

The notable researcher, who was on his third cocktail at 5:45 pm, after presenting in two sessions earlier, and listening to only one session that day (because it was a colleague in his university department) said, "What are you talking about? It's FREE public education. I mean, ya know, taxes, but it's FREE." The word seemed to have a whimsical feeling in his mouth almost like the word "weeee."

It was not entirely clear to the NEB if the notable researcher (who was also notably inebriated) understood or even considered the controversial point that the NEB posed. "By definition, that's not free. It's just differently funded." The NEB quickly found a definition of "free" in an online dictionary just to be sure. The projection of the definition flashed in front of its fake human face in infrared lighting so the *homo sapiens sapiens* (HSS) in the room could not see it. "Free: not costing or charging anything ... *free school*" (Merriam-Webster). Oh, so that's why the HSS said that statement. The dictionary even seems to believe that schools are free, which the NEB thought was weird since nearly all students of color attend schools that are underfunded throughout this country that constantly boasted of itself like an insecure teenage boy. The language usage seemed sloppy, but this was a pattern that happened fairly consistently ... never mind the poor spelling of words.

The HSS professor looked at the NEB and paused for a moment. "Yea, you're right!" he exclaimed and finished his cocktail. "Why do we call it 'free' instead of 'publicly funded'? That's totally inaccurate." The NEB assessed the HSS professor to be at a minor level of inebriation with a maximum amount of mental plasticity.

The HSS professor seemed to go into car salesman mode: "But, we live in a capitalist society, ergo it goes that little de facto capitalists we make. Yes, even in science. Or, maybe on some level because of science, right?" He was a little surprised by his own admission. "Was I loud just then?" The HSS professor had become very well known in recent years and made a second career with keynote speeches, speaking engagements, and large corporate-funded grants that his university boasted about on their web page and he boasted about on his curriculum vitae and others boasted about when they introduced him prior to him giving a speech that they paid him for. The NEB made a face gesture that was deliberately ambiguous. "Ok, good," said the HSS professor. "I can't believe that I just said that." It was almost like he broke character or forgot his line. "Well, maybe I can. I am certainly doing a lot better financially these days," he whispered rather deviantly, like a kid who just realized that he got away with stealing a pocketful of candy from a convenience store.

The NEB already knew the answer, but the HSS professor was always happy to hear the affections of an adoring academic, "I know. What you're saying seems to be really well-received by so many in the field, right now." Fortunately, the NEB species biologically has no gag reflex.

In a moment the NEB froze, scared that one of the HSS in the room called to it by its actual name. Then, the NEB reminded itself for the 42nd time that its real name was also pronounced very similarly to the acronym of that conference organization. The NEB hummed a few notes of the D minor scale to get back in tune.

"So, here's the thing. Here's what people in our profession don't realize about all of this."

"That it's all male cow manure?" asked the NEB with a perfectly straight face.

The HSS professor replayed that statement in his head. "Ha!" He chuckled some more, stopped the waiter's forward motion, ordered another 14-dollar hotel lobby beverage, and stated, "It's like a balanced chemical equation. If you say STEM or STEAM enough times, talk about future jobs, and call it 'equitable,' you can cash in, too." He paused, carefully looked around the room for possible competitors, and turned back to the NEB dressed as a human, "I'm White, right? So are most academics in science education, aren't they? I genuinely want all people of all backgrounds to come into STEM fields. I genuinely do. But, when you look at it through critical eyes, you know that this is just another iteration of White dominance. I'm not saying that I want that. I don't. But, the STEM fields are not truly open to diversity, social justice, cultural criticality … Ooooo, I'm not even sure what that means, but I like the sound of it. It's all whitewashing and capitalist at the end of the day. Kids are being treated like 'pre-employees' at best. On the one hand, I love STEM and science education, but on the other hand, I know that they are both just about jobs. It's been that way since the U.S.'s response to Sputnik. It's dehumanizing, and I feel dehumanized every time I talk about it like that."

The NEB said, "You seem conflicted." The NEB learned a simple counseling technique and utilized it at that moment to see what response it might elicit from the HSS professor.

"Yeah. I am. I have talked about STEM education as a form of equity for so long that I actually started to uncritically believe it. But, here's what really happens. Someone poor or someone diverse does better in school, does well in math and science classes, majors in some kind of engineering in college maybe with some scholarships, and then they are happy to go work for Exxon or Bayer. Can you see the problem there, maybe? It's not okay. If they go to Exxon, they become a scientist who helps add more carbon dioxide to the atmosphere or they work for Bayer and create pesticides that kill off primary pollinators like bees and create colony collapses. Not exactly the best uses of STEM, but at least they have an upper middle class paycheck, right?" He sighed. "Social justice, everybody! High five yourselves. Job well done. Equitable Anthropocene, right there!"

He paused again. "Sorry, I don't *normally* talk like this. I *normally* play the buttoned up, self-important professor part much better." He paused and looked at the NEB as if he told a pun and was waiting for the NEB to respond. "Ah? See what I did there?" The NEB saw a flash of Fozzie Bear in the HSS professor. "Get it? 'NORMally' and 'play the part?' That's what we are all expected to do here, not that we all do, of course. When you start getting more attention, you pay more attention to what sells rather than what you really think, apparently." He looked disgusted at himself for a moment but became instantly aware of the colleague (who really wasn't his colleague) looking at him.

"So, you're saying that money and prestige are keeping you from saying more openly to your colleagues what you're really thinking?" The NEB assumed this was the moment to finally ask that direct, critical, and personal question. The HSS professor nodded his head and sighed.

"Yes." The HSS professor recoiled a bit. "I don't know." He sighed, again. "Probably." The HSS professor took a breath, began to speak, but then paused to consider what he was about to say. "The history of science is not neutral. There is a long story, since at least Francis Bacon, of scientists performing as showmen for profit. Bacon did it for the king, and many others did it for people with power and wealth (Merchant, 1980). In that sense, scientists are not unlike many artists throughout history. Many artists produced art for nobility, the Church, and very wealthy people largely because that was their source of revenue, as well as notoriety. Similarly, scientists throughout the ages have made discoveries and created technologies that connected to their own personal gain in the form of profit and notoriety. In that sense, many of us are greedy servants to the wealthy class, right now."

The NEB quoted Max Horkheimer and Theodor Adorno in the *Dialectic of the Enlightenment* (2007), "What human beings seek to learn from nature is how to use it to dominate wholly both it and human beings" (p. 2).

The HSS professor came back with more Horkheimer and Adorno, "The earth is radiant with triumphant calamity" (p. 1). The HSS professor continued, "I think I and so many others in science education may be carrying on the tradition, sadly. Why are there so many conference presentations on STEM and so few about sustainability in science education, no less? The International Panel on Climate Change has been saying for years that we have only until 2030 to make change if we have any hope of staving off the catastrophic effects of global warming, but here we are at a conference where STEM for corporate interests dominates while science education for sustainability is still not a full-throated effort on our parts. We care far more about diversity and equity for STEM, which really means just having a more diverse set of STEM employees that ultimately funnel more wealth to the already super wealthy, rather than putting all of our efforts into reconfiguring science education to aim towards sustainability. Maybe it gets a couple of weeks per school year or is a discussion on Earth Day, but we resort back to STEM for profit as the default. I'm doing it, too. Why? Are we really so trapped in greed motives that we can't operate in a different mindset? Are we so trapped in Western thinking that we are inevitably just going to destroy the planet?"

"Are *homo sapiens sapiens* really just *homo greedyus*?" asked the NEB. The HSS professor snickered. "It seems like there's a tokenizing of people of color in a White-dominated STEM culture in order to justify that domination. It seems like an attempt to wallpaper over white walls where the foundation ultimately remains unchanged. Have you made a name for yourself, in a way ... probably not intentionally, though ... doing some of that wallpapering?"

The HSS professor stared deeply into the glass in his hand now less filled with alcohol. "I wish I could say that I hadn't."

The NEB's curiosity about the HSS professor's thoughts was intensifying. The NEB wanted to know how important aiming toward sustainability was to the HSS professor. The NEB continued, "It seems to me that sustainability or its current failings are more of a cultural and values problem than they are a science problem. On the one hand, greed seems to be driving the show in science education similarly as it did throughout the history of science in many cases (Merchant, 1980). Greed is ideologically tied to freedom for Western thinkers, right? They may not say it that way, but personal profit is a major motivator and a measure of success, and they want the freedom to pursue it. And the portrayal of pursuing that greed is framed as a kind of freedom in this society. To that end, the framing of science, science research, and science education is heavily influenced by this culture of greed. Despite knowing pretty well what to do scientifically about climate change and having strong inklings of how to proceed sustainably as a society, *homo greedyus* views STEM with profit in mind, which is then inherent in how equity and diversity are viewed." The NEB was waiting for pushback.

"Well, I think maybe Chet Bowers (2006) had it right by focusing on it as a cultural issue rather than a species issue, but I certainly take the point. If you mean *homo greedyus* as a tongue-in-cheek way of describing members of the human species enacting a culture that is oriented around mindsets, values, and practices of greed, then I can agree with that." The HSS professor was nodding his head as he spoke. "The death grip of greed in the long history of humans is relentless, especially in industrialized, really post-industrialized societies. It manages to colonize our minds just as it has done for hundreds, if not, thousands of years. I'm aware of it, and yet I'm still complicit way more than I even recognize." He was looking into the distance as he spoke, but did not seem to focus on anything.

The HSS professor continued, "STEM/STEAM education is constructed in the tradition of a greed ideology with the veneer of a meritocracy packaging to make it all shiny and equitable."

"Sometimes I feel like I'm from another planet," said the NEB as it chuckled to itself. "The United Nations' IPCC said that 2030 is about as long as humanity has to make any real changes to slow down the catastrophic effects of climate change. It's a growing calamity right before our eyes. And, yet, when I come to these conferences, read the journals, or look at the standards, the dominant message does not connect. If I didn't know any better, I would say that this is a modern-day evolution of centuries of colonization and a pro-colonization mindset, which is really to say greed and domination. Climate change and sustainability are every bit a STEM/STEAM-based set of issues, content, and practices. Yet, many 'experts' in science education still seem to be ensnared or even willing soldiers of the colonization of corporate, for-profit STEM/STEAM. When climate change and sustainability are actually taken up, they rarely rise above pro forma performances for each other."

"Perhaps, we are more ensnared than we are willing soldiers. I wouldn't call myself a willing soldier of corporate greed, but I am realizing how entangled in it my pro-STEM/STEAM discourse tends to be. I think we are tricked discursively by a 'good guy syndrome' or something to that effect." The HSS professor stared off into the distance, again, but seemingly came rushing back. "'We're the good guys' is really holding us back. As we said before, science is neither inherently good nor inherently bad. The 'goods' and 'bads,' so to speak, are built into it by the builders and maintainers. Humans have done plenty of damage using science and technology throughout history. Science and technology are what we make of them. Humans, led by greed (and deliberately cloaking it in discourse of 'progress, opportunity, and advancement'), used STEM to create a fossil fuel-dependent industry, which was incredibly exploitative. It employed and still employs people in ways that exploited them, and it exploited the environment, right? Science, technology, engineering, and mathematics were all used to that end. STEM is currently used today to continue exploitation and destruction. STEM is used to create pesticides that kill off bee colonies. STEM is used to create war technology. STEM is partly behind school violence in gun technology. STEM is used in eugenics research. STEM/STEAM isn't all bad, of course, but it's not all good, either. Science education and STEM/STEAM education have a driving discourse that herofies science, which acts as a form of indoctrination in schools if teachers leverage it that way."

The HSS professor looked at his cell phone and saw that it was 6:23 pm. He was giving a dinner keynote at 6:30 pm entitled: *STEAMing into the Future: How STEAM Education Programs Create Equity and Social Justice Opportunities.* His 45-minute speech was going to be recorded in the multi-million-dollar grand ballroom, in the multi-million-dollar hotel that was part of a multi-billion-dollar hotel business, and the focus of the speech was the claim that STEAM education addresses equity and social justice (just to reiterate a point of irony). The room resembled a small mega church but with a buffet line. STEAM was the gospel. The HSS professor had become a cardinal. "It was quite the spectacle of hegemonic hubris," the NEB wrote in its infrared notes.

"Oh my goodness! I need to go. It was really nice running into you, again. It's been so long. Let's find a time to talk in a couple of months or so." The HSS patted his shirt pocket to make sure that the thumb drive with his presentation was still there, which it was. "Take care!" With that, he darted over to the ballroom and was immediately greeted by the session moderator, who seemed to be trying to play off that he was not a little starstruck by the keynote HSS professor. They disappeared into the very busy, dimly lit room.

The NEB glanced at the waiter walking by with a tray full of empty glasses that were previously all filled with overpriced alcohol. The NEB looked into the eyes of the waiter and wondered how often and how deeply humans think about their societies' systems of control, limitations, and exploitation, especially about how it affects them individually. Was everything just so normalized and reified that oppression was not even seen any more in these societies? Did they realize that from at least tod- dlerhood through elderhood there was a very powerful group profiting that used social constructs and personal insecurities to create exploitative labor and addictive consumerism? The NEB wondered further if humans— who were exploited Earth day after Earth day, Earth year after Earth year—considered that only one social construct (money, which was a man- ifestation of greed) was at the root of every social and environmental issue of dominance and exploitation. Would humans ever drill down far enough and see how they filter nearly all of their thinking through money? The NEB added to its infrared notes, "Money, which is to say, the desire to exploit in the form of a socially constructed symbol, is the greatest mental trap of humanity and will likely parallel the outcomes of a planetary impact by a comet if the patterns remain undeterred." From the NEB's perspec- tive, based on its home planet where societies banded together to elimi- nate exploitative practices, Earth seemed excessively and preventably stressful, but humans showed very few signs of actually changing. The NEB, who was a professor on its home planet, was happy that this portion of its ethnographic data collection was done.

PART 2: SCIENCE EDUCATION TOWARD THRIVING AND *HOMO THRIVEUS*

It took a few moments, but the video call app opened on the laptop. The app was acting a little glitchy, a pet peeve of the NEB. "Why is tech on this planet so inconsistent? It's so annoying," the NEB thought. A notification appeared on the screen asking for a participant to join. The NEB accepted.

"How are you?" asked the voice on the screen. She was using a fake background in her video that made her appear like she was in the captain's seat on the bridge of the U.S.S. Enterprise from *Star Trek*.

"Live long and prosper," exclaimed the NEB, who wished that it used the pineapple under the sea background. The NEB morphed into and assumed the form of a White, female graduate student who was in a science education doctorate program. The video call was supposed to be an interview for a smaller paper for a course that looked at contemporary thinking in science education. "Thank you so much for taking the time for this interview. I really appreciate it. I also love the arguments that you make in your publications, so it is really nice to meet you and talk with you." The NEB was not an actual human or even a graduate student on its home planet, but of all the HSS written work that it read regarding science education, this HSS professor seemed to be unknowingly aligned with the core philosophies and practices on its home planet. The NEB was very curious why this HSS professor seemed to be at that point in her thinking, as much of the rest of that field of thinkers just did not seem to get there, yet.

"My pleasure. How can I help?" There was a genuineness about the HSS professor.

"I guess that I am most intrigued about how you arrived at your arguments, in general. It's not just that I tend to agree with them, but it's more that I do not see others really writing about this perspective, at least not at the depth that you've gone." The NEB had read many articles and books about science education and the very popular rebranding in the form of STEM/STEAM education, but so much of it rang hollow for the NEB, especially given the history of its home planet. Coming from a planet where life once teetered on the edge of existence because of the powerful elites who used to vehemently defend a system of exploitation that favored only them, but living in a time after that turmoil was fairly resolved, the NEB wanted to see more closely how those dynamics were playing out on Earth in real time. Much had been written about the Great

Peaceful Transition on its home planet, but mostly in retrospect. To potentially observe it in real time, or at least to talk directly with HSSs who might be forming the base rationale for their own transition, could provide some deeper insight about how new thinking emerges and grows in a sustained way.

"I wish that I could have some of the little, fleeting thoughts over the last decade that have culminated in my thinking, today. I'm not intending that in an egocentric way at all," said the HSS professor, "but I find it fascinating to know that, as just one individual, I have had years of thinking that somehow formed a fairly cohesive argument at some point. I imagine that others have that kind of punctuated disequilibrium happening regularly, too. It's a difficult thing to pin down and describe, now that you've got me thinking about that." The HSS professor was seemingly increasing with intrigue as she talked more about it.

"Well, if we were to break the facade a bit, and try to isolate some of those seemingly random thoughts, are there ones that you can look back on that seemed to punctuate your disequilibrium?" The NEB could not tell if the universal translator asked the question in a nuanced enough way.

"Uuuummm, hmmmm..." the HSS professor shared profoundly. "I guess I probably did what I imagine a lot of people do when they are an angsty teenager. I questioned the hell out of everything. I felt like the way we were living life was so totally off and so totally below our greatest potential as a relatively advanced species (at least on this planet), that I just kept picking apart our motives for doing things. As a teenager, I loved old school punk, but the giant spiked hairstyle was long gone by the time I hit my teens. I loved how pissed off that music was. I was pissed off, but I also had this deep appreciation of what it meant to be human, or at least I kept trying to figure out what that was. I still am, of course."

"What do you mean by that?"

"Yeah, I had a feeling that you were going to ask me that as soon as I said it," she laughed. "It's weird to me that academia doesn't seem to go here very often. I think academia wants to portray itself as organized and confident, but life just feels so disorganized, vulnerable, and raw most of the time, at least to me. I don't even know why I publish because I never feel organized or confident about what I'm writing about." The HSS professor paused.

"Well, I can say this. Your writings inspire me to remain open and humble as someone who wants to be a thinking thinker." Again, the NEB was unsure about the universal translator's ability to provide enough nuance.

"That's really all I'm aiming for here. I just want to be a thinker who thinks … critically questions, doesn't stop asking why, stays vulnerable, and connects with the rawness of life, at least as I feel it within and around me. That's where much of my thinking and writing comes from. I'm actually surprised every time something gets accepted for publication or for a conference." She rolled her eyes.

"Conferences?" The NEB knew that the HSS professor stopped going to science education conferences eight years ago.

"There aren't a lot of vulnerable, raw presentations at science education conferences," she laughed hard at that one. "But, there really should be. Science isn't about performing personas or getting all flashy with the newest trend, and it certainly shouldn't be involved in propping up globalization, climate change, or other destructive practices. As much as I love science education is as much as it depresses me. As a child through my young adult years, science was a way for me to get closer to nature, but at a certain point, I began feeling like science was very much about keeping nature at a distance so that it could be observed and exploited. The very thing that gave me an oasis at my soul was now becoming a profound source of sadness. Science, well, really STEM was destroying the planet, exploiting people, and continuing colonization, and somehow I was expected to be on the science education bandwagon like a cheerleader. Like I just told you, even as a kid, I could never be that. I had to question everything, especially if other people were banding together around an idea and trying to sell it. I hate when things get popular. I can't function in that space."

"Are you a contrarian by nature?"

"I'm a skeptic more than a contrarian. If someone else is jazzed up about something, that's fine for them. I'm very filled with energy and passion for life. This is a pretty damn amazing thing, life. It never fails to intrigue the hell out of me, but I need to ask why, and poke around doing my own evaluations of it. I do it even more if someone agrees with me about something, especially science education. I'm very aware of the craze of paradigmatic thinking, and that scares me, constantly." She seemed to have an awareness of the of level how much she was sounding different than what she perceived to be the thought culture of her field. "Does any of this make sense?"

"Well, if I agree, are you going to be uneasy?" The NEB laughed.

She laughed, "Definitely."

"So, how does all of that manifest into articulating a framework of thriving in the context of science education?" The NEB was very aware of key historical figures on its home planet who started the thought revolutions that led to generations of thinking, rethinking, and unthinking that eventually broke the stranglehold of the powerful elites. The NEB was sure that the kind of unthinking that this HSS professor was doing was similar to the historical figures on its home planet.

"I think for me it just came down to seeing the extent to which exploitation was occurring in science education. It was never more clear to me than when the wave of popularity grew around STEM education. At first, I thought that the idea of integrating science, technology, engineering, and mathematics was a great idea, but then my inner skeptic wondered about the underlying motives, which, as it turned out were openly expressed. Jobs. That's it, people. Nothing more to see, here. Jobs. I was disappointed in myself. How could I have been naive as to think that STEM education was for anything but the greed of the wealthy, dominant class? The wealthy, dominant class has only been colonizing the planet for 500 years in some form of capitalism, and for thousands of years in feudalism, kingdoms (that is domination by a king), imperialism, and any form of takeover that they can get away with. Now, we just call it 'globalization.' Why would that stop, now? Or, at the very least, why wouldn't the academic class not continue to support that motive since they are paid to provide the thinking in order to continue respective contemporaneous practices of satiating greed? They, too, benefit from it. They go on their little speaking tours, increase their status, and get groupies or whatever. I have had enough of that. That, all of that, is greed to me. Education, more broadly, suffers from the same underlying motive. Greed tells us to tell the kids that social justice means getting an upper middle class job for four decades of their life, making someone else richer." The other HSS professor at the conference made a very similar point. Could it be that professors on seemingly different ends of the spectrum were starting to merge their analyses much like what happened on the NEB's home planet?

The HSS professor continued, "Science education should be about caring for the planet, figuring out ways to do that better, and excavating out root values that stall us from doing these practices that would hopefully be more sustainable and help us thrive as a society … and not just be at the beck and call of the wealthy class who ensnares us with their mediocre salaries for most of our lives. One of the things that always bothered me, even as a middle schooler in science class, was that science pretended to not be of humans. Like, somehow it was pure, neutral, and free from toxins. Science/STEM/STEAM is all human based, and its outcomes are as good or as bad as we make them. We can destroy the planet with pollution, greenhouse gasses, and nuclear weapons, or we can figure out how to be a deeply thriving species that is every bit a part of nature. We love to be selfish and destroy things for our own individual benefits, usually for money. Why are we so horrible? These are the very angsty questions that fuel my work, still. Why do we care so much about participating in destruction when creating a thriving planet is so much better for all of us? Are we incapable of core remediation once greed sets in? For crying out loud, Charles Dickens was writing about the greed of Ebenezer Scrooge over a hundred years ago. Does humanity need a bunch of ghosts to give us an existential crisis? I'm clearly careening off the cliff. Let me come back to science education and STEM." She took a breath.

"No, seeing the interconnections across the silos is what I see in your work. It really does inspire much thinking and critical questioning. It's humanizing, and you're right, it's outside of the performances of science education personas that are expected and normalized." The NEB also knew that the historical thinkers from its planet started with integrative thinking across contexts and content. They explored big concepts together rather than the usual recommendation of nearly every dissertation advisor to avoid broad topics. It was the joining of broad topics that helped generations redesign their thinking and practices. "If I had to guess, I would say that the angst and the integrative thinking that you do seems to work together as the background for the framework of thriving in science education?"

Her eyes seemed to light up a bit. "I don't know if I've ever put it quite that way, but that concise description definitely resonates with me."

"I've read about the framework many times, and each time I do, I have a whole new set of connections in my thoughts," the NEB was pulling from its own historical accounts on its planet and restating it here to see how true that might feel for the HSS professor, "but can you explain it in your words, now?" The idea with this question was to see if the HSS professor would continue to grow in her thinking about her framework.

"Well, sure. It's a good exercise for me to say it out loud from time to time." She chuckled. "First, science education as we teach it today is still very much a response to the Sputnik panic. Science education is taught mostly as form and function of the universe. It's nuts-and-bolts city. The

Sputnik panic created science as a character or a package for children that was essentially always right, had good motives, and advanced its society with disregard to its actual practices. It's nearly devoid of anything social except for the constant social desire to make science sound wonderful. There's a real problem with avoiding the social aspects of science. Science that ignores exploitative motives and values is sexist, classist, anthropocentric, and racist. How can I make that claim? Well, to ignore or tolerate racism is also racism. To ignore or tolerate exploitation is also exploitation. So, if science education is packaged in a way that ignores, omits, or tolerates exploitation, then it is exploitation, too. Science that is critically connected to its social aspects has a much greater potential for equity- and justice-oriented outcomes. The scientific process remains a very high bar for discovery and achievement, and I, like so many, am a fan, but I'm also nuanced. Science and STEM has been a terrible weapon (do I need to include eugenics?) and a destroyer of the planet as much as it has the ability to support life. That is the ground floor of the framework."

"Where we go from there is simply a comparison of science toward greed/exploitation or science towards thriving/undoing exploitation. If we are telling ourselves that STEM for corporate jobs is equity, that's straight up subjugation and propaganda. A science education and STEM education movement should absolutely be redesigned around sustainability, non-exploitative endeavors, and understanding the integrative conditions of thriving. We can use science, and, thus, teach science as a way to vastly improve life on the planet if we are doing so with social aspects at its side. The Sputnik panic design really needs to go away completely and be replaced by a framework that moves us towards creating sustainable societies. To make a long story short, science education (and science, itself) should be in service of creating thriving, sustainable, socially just societies, but it is currently doing the opposite by being in service to creating tech-based workers who are primarily there to funnel wealth to the already super wealthy." She described a fundamental shift in science education that became integral and foundational to all of education on the NEB's home planet many generations ago.

The NEB jumped in excitedly, "It's like if you lived your entire life on the Death Star from *Star Wars*. As a kid, you might think that becoming a Stormtrooper is a great entry-level job and that becoming a commanding officer, or better yet, a Sith Lord is a form of ultimate success, but all you're really doing is contributing to the operation of a weapon that destroys planets for the sake of the Empire. In the case of actual humans on Earth, they are rising in the ranks of planet-killing corporations but focused on increasing their salaries and status. Yet, they call that social justice."

"HA! Yes, that is such a great analogy!" The HSS professor laughed to the point of a small coughing fit. "People are really not seeing that, though. And, to be honest, they are not going to see it if it is just in the form of critical analysis, either. Some do, but the mainstream thinkers won't. They will need to have an entire replacement paradigm up and ready to go in order to move them in that direction, and it will need to be 'soft' enough in its criticality so that it does not threaten them because that will prolong their transition from a greed paradigm to a thriving paradigm." The NEB could hear much frustration in the HSS professor's voice.

The HSS professor continued almost immediately, seemingly in order to avoid getting overly frustrated, "So, that means that we need to figure out what we mean by 'sustainability' and 'thriving.' Both are terms that have integrative meanings, which is to say that they are more than just one context. The most basic framework would mean that well-being (overall state of inner balance at the individual level) combined with sustainability (ecological, social/cultural, and economic conditions at the community and societal levels) lead to a state of thriving (beyond just surviving, a reciprocal state of healthy well-being and balanced community/society). These will probably always be moving targets to some degree, especially as we continue to learn more about them, but a redesigned science and STEM curricula will be important parts of this transition. Ultimately, we need to redirect our vision from creating STEM-based workers to creating thriving communities, and we need science and STEM education on board with that, now."

"It seems so overly obvious that this should be the primary effort, and probably should have been decades ago," said the NEB with exhaustion because that was its planet's history of transition. "We should be moving from an outdated, outmoded Sputnik-era science education to a science experience that helps create sustainable, thriving societies. Why do you think that there is not a rapid shift in this direction, right now?" The NEB wondered why societies wait so long to shift, especially this one since, quite literally, this society already has everything that it needs to shift … minus a replacement mainstream mindset focused on thriving together. Why was greed so important to hang on to as this society's core ideology? Back on the NEB's home planet, breaking that grip was crucial to remaking all aspects of living much more peacefully and sustainably.

The HSS professor answered immediately, "People project a preferred version of reality in front of themselves every minute of every day. That reality projector is fine tuned to their individual preferences and creates a narrative that aligns with those preferences. The more that their way of life is threatened, the more that they reinforce that version of reality to themselves until it shatters completely, leaving them to rebuild a version of reality for their reality projectors. The important point there is that the actors project a version of reality to themselves and live by it. They tend to double down on it many, many times before finally giving up, if they give up at all. It's a real loss to be in a society so self-limiting and often so ready to defend its version of reality despite its very observable failings in real time. I am sure that the threat of not having money and status is a major insecurity for many, and it keeps them from opening up to other reality narratives. The question remains of what will make their reality narratives of selfishness, greed, and exploitation finally shatter so that they can be more open, vulnerable, and willing to care for others and nature. I think that is the primary question of our time. Will it be a plague? Will it be the collapse of ecosystems? Will it be an economic disaster? Will all of them occur together? It is pretty obvious to me that our society is heading down that combined disaster path, but wouldn't it be lovely if we got out in front of that? Wouldn't it be so much better if we started making changes now that could significantly mitigate that impending eco-social disaster that is highly probable? Wouldn't it be great to teach about a science that is capable of being a major player in the mitigation and remediation of disaster, and, better yet, a science that moves us towards long-term thriving? Again, this seems obvious to me, probably you, and probably many, but why are so many more just sitting idly by and not putting pressure on the system to change? We are not powerless even in the current structure."

"Agreed. Perhaps, we are searching for a different route than was taken in history?" The NEB knew quite well that on its planet, change occurred as the collective consciousness changed. In fact, the NEB is a faculty member in that department at its university: The Department of Collective, Integrative Consciousness. "Maybe, that is where long-term change always resides."

"Maybe." The HSS professor sighed.

"Well, thank you very much for your time. You've given me so much more to add to my thinking," said the NEB.

"That's the highest honor anyone can give me. Be well," said the HSS professor.

"You, too," said the NEB. That was the NEB's last data collection effort for its study.

PART 3: THE FINDINGS

The NEB interviewed 217 HSS professors, teachers, and administrators and observed over 1000 of them in various settings (such as conferences, college classrooms, department meetings, and in calls, as well as over drinks at different hospitality establishments) over a three-(Earth) year period, but the HSS professor at the conference at the hotel lounge and the HSS professor on the last video call seemed to encapsulate and illustrate the range of the collective thinking among science education professors, administrators, and teachers. There was either a real attraction to sticking to the for-profit thinking of STEM education (reinforced by grants, awards, and recognition) or an almost aggressive pull away from the for-profit paradigm in favor of a more sustainable and thriving exploration of science and STEM. The humans of Earth were very much the cause of their own suffering, but they were seemingly (hopefully, temporarily) stuck in a moment of real tension: Do they still seek profit as a way to buffer the failures of their efforts, or do they address their failings head-on, which presented as a seemingly impossible task, in their hubristic eyes? Many humans spoke about themselves as a superior species, which always made the NEB chuckle, but unfortunately, as the NEB wrote in its initial report, "that was part of the problem because it prevented them from seeing themselves as a problematic, invasive species." The NEB's qualitative coding of data kept leading back to two ways of framing science education (and STEM education): Greed-oriented (individually oriented) or thriving-oriented (cooperatively oriented). There seemed to be a growing reluctance to stay with the current dominant frame of for-profit reasoning, but even as the core of it became wobbly over time, it was the direction that those with the most power chose unanimously. It seemed like there was a fear of letting go of the for-profit frame since it had been the foundation of science and eventually STEM education for six Earth decades (since Sputnik). Teachers, administrators, and professors of science and STEM education are a self-selecting group of HSSs. Most went into those positions initially thinking quite positively about science and STEM within the for-profit frame, but that collective confidence did seem to be dwindling. In contrast, the second group of critical pushers were often frustrated emotionally and lost confidence in their colleagues despite their own willingness to keep pushing toward a replacement paradigm of cooperative thriving.

The NEB wrote as its conclusionary statement, "If Earth history parallels the history of [the NEB's home planet], there will be the coinciding factors of an increasing collective, integrative consciousness coupled with an unfortunate set of circumstances that HSSs experience, which may change their levels of collective willingness to become more open to fundamental change. The more beneficial path would be to make changes prior to the catastrophic events, of course, but the current trends do not appear to arc strongly enough in that direction at this time."

REFERENCES

Bowers, C. A. (2006). *Revitalizing the commons: Cultural and educational sites of resistance and affirmation*. Lexington Books.

Horkheimer, M., & Adorno, T. (2007). *Dialectic of enlightenment* (G. S. Noerr, Ed., & E. Jephcott, Trans.). Stanford University Press.

Merchant, C. (1980). *The death of nature*. Harper & Row.

CHAPTER 13

Practices of Care with the Anthropocene: Scenes from the 2019 Nebraska Flood

Susan Naomi Nordstrom

A VIEW FROM A BODY

In July 2019, I drove along Interstate 29 in Western Iowa (United States) toward my childhood home in Nebraska for a brief vacation and with a compelling curiosity to see the aftermaths of the March 2019 flood that I had followed from my current home in Tennessee. Fields that should have been filled with corn and soybeans had become lakes. Some of these lakes had beach-like banks made from Missouri River sand that migrated along with the water from several miles away. Center-point irrigation implements sprung forth from the lakes. A tractor became an island in one lake. Bloated silos, some with corn and soybeans spilling from them, stood like volcanic islands in some of the lakes. A few homes sat in the middle of lakes, their decks now more like docks waiting for boats that would never arrive. Exits that punctuated the drive became waterways, closed due to water still covering the roads. Percival, Iowa, always a marker of "almost

S. N. Nordstrom (✉)
University of Memphis, Memphis, TN, USA
e-mail: snnrdstr@memphis.edu

S. Tolbert et al. (eds.), *Reimagining Science Education in the Anthropocene, Volume 2*, Palgrave Studies in Education and the Environment, https://doi.org/10.1007/978-3-031-35430-4_13

home," was flooded. I would soon learn that citizens of Percival were negotiating whether or not to rebuild the town. The flood happened in March 2019. It was July.

Prior to my July trip, I had followed the flood with great interest, as my parents live in what was called one of the Nebraska Islands. Several towns became islands, surrounded by floodwaters that choked off all road access in and out of the town. Our conversations during spring 2019 frequently focused on the flood. I followed events via the *Omaha World Herald*, the state's most widely read newspaper, and social media. Still, I could not anticipate what it would be like to see the devastation, to walk on river sand that was not even close to the riverbanks, to see how river ice chunks made trees into postmodern sculptures, and to find debris—a shoe, lawn furniture, a door, and so on—recreated into river-made art installations.

Donna Haraway (2008) wrote that we are all "ordinary beings-in-encounter" (p. 5) flowing together to create "unpredictable kinds of 'we' " (p. 5). Within the posthuman literature there is a call for these kinds of "we's" that contingently unite humans, nonhuman animals, plants, trees, and all that inhabit the earth in the precarious Anthropocene. Those flood waters did pull me into a series of we's that stretched across time and space among humans and nonhumans. Those we's created water-like movements of a particular conceptualization of affect-rich care as an ongoing and situated and ethical practice in the Anthropocene (Puig de la Bellacasa, 2017). This chapter is about "the urgency of the Anthropocene, Capitalocene, and the Chthulucene [and how they] demand a kind of thinking beyond inherited categories and capacities, in homely and concrete ways" (Haraway, 2016, p. 7) and how care shifts in that urgency into affective movements. I write this chapter from a "view of [my] body, always a complex, contradictory, structuring and structured body" (Haraway, 2002, p. 683), a body learning to live in the Anthropocene. Following Gough's (2006) call for more literary and artistic modes of representation in science education, I write to break free of the arborescent ways I was taught to think about science toward an affect-rich account of learning to live in the Anthropocene that is grounded in situated knowledges and affective partial connections of watery we's (Haraway, 2002). In this chapter, I follow the work of the "we's" that the flood collected and the affective practices of care that flood continues to generate. To do this work, I first situate the 2019 Midwestern floods, with a focus on Eastern Nebraska. Then, I share how I put Puig de la Bellacasa's (2017) conception of care to work by examining moments that created different affect-rich currents in practices of care.

THE FLOOD

The March 2019 flood was a result of extraordinary 2018–2019 fall and winter weather events. During fall 2018, areas of Nebraska experienced significant rainfall that saturated the ground that would soon freeze over the winter. While the early winter (late December to early January) was relatively mild, it was cold enough to keep the water-saturated ground frozen. January 2019 began with high temperatures that chased records. As the month went on, the weather marched along with a comfortable normality, cold but not too cold with a few mild days. Toward the end of the month, near-record cold temperatures dipped well below freezing. The first week of February offered a rush into near-high record temperatures followed by a return of at or below freezing temperatures. This cold air was accompanied by weekly precipitation of either ice or snow. This is called a blocking pattern in weather. Blocking patterns happen when fronts cannot move and are blocked in by other fronts. February became a month of brutally cold weather, the coldest February on record. In Omaha (approximately 30 miles east of my hometown), a record 27 inches of snow fell on the ground. Nebraskans were blocked in by winter. Each lashing of winter weather beat down hopes for an early spring into a snowy-icy pulp piling up on frozen ground.

A National Weather Service tweet on February 28, 2019, predicting the first week of March's weather in Nebraska joked that March would come in like a penguin rather than a lion or a lamb. Meteorologists predicted temperatures well below the normal lower 40s Fahrenheit (15–35 degrees lower). The second week of March was no different. Below-normal temperatures further hardened the frozen ground that was covered with snow and ice. These temperatures also further crystallized river ice. March marched in like a penguin over the fifth wettest fall-winter in 124 years and the eighth coldest February in 124 years (Gaarder, 2019).

From March 12 to 14, a battle between winter and spring was happening in the atmosphere. The battle began over Arizona and then moved through the atmosphere over Colorado. Pulling further strength from passing over the Rockies and into the plains of Western Nebraska, the battle intensified over Nebraska and other Great Plains states. According to the *Omaha World Herald* (Gaarder, 2019), the satellite imagery showed what looked like a land-based hurricane. The air pressure dropped 24 millibars in 24 hours. One to three inches of rain fell over Eastern Nebraska. A blizzard overwhelmed Western Nebraska with snowfall of a foot and 89-mile-per-hour winds, and shut down parts of Interstate 80, the only interstate roadway in Nebraska. Emergency Management Services sent warnings to Eastern Nebraska about the potential for flooding as rain fell on frozen ground. They predicted that the water would move across the frozen ground and find its way to ditches, creeks, and rivers that would then flood.

In the early morning of March 14, the Spencer Dam on the Niobrara River in Northeastern Nebraska failed. Built in the late 1920s, the hydroelectric dam sought to harness the power of the confluence of the Niobrara River and Missouri River. Following the dam's failure, an 11-foot wave of water was released that took out the river gauges. Also, on March 14, the ice jams jostled their way out of the Elkhorn and Platte riverbanks, the rivers that border my hometown. The ice from the river moved out of the banks and slid across the frozen ground like glaciers and freed the water from the banks. By March 15, the Platte River crested at 12.65 feet, the highest on record. The Elkhorn River set a record of 24.11 feet, 5 feet above the previous record. The Missouri River that forms the border between Nebraska and Iowa had it third highest crest of 31.12 feet on March 18.

On March 19, the Nebraska Emergency Management Agency declared emergencies for 74 counties, 83 cities, and 4 tribal areas. The human population affected by the floods was numbered at 1,735,635 or 95.03% of the state's human population. Approximately 75% of the land in Nebraska was affected by the flood. The Nebraska road conditions website map had so many roads marked red to denote closures that the map looked like red veins in a body. Nebraska's roads, its veins, had been opened and blood water poured forth from them. The water took out roads and bridges all over the state. Main streets of small rural towns became canals for the rivers and creeks. Homes destroyed. Farms destroyed. Businesses destroyed. Infrastructure carried away by water. Entire towns and villages swept up in the floods. As early as March 21, 2019, the damage of the flood was estimated to be 1.3 billion US dollars, which included infrastructure damage as well as crop and cattle loss (Schwartz, 2019). All the critters of Nebraska relationally came together in a water-filled we, whether they wanted to or not.

WATER CARE

Everything, it seemed, belonged to the water. Online videos and photographs made my home state look like it had given itself back to the Western Interior Seaway of the Cretaceous Period. So much water. So many people displaced, the amount of financial damage seemed to keep growing, towns erased from the map, rebuilding efforts, and … and … and …. The water seemed to be " 'tearing down, dancing over, laughing at' our efforts to restrain it" (Schneiderman, 2017, p. 17). With my phone pressed to my ear during my weekly phone conversations with my parents, I listened to their updates about living on an island, with every road out of the town flooded or ruined by the Elkhorn and Platte floodwaters. I read newspapers online. It seemed unfathomable. And it still does. There is a degree of incredulity about these events. The water washed away moralizing, human-centered approaches to care. The water opened space to consider "that a politics of care engages much more than a moral stance; it involves affective, ethical, and hands-on agencies of practical and material consequence" (Puig de la Bellacasa, 2017, p. 3). The water washed away the "shoulds" that often populate moral stances and brought about watery entangling flows collecting people in "we's" in which we can only work with entanglements.

As the Anthropocene collects us into unanticipated we's, Puig de la Bellacasa's (2017) care is useful to think with. Her conceptualization of care disrupts the moral imperative of care or a sense of care that can be centralized within a subject or structure of power (Puig de la Bellacasa, 2017). Instead, she argues for a relational and situated care that moves between humans, nonhumans, power structures, and other entangling forces in a world that is making and remaking itself. Consequently, care is an ongoing practice in entanglements and cannot be predetermined. Care is generated within relations in entanglements, no matter the time-space distance, and pulls us into a series of we's that require an ongoing practice of care to live as well as possible in the unpredictable Anthropocene.

As the floods happened and the ongoing aftermath of the floods, the practices of care became flood-like, ebbing and flowing as river levels rose and fell, river water flooded and receded. This flooding sense of care became "continuously reenacted in inseparable entanglements between what is 'personal'—how one individual is affectedly engaged in attachments—and what is 'collective'—a web of compelling relations, with humans and nonhumans, included in a community of practice in situations" (Puig de la Bellacasa, 2017, p. 166). While care has always been generated by the relations between myself and the humans and non-humans of Nebraska and has shifted over time and space, that sense of care became something otherwise with the floods. The practices of care became curious as there were elements that "shifted [my] priorities" (p. 167). In the following sections, I offer movements of this sense of care.

#NebraskaStrong

In the immediate aftermath of the flood, the hashtag #NebraskaStrong gained popularity on social media. The hashtag offers a way to consider how some Nebraskans stayed with the trouble of the floods (Haraway, 2016). Haraway wrote:

> In fact, staying with the trouble requires learning to be truly present, not as a vanishing pivot between awful or Edenic pasts and apocalyptic or salvific futures, but as mortal critters entwined in myriad unfinished configurations of places, times, matters, meanings. (p. 1)

The hashtag materialized watery we's that the flood collected and how people and nonhumans configured themselves in the moment so as to "live and die well with each in a thick present" (Haraway, 2016, p. 1).

The realities of the we's materialized from the hashtags are ongoing and responsive to the always already configuring nows that collect humans and nonhumans. In this way, the hashtag actualized the care work of the we's. Numerous photographs show efforts such as sandbagging, rebuilding, and others that featured Nebraskan humans coming together to live better in the aftermath of the flood. For example, while two men were cleaning up flood debris on their land, they found a beer fridge that had traveled four miles from its home (Salter, 2019). After a long day of work, they cracked open a couple of beers and took a selfie that then went viral. Not soon after, the owners of the fridge contacted the men and told them about its four mile journey. The beer fridge has since been returned to its original owner. Simply put, #NebraskaStrong was never about some past or future, it was about the now that was configuring itself and the we that the now collected. In this way, the hashtag became a way to focus on a particular kind of care in action.

#NebraskaStrong is an ongoing-everyday care that is grounded in helping others. The hashtag provides a way to see how the collected and contingent we's put into practice an everyday care. Puig de la Bellacasa (2017) wrote, "Focusing on everydayness, on the uneventful, is a way of noticing care's ordinary doings, the domestic unimpressive ways in which we get through the day, without which no event would be possible" (p. 117). In many ways, the tweets using the hashtag became a way to materialize how the "we's" got through the day. Nebraska Strong care is an everyday practice that is "about relating with, and partaking in, worlds struggling to make their other visions not so much visible but possible" (Puig de la Bellacasa, 2017, p. 118). In other words, #NebraskaStrong is not so much about the everyday as much as it is about what possible worlds assemblages of humans and nonhumans could create. The careful actions materialized in the hashtag generate a particular Nebraska Strong practice of care. Such a practice is ongoing (as flood relief was still happening as I wrote this chapter in December 2019) as it creates careful actions that are about getting through the day.

Haptic Flows of Technotouching

The #NebraskaStrong hashtag along with newspaper articles and weekly phone calls home became a way for me to touch, but not understand, these practices of care happening in Nebraska. As a non-resident of the state, I watched from afar and did not directly participate in these practices of care. I could only listen from hundreds of miles away to these practices on social media, online newspapers, and to the stories my parents told me. I could not fully grasp the flood and these care-filled practices. There was no way for me to know what it was like to live on a flood-made island as my parents did and directly engage in these practices.

Puig de la Bellacasa (2017) might call this technotouching, or technology developed for and/or used for sensorial immersion that does not "transcend human limitations" (p. 104). While I could not see the floods firsthand as they were happening nor the immediate aftermath, I was still able to touch them through technology. Technotouching became a way for me to co-create meanings and engage in practices of care. Puig de la Bellacasa wrote, "Touching technologies are material and meaning producing embodied practices entangled with the very matter of relating-being. As such, they cannot be about touch and get, or about immediate access to more reality. Reality is a process of intra-active touch" (p. 114). In other words, just because a human can haptically engage from afar does not mean that they get "it," whatever it may be. Technotouching does not collapse space and distance. Nor does it render meaning as something transferable in a one-to-one correspondence. Rather, technotouching generates a distributed reality through space and time in which multiple situated meanings are created in an assemblage-like fashion. In this instance, an assemblage of technology, humans, nonhumans, floods, and so on co-created a reality and multiple situated meanings.

This is a speculative reality that is "an invitation to participate in its ongoing redoing and to be redone in the process" (Puig de la Bellacasa, 2017, p. 117). In so doing, this reality offers no guarantees of sure knowledge or that knowledge can be enhanced. In the twists and turns of technotouching, one can only be undone, affected, by what one touches.

Technotouching took my breath away with its continual waves of affect that undo me and render knowledge situated, contingent, and shifting. I could not believe my eyes as they haptically touched the floods as I watched the Nebraska Educational Television (Kelly, 2019) documentary about the floods, *And the Floods Came*. The amount of damage. The cost of that damage. The stories. The photographs. Farmland becoming lakes. The resilient beauty of the practices of care happening in my state. And, it went on, even into December 2019, when I wrote this chapter. Each reality of the flood that is generated over space and time is undoing itself as it undoes me. The incredulity of it grabs me and pulls me under and constantly shifts meanings.

Thinking-Being with Care

As I noted in the introduction, I had the opportunity to see the flood damage and flooding in July 2019. Even when I could physically touch the flooded areas in person the flood overwhelmed me. Proximity to the floods granted me very little, only more co-created realities of undoings. How does one think when each moment creates new arrangements in such urgent times? One can only speculate.

Puig de la Bellacasa (2017) wrote:

> We learn that to speculate is also to admit that we do not really know wholly. Though there are indeed many things that knowledge-as-distant vision fails to feel, if touch augments proximity, it also can disrupt and challenge the idealization of longings for closeness and, more specifically, of superior knowledge in proximity. (p. 117)

Prior to my trip in July 2019, I idealized proximity. If I could just see the damage with my own eyes, then I might be able to make sense of it. Someone else's stories about the floods might spark an insight that might make me be able to say something. If I could walk the flooded areas, then I might create some sort of touch-knowledge. I fell prey to post-positivism's supposed guarantees that proximity would grant me knowledge. It did not.

If anything, proximity generated more speculative thinking. There was no way to know this flood. I could only think with it, walk with it, story with it. During the visit, I took many notes and many photographs that followed the ever-generating speculations, never catching up to them. The vast majority of these notes and photographs became the evidence of listening to the flood, to its resonances in both real and produced ways (Haraway, 2002). The flood invited me to listen to its resonances (Haraway, 2002). Listening to resonances became acts of humility toward something of such magnitude that I cannot understand, of partial "connections and unexpected openings that make situated knowledges possible" (p. 684). Assemblages, such as the 2019 Midwestern Floods, ought to bring about humility that only comes with the recognition of our interdependency, not only with other humans, but with nonhumans, the environment, and so on. In such a humble interdependency, listening to resonances is critical. Resonating partial connections and the interdependency of those connections invites us into the fray and asks for "accountability and responsibility for translations and solidarities" (p. 684). Acting is done with caution as it results from examining the multiplicities of the assemblage in which we find ourselves and creates uneasy, partial, and shifting solidarities.

Nebraska Strong Onto-ethico-epistemology

The realities that are distributed in technotouching materialize frictional ways of knowing. Puig de la Bellacasa (2017) wrote:

> Thinking with care also strengthens the notion that there is no one-fits-all path for the good. What as well as possible care might mean will remain a fraught and contested terrain where different arrangements of humans-nonhumans will have different and conflictive significances. (p. 220)

Put another way, thinking with care is about friction, negotiating that friction, and creating practices of care that work within that friction to live better.

This friction materialized in conversations with my parents. It was easy for me to be at home in Tennessee on the phone with my parents and say, "Well, all the research is pointing to stronger storms. I can't say I'm surprised that the flood happened." Or, "This is the climate crisis in action." Or "We should be prepared for these kinds of disasters to happen at any time." Any time I uttered these statements, I could sense my parents' frustration through their silence—their unspoken way to say, "You've gone too far." I relied upon a scientific knowledge over an affective knowledge. I offered moralizing statements in the face of constant uncertainty that undermine these statements. My parents' knowledge of the flood lived in the realm of affect. It was not until my trip to Nebraska in July 2019 that their sense of knowledge came into being for me.

On one of the last days I was there, my father drove me to the area of town that borders the Platte River. We first drove through the state recreational park outside my hometown. I knew the roads of that park, the lakeside beaches that I swam in as a child, the lakes where the geese and other waterfowl tended to congregate, and the areas where certain birds nested in the spring and summer. The flood changed the park. Roads were damaged. Individual lakes that had been separated by picnic areas and roads were now joined together. Two lakes that bordered the state park joined with others not previously in the state park. The water created new borders. The water asked me to learn how to navigate the park again.

Then, we drove on Ridge Road, which provides access to the Platte River. The road greeted us with a sign that warned visitors that they entered at their own risk. We drove slowly so I could see how the water created new landscapes. I stepped out of the car to take photographs. Unbeknownst to me, a group of deer were feeding next to a house made uninhabitable by the flood. I heard warning grunts and looked up and saw them. It seemed like I had offended them somehow. A human presence in an area that they now appeared to claim as their own. I softened my gaze as I looked at them in an attempt to communicate that we could be together without fear of each other. Soon after exchanging looks with me and each other, they ran off into what nature had reclaimed.

I carefully walked along the road in river sand. I studied trees bent at angles that continued to grow. I touched the bark to see where the river ice made them bend. I noticed how the river left debris—lawn chairs, doors, a shoe, outdated wallpaper, and so on—scattered across the area. In a couple of areas, toads once camouflaged by the sand hopped from the sand as I disturbed their area. As I walked, waves of affect flooded me and hopped along with me. I listened. I let those waves wash over me again and again. I returned to the car. Silent. My father and I did not talk the rest of the way home. I tried to "write to be in the reverb of word and world" (Berlant & Stewart, 2019, p. 131) when we arrived home. But, the reverb of the affective floods was too much.

The next day we went to Desoto National Wildlife Refuge, a refuge that borders the Missouri River and is approximately 30 miles away from my hometown. As we walked into the visitor center, cliff swallows, a bird uncommon to the area, populated the eaves of the center. We learned that the cliff swallows had come en masse to the area because of the number of mosquitoes living in the flood waters. Mosquitoes are cliff swallows' primary source of food, and the flood provided fertile ground for mosquitoes to proliferate. We also learned that migratory birds that have come to this area for centuries diverted their paths because of the floods. Carp populated the flooded areas, and bow fishers (people who use a bow and arrow to fish) captured truckloads of them. Sometimes catching two fish with one arrow.

I walked a path at the refuge and was once again met with toads hopping with me, just as they did the evening before. Water had taken some of the paths, requiring new paths to be made. The water did the same for roads. The water created new ways of knowing and being for so many critters.

As we drove through the park, a snapping turtle slowly made its way across the road. We stopped the car and got out to watch the turtle. The turtle stopped as the humans stared at it, trying to guess its age, wondering about where the turtle had been and where it was going. We also talked about how the turtle had lived this flood and wondered how its life had changed. Slowly, ever so slowly, the turtle made its way across the road. I wondered what it thought of these bipedal mammals chattering on about it.

The flooding reverberations overwhelmed me again. Puig de la Bellacasa (2017) wrote:

> These engagements do not so much entail that knowing will be enhanced, more given, or immediate through touch than through seeing; rather, they call attention to the dimension of knowing, which is not about elucidating,

but about affecting, touching and being touched, for better or for worse. About involved knowing, knowledge that cares. (p. 118)

The everyday practices of driving with my father and visiting a wildlife refuge center I had been to countless times in the past affected me differently. As the flood enveloped me into a watery we, I realized that how I had known the flood in the past was based on an impersonal, scientific view from Tennessee. The we called me to the affect-rich flood that created so many we's I had technotouched from afar. The everyday practices are about being affected, being touched by others, and, in turn, affecting and touching others generated affective and caring ways of knowing. Nebraska Strong not only became a series of practices of care, it became an affect-rich way of knowing and being within those practices of care.

Unruly Nebraska Strong Solidarities

The hashtag #NebraskaStrong and the practices of care materialized within that hashtag disrupted any sense of moral imperatives and centralization. Analysis of the hashtag, a hashtag that was widely used for some time, and following it as it has moved across space and time demonstrate how the affect-rich practices of care materialized in the hashtag resist centralization and moral imperatives.

The state of Nebraska through the Nebraska Emergency Management Agency and other public and private partnerships borrowed the hashtag #NebraskaStrong and named their recovery efforts as "Nebraska Strong." The government website offers a secure way to donate funds to flood victims and links to resources. Furthering the idea of Nebraska Strong is the University of Nebraska Public Policy Center. This center is offering mental health counseling and assistance to community rebuilding across the state. The use of the hashtag by a policy center and the state government can be viewed as a way to centralize and normalize the ethical work of the "we's" coming together into a moral imperative, which Puig de la Bellacasa (2017) warned against.

The hashtag remained active on Twitter as people continue clean-up efforts among other activities. Analysis of the hashtag from late spring 2019 and through 2020 materialized a new series of "we's" as people continue to find themselves caught up in further relief efforts that are not associated with the aforementioned agencies. These series of "we's" disrupt any efforts of centralizing and normalizing practices of care.

These series of "we's" were recently celebrated by the *Omaha World Herald* when it named the Flood Helpers as Midlanders of the Year (Duffy, 2019). In the past, this award was given to individuals rather than collectives. However, the 2019 Midlanders of the Year focused on those who practiced Nebraska Strong ontoepistemological practices of care. A December 22, 2019, article featured numerous photographs of flood relief and stories of how these people practiced affective care. The news section of the hard-copy paper featured two full pages that described this care work across the state. In brief (no more than five sentences) descriptions of this care work, "we's" of all Nebraskans (human and nonhuman) were celebrated. Nebraska Strong, indeed.

Affect-rich care work in the Anthropocene cannot be regulated or normalized. It will always be generated responsively. This response is never to the Anthropocene; it is always with it. As the climate crisis continues on, affect-rich care practices are generated with it as we continue to learn to live in crisis. These series of "we's" must be inclusive as we think, be, and care with the Anthropocene. In other words, these "we's" must "think together anew across differences of historical position and of kinds of knowledge and expertise" (Haraway, 2016, p. 7). Simply put, we must think together collectively, recognizing the wide ranges of expertise from all as we respond with the Anthropocene without falling prey to the god-trick (Haraway, 2002). This unruly work is always already imperfect, for the human elements of the we's must recognize that the effects of the Anthropocene many times further displace those people not historically centered. The Anthropocene is not a singular experience for all people. We must listen to the resonances without "appropriating the vision of the less powerful while claiming to see from their positions" (p. 679). We must work to see how all our stories of the Anthropocene "join with another [without] claiming to be another" (p. 681) in uneasy, shifting solidarities of collective we's. Such a solidarity moves beyond the moral imperative of creating a good life for all as we live and die in the Anthropocene.

Interestingly, Nebraska's former state motto was "The Good Life." For many Nebraskans, the question became "For whom is this good life?" as this good life seemed to be grounded in White, middle-class ways. Perhaps answering that question, the state recently changed its motto to "Nebraska. Honestly, It's Not for Everyone." Likewise, Nebraska Strong can be similarly questioned. Nebraska Strong features predominately white people at the exclusion of the Latinx population, Black Nebraskans, the Indigenous Tribes, and the White working class who also call Nebraska home, despite these populations living in some of the hardest hit areas of the 2019 flood. Where are their voices, their historical positioning, and their knowledge in Nebraska Strong? How might their embodied and situated knowledges have generated a stronger Nebraska Strong, a strength that draws from humble listening to resonating partial connections of "living within limits of contradictions" (Haraway, 2002, p. 684) of the flood?

Both the presences and the absences in the we's that constituted Nebraska Strong point toward a practice of care that moves beyond moral stances to "affective, ethical, and hands-on agencies of practical and material consequence" (Puig de la Bellacasa, 2017, p. 4). Affective stories, such as this one, have the potential to suggest an interdependent onto-ethico-epistemological state in which we (all humans and nonhumans) are relationally living and dying with each other in the Anthropocene. Relational "both ands" create space for humans to consider both presences and absences of stories, how those stories are situated in dense entanglements, and how those stories are both practical and material. Such an affective care is about staying with the trouble (Haraway, 2016) of interdependence. It is about making kin with diverse people and nonhumans as we learn to interdependently live and die together, attending to the resonating affect-rich situated stories of living with the Anthropocene (Haraway, 2016). Staying with the trouble (Haraway, 2016) becomes the moral imperative (if one can call it that) as the Anthropocene continues and will continue to collect us into unanticipated "we's."

SCENES OF HOME

In writing with the everyday—from technotouching from afar to touching in proximity—I hope to illuminate practices of care that are affective-rich messes. These frictional messes are never easy. And, I suspect more will happen (have happened, are happening) in the Anthropocene. We write to reverb with the world (Berlant & Stewart, 2019). We write in affect-rich assemblages that sometimes stun us in silence as we listen to, become with, the earth.

As we write these accounts that hopefully do the work that Haraway (2016) asks us to do, to write stories with matters that matter, the "I" who writes is always already a "a mass of reactions vaguely jarred into being at the glimpse of a method or a thought. Just trying to catch up with whatever's happening" (Berlant & Stewart, 2019, p. 123). So as we examine the knots, descriptions, ties, and matters (Haraway, 2016) that create the

"I" that is just trying to keep up with affects that always already over-whelm us in the Anthropocene, "we write to invite and to goad, to bring the weight of scenes home" (Berlant & Stewart, 2019, p. 131).

The reverberations of this flood continued to make we's well into December 2020, nine months after the March 2019 flood. As the climate crisis continues and provokes more powerful storms, I suspect that all crit-ters will continue to form new we's that demand different practices of care. These practices of care, like Puig de la Bellacasa (2017) notes, are entangled, material, and difficult. These practices do "somehow own us, we belong to it through the care that has attached us" (p. 167). In other words, climate crisis creates we's and practices of care that own us; we do not own them. They destroy moralizing stances. They entangle us and ask us to attend to resonances from all critters while adopting practices of care that get to work in those resonating entanglements. Practices of care have the potential to link us together, forever change us, and work toward more better lives.

The practices of care that I technotouched, engaged with, and ulti-mately owned me (Puig de la Bellacasa, 2017) with the March 2019 flood created eddies and flows in my life. The more post-positivist understand-ings of the flood that included weather patterns, river measurements, the numbers of the critters affected by the flood, and the financial damage from the flood provide little to help me think with the flood. It is easy to stand on a river bluff that overlooks the river in which the flooded waters will pass as they make their way to the Gulf of Mexico and offer up these numbers as the only explanations. These explanations negate the move-ment of affect and the ways in which these movements generate practices of care, which I hope to have illustrated in this chapter.

In the knots, descriptions, and ties (Haraway, 2016) that the flood gen-erated and, in so doing, created a series of we's, I have tried to write a scene from my home state (Berlant & Stewart, 2019). The affects gener-ated by listening to the earth and attending to the we's it generated shifted me and created new practices. These new practices owned me and shifted my thinking into the powerful realm of affect and what affective knowl-edge can do in the Anthropocene. For, when we listen, really attend to the all the critters—both humans and nonhumans—maybe we can begin to make better possible worlds. I am reminded of the flight shifts of migra-tory birds that happened in March 2019. I wonder how I might be changed by the realms of affect that blow like atmospheric winds, shifting onto-ethico-epistemological patterns and what those new patterns will demand of me. It is the only thing I can anticipate. There are no predictions. No meteorological models can predict how practices of care and affective ways of knowing and being within those practices will collect me. A different scene of home will always materialize in the Anthropocene.

REFERENCES

Berlant, L., & Stewart, K. (2019). *The hundreds.* Duke University Press.
Duffy, E. (2019, December 22). Flood helpers named Midlanders of the Year. *Omaha World Herald.*
Gaarder, N. (2019, March 17). Record snowfall, 'historic' bomb cyclone are forces behind Nebraska floods, blizzard. *Omaha World Herald.* https://www.omaha.com/weather/record-snowfall-historic-bomb-cyclone-are-forces-behind-nebraska-floods/article_b7b6547d-d4d2-5363-ad64-1142f87a513a.html
Gough, N. (2006). Shaking the treed, making a rhizome: Towards a nomadic geophilosophy of science education. *Educational Philosophy and Theory, 38*(5), 625–645.
Haraway, D. (2002). The persistence of vision. In N. Mirzoeff (Ed.), *The visual culture reader* (pp. 677–684). Routledge.
Haraway, D. (2008). *When species meet.* University of Minnesota Press.
Haraway, D. (2016). *Staying with the trouble: Making kin in the Chthulucene.* Duke University Press.
Kelly, B. (Writer & Producer). (2019). And the floods came. [Television documentary]. http://pivotal.netnebraska.org/nebraska-floods/
National Weather Service [@NWSOmaha]. (2019, February 28). Cold start to March is in the forecast across the area. And it's not just a little cold, it will be around 15 to 35 degrees below normal the first week of March! Brrr! #newx #iawx [Tweet]. *Twitter.* https://twitter.com/NWSOmaha/status/1101150118450028544/photo/1?ref_src=twsrc%5Etfw%7Ctwcamp%5Etweetembed%7Ctwterm%5E1101150118450028544&ref_url=https%3A%2F%2Fwww.omaha.com%2Fweather%2Fomaha-can-expect-one-of-the-coldest-starts-to-march%2Farticle_3ccd17c2-bc4e-54fa-a9ab-862f973e34bc.html
Puig de la Bellacasa, M. (2017). *Matters of care: Speculative ethics in more than human worlds.* University of Minnesota Press.
Salter, P. (2019, March 19). 'A gift sent from the heavens': Magic beer fridge found in flooded field. *Lincoln Journal Star.* https://journalstar.com/news/state-and-regional/nebraska/a-gift-sent-from-the-heavens-magic-beer-fridge-found/article_2af5bd7f-7810-57ec-bcd2-d5ace989d611.html

Schneiderman, J. S. (2017). The Anthropocene controversy. In R. Grusin (Ed.), *Anthropocene feminism* (pp. 169–198). University of Minnesota Press.

Schwartz, M. S. (2019, March 21). Nebraska faces over $1.3 billion flood losses. *National Public Radio*. https://www.npr.org/2019/03/21/705408364/nebraska-faces-over-1-3-billion-in-flood-losses

PART IV

Science Education for a World Yet to Come

CHAPTER 14

Science Fiction, Speculative Pedagogy, and Critical Hope: Counternarratives for/of the Future

Brittany Tomin and Ryan B. Collis

INTRODUCTION

Tom Moylan (2021) begins *Becoming Utopian: The Culture and Politics of Radical Transformation* with a survey of the current moment: "It's not yet the worst of times, but things are worse every day. It's far from the best of times. Harm abounds everywhere" (p. 1). Expanding on this by mapping interrelated webs of ecological destruction, the boundless reach of capitalism, ongoing wars and human suffering, and political division further deepening discrimination and injustice, Moylan convincingly lays the groundwork for a critical realization: we have, perhaps more so than ever before, a desperate need for new stories, different orientations, and

B. Tomin (✉)
University of Regina, Regina, SK, Canada
e-mail: Brittany.Tomin@uregina.ca

R. B. Collis
York University, Toronto, ON, Canada
e-mail: rbcollis@my.yorku.ca

S. Tolbert et al. (eds.), *Reimagining Science Education in the Anthropocene, Volume 2*, Palgrave Studies in Education and the Environment, https://doi.org/10.1007/978-3-031-35430-4_14

alternative ways of being in the world that can catalyze transformation through critical hope. With similar sentiment, Matthew J. Wolf-Meyer (2019) in *Theory for the World to Come: Speculative Fiction and Apocalyptic Anthropology* suggests that speculative storytelling can be an integral tool in envisioning possibility beyond—and perhaps *through*, or within—calamity. This is an especially important offering in the times in which we live: times characterized by a kind of uncertainty wherein we are prone to turn inward and focus on our own *individual* futures because the prospect of collective continuance feels too tenuous (Rubin, 2013); times in which eco-anxiety runs rampant (Ojala, 2016; Pihkala, 2020) and ecological grief abounds (Cunsolo & Ellis, 2018), often unaddressed and disregarded; times where narratives of climate change in particular are decidedly and perpetually negative (Kelsey, 2016) and hope teeters on the edge of impossibility; and in times where the very institutional systems and structures of the Anthropocene/Capitalocene seem to encourage—and *demand*—that all humans stay the course in spite of the unfaltering truth that the "kind of thinking that created today's global turbulence is unlikely to help us solve it" (Moore, 2016, p. 1).

Accordingly, in this chapter, by centering science fiction (hereafter SF) and speculative storytelling toward a broader speculative pedagogy, we take seriously a couple of key assertions: first, as Donna Haraway argues, it "matters what thoughts think thoughts. It matters what knowledges know knowledges. It matters what relations relate relations. It matters what worlds world worlds. It matters what stories tell stories" (2016, p. 35). As we expand upon below, a pedagogical orientation rooted in complex understandings of interwoven pasts, presents, and futures grounded in storytelling must be attentive to the power of stories, and how the speculative stories we tell in the present help us tell new stories of the future while simultaneously reorienting us to ever more complex understandings of our interrelatedness *in the present* beyond such binaries as utopia/dystopia, progress/stagnation, hope/fear. The past, present, and future operate reciprocally, past and present informing future and future informing present action and rearticulations of the past. In this, speculative stories about the future can help us transform the present. Building on this capacity of stories to bring us into new relations with the world across time, wherein stories help us trace how "ways of life are to a large extent, manifestations of concepts—of ideas they foster and the possibilities of action they afford, delimit, and rule out" (Crist in Moore, 2016, p. 24), speculative, future-oriented storytelling and speculative pedagogy can offer us new *modes* of thinking and new paths forward beyond paradigms of capitalist continuance and/or totalizing destruction.

Further, given that "fictional worlds are not just figments of a person's imagination" but rather, "circulate and exist independently of us and can be called up, accessed, and explored when needed" (Dunne & Raby, 2013, p. 71), this chapter considers storytelling—through engagement *and* creation—as a critical tool that is vital in thinking and acting differently in the world. Moving beyond feelings of inevitability with regard to the future through the concept of *future-making* (Montfort, 2017, p. 4), how might we imagine new possible, probable, and preferable futures (Bell, 1998; Kelsey, 2016) through which we can uncouple ourselves from the seemingly inevitable, forced imaginary of ceaseless, anthropocentric, capital-driven destruction of all life? If future-making—an orientation toward the future imbued with a sense of agency through pairing action with imagined future possibility—can help expand the anti-utopian limits we impose upon ourselves (Olin Wright, 2010, p. 23), can speculative pedagogical approaches rooted in stories help us imagine beyond neoliberal notions of progress, change, innovation, and productivity? Can they expand the limits of our utopian consciousness and ability, while we remain within the realm of uncertainty? Inspired by Haraway's *Camille Stories* from *Staying with the Trouble: Making Kin in the Chthulucene* (2016), which exemplify what it might mean for "science fiction and science fact [to] cohabit happily" (p. 7), in this chapter we therefore consider the interdisciplinary potential of science fictional and speculative stories to aid us in inhabiting this cohabitation in environmental and science education contexts: what might a centering of SF and speculative storytelling as a speculative, pedagogical orientation offer us as a means of *rethinking* science education? How might we carve out space for fiction and fact to meet? Drawing also from a collaborative, speculative world-building project conducted by one of the authors, this chapter provides a conceptual overview of SF and its uses in and beyond education as a mode of thought, examines potential roles the concept of the future might play in reshaping how we approach environmental and science education speculatively, and concludes by mapping out the parameters of a speculative pedagogy; an orientation toward teaching and learning grounded in an openness to "what if?" questions that can support collective disruption, critical hope, and the creation of counternarratives of and for the future.

Science Fiction and Critical Hope

Stories of future destruction are now enshrined in our collective consciousness: the inevitable extinction of non-human species, the tireless and unstoppable expansion of corporate power and consumption, and the irreversible warming of the earth, to name a few. The prevalence of this genre has normalized the assumption that humanity is on an unalterable collision course and that there are no alternatives. Sarah S. Amsler (2015) in *The Education of Radical Democracy* describes this as a crisis of hope, a foreclosure of possibility wherein change feels impossible and, even if it were, "people are not even sure that making change would make a difference" (p. 21). This sense of helplessness has significant implications for education, and like Amsler, many working on eco-anxiety and ecological grief turn to complicated expressions and possibilities of hope in our contemporary moment of existential turmoil in search of paths forward. For example, in considering the affective experience of climate education on youth, Maria Ojala (2016) advocates for a pedagogical approach "that focuses on a critical hope that is based in an acknowledgement of the negative, a positive view of preferable futures, the possibility of societal change, and that is related to concrete pathways toward this preferable future" (p. 42). Resisting the neoliberal impulse to "privatize hope" (p. 46), Ojala emphasizes the importance of teaching toward the undecided nature of the future and argues that "to face the negative is a starting point for constructive hope" (p. 51). A *critical* hope, positioned in this way, does not lead youth away from the truth of our current moment but, rather, carves out space for new stories of possibility to emerge from darkness. This is echoed by Kelsey (2016), who, in the context of environmental education, asserts that "we need to recognize the power of narratives, and help the children and communities we engage with to recognize them too" (p. 32). She goes on to argue that the "stories we tell ourselves shape how we live and what we believe to be possible," positioning narratives of inevitable destruction as profound obstacles thwarting meaningful change. There is a need for the kind of radical hope that Rebecca Solnit declared "is not a lottery ticket you can sit on the sofa and clutch, feeling lucky. It is the axe you break down doors with in an emergency" (Solnit in Gannon, 2020, p. 19). If clinging "to a narrative of doom and gloom that leaves its most vulnerable constituents frightened and disempowered clearly needs to change," and "the narrative needs to be changed in a way that does not create ethical tensions around the issue of raising false hope" (Kelsey,

2016, p. 28), how might SF and speculative storytelling—as a space of intermingling fiction and fact—offer space for this renarrativization?

In this chapter we therefore offer speculative storytelling—and SF in particular—as a path toward meaningfully addressing the pervasive and paralyzing nature of dystopic worldviews and as a pedagogical orientation through which we can imagine *otherwise* in science education (and elsewhere) alongside our students without abandoning the world as it is. Critical to placing speculation at the fore of a pedagogical orientation of openness toward the future is navigating the parameters of speculative fictions as distinct genres with specific narrative capabilities. Importantly, while the umbrella of *speculative* fiction encompasses myriad genres—SF, fantasy, horror, weird fiction, etc.—here we focus on SF because, as Istvan Csicsery-Ronay Jr. suggests, "sf has come to be seen as an essential mode of imaging the horizons of possibility" (2008, p. 1). SF, as a largely future-oriented genre, has a long history of reaching toward otherness as it fictionally grasps at the contours of change (Campbell, 2019). Although commonly associated with its tropes—robots gaining sentience and threatening the end of humanity, an isolated scientist overwhelmed by their own ambition and driven to destructive innovation, or a crew of misfits on a long-haul interstellar voyage to galaxies unknown—SF is particularly unique in its capacity as a genre of difference wherein the future in treated as a "locus of radical alterity to the mundane status quo" (Freedman, 2000, p. 55). By catalyzing comparisons between unfamiliar futures and our own *seemingly* familiar present, SF has routinely been mobilized toward envisioning myriad possibilities and used as a means through which we might answer critical questions about how to live meaningfully together in the present.

In her critical look at SF for children and teens titled *The Intergalactic Playground* (2009), Farah Mendlesohn begins with a structural articulation of what makes a text SF: "dissonance, rupture, resolution, consequence" (p. 10). This framework is helpful in beginning to map out how a generic structure might be translated into a pedagogical orientation and a means through which we might work with young people to think *science fictionally* or speculatively about future possibilities using fiction. In SF, drawing from Darko Suvin's (1979) *The Metamorphosis of Science Fiction: On the Poetics and History of a Literary Genre*, Mendlesohn explains how science fictional stories create dissonance and rupture through the introduction of a novum—a "new" catalyst of change around which a story revolves—and the experience of the reader as they begin, through comparison and cognitive estrangement, to see their own present as fundamentally strange. In their ground-breaking anthropological examination of scientific laboratories, Latour and Woolgar (1986) used this approach of making "strange those aspects of scientific activity which are readily taken for granted ... to dissolve rather than reaffirm the exoticism with which science is sometimes associated" (p. 29). This "strange making" provides a new vantage point from which to observe and understand that which otherwise might be automatically taken as accepted. As Joseph W. Campbell (2019) further describes, SF has the potential to help us see our empirical environment from a critical perspective because, immersed in a fictional world that might look like ours but is clearly not *our* world, all our assumptions about the fictional future (and, by extension, our empirical present) must be called into question. Returning to Mendlesohn's framework, "in the 'full SF story,' the resolution is not the end of the story, it is the beginning, for sf stories are about *change* and *consequence*" (Mendlesohn, 2009, p. 12). Taken together and following this structural grammar of SF, we can see how SF functions as an opportunity to engage with difference by destabilizing what is familiar. In SF stories, we see a world alike but different from ours and, in turn, see our own world *differently*. In the process of coming to understand the new world we've been presented with, we are then led to see the possible consequences of change. If, as Campbell asserts, the "social critical work of science fiction is to bring the reader to the encounter with the other" (Campbell, 2019, p. 67), how might we mobilize SF toward pedagogical encounters with myriad others and with alternatives beyond the widely accepted narratives that perpetuate overwhelming ecological grief and enforce helplessness? How might SF facilitate our ability to "transgress, or disrupt, deeply held and taken-for-granted norms, norms that are at the roots of oppression and unsustainability," and imaginatively chart paths toward "acting in surprising, creative, and boundary-crossing ways"? (Ojala, 2016, p. 43).

Science (Fiction) Education

Although underutilized and often dismissed as unserious (Westfahl & Slusser, 2002), the use of SF in science education specifically is not a new or novel phenomenon. The 1978 Modern Language Association special session panel titled "Teaching Science Fiction: Unique Challenges" between Gregory Benford, Samuel Delaney, Robert Scholes, and Alan J. Friedman, published in 1979, serves as an early example of discussions on the use of SF to motivate students to pursue science, echoed in later work on how SF can be used in science education to deepen student engagement (Oravetz, 2005; Singh, 2014; Smith, 2009; Subramaniam et al., 2012) and teach students about the nature of science and scientific innovation (Hasse, 2015; Reis & Galvão, 2007). Taking an interdisciplinary approach further expands on the promise of SF, not just as a generator of scientific interest but as a tool through which change—scientific, technological, political, or social in nature—can be thought of differently. From Noel Gough's (1993) early work exploring the intersection of science fictional literary form and science and environmental educative storytelling, to more recent work using science fictional and speculative storytelling with youth to reimagine community, education, and the world (Mirra & Garcia, 2020; Toliver, 2021; Truman, 2019) and engage in futures talk (Priyadharshini, 2019), SF and the act of *speculation* can be a powerful tool that can help young people articulate their hopes and fears, resist dominant epistemological and ontological narratives, and see themselves in the construction of the future. Building on this latter application of SF, not just to generate interest in science or on any other subject but rather to use SF and speculative storytelling to reorient how youth conceptualize their relationship with the present, the future, and the potential for change in increasingly turbulent times, the remainder of this chapter focuses on approaches to science fictional storytelling and how we might use science fictional and speculative storytelling to construct counternarratives of possibility.

Collective Speculation and the Collaborative Building of Worlds

Exemplifying the construction of speculative counternarratives of the future, in the spring of 2019, one of the co-authors of this chapter engaged in a speculative world-building project with a group of high school students in a secondary English class (Tomin, 2020), using SF and collaborative world building (Hergenrader, 2019; Tuttle, 2005) to envision future possibilities focused on the city in which the research occurred. Over the course of two months students explored examples of SF and elements of science fictional storytelling, concurrently engaging in conversations about their hopes and fears for the future and expanding on connections they were beginning to make between their understanding of the present moment and how that understanding informs what futures they envision as being preferred, probable, and possible (Bell, 1998). This work emerged as students began to consider how SF authors construct the worlds in which their stories take place, focusing on the extrapolative process through which authors critically expand upon their present to imagine radically different futures. Alongside extrapolative storytelling, students examined self-selected and collaboratively-engaged SF texts in the context of world building—the process SF authors and creators follow when creating the worlds in which their speculative stories occur. Building off this exploration, in the latter half of the study students embarked on the *Toronto 2049* project: a project that used science fictional storytelling and world building to imagine the future of the city they lived in and the world in which it might exist.

Toronto 2049 took the form of a collaborative wiki-style catalog of the future, and students were asked to build the world from broad to specific, identifying enduring problems of concern to them alongside their understanding of the present and using extrapolative processes to envision possible futures together: based on the present, what did they think the city would likely look like? What did they want it to look like, and how did they want it to *feel* to live in that future? What did they fear? Over the remaining four weeks of the study students used in-class time to discuss contemporary issues spanning reproductive and sexual health for teens, climate change, poverty and food security, medical care, education, religious freedom, automation, and myriad other topics. As the world they built became more populated—co-editing each other's wiki entries, researching topics central to their imagined future, balancing dystopian and utopian imaginaries alongside their hopes and fears—they were asked, as their final entry to the project, to write from within the world they had built. Would this be the preferred future for everyone? Would people experience the future they had imagined in the same way? If this represented their personal preferred future, what do they imagine might have had to happen between their present and 2049? How might that future-history map onto a lived life? As the culminating contributions these students would make to the project, the entries reflected multiple formal planning discussions wherein students drew from their own personal experiences and intersectional insights on contemporary issues but also from informal collaborations through which students were able to learn from others who held often different hopes and fears for the future than they did. The diversity of first-person perspectives reflected this variety of perspectives, spanning the narratives of a young university student in an illegal, "internatural" relationship with a robot peer, to a teenage girl who attends school entirely online and has never met any of her peers in person, to a fashion designer hosting a showcase of a biodegradable clothing line on a flooded city street, to a child asking her father what it was like to live in single-family homes in a society that has long since abandoned them.

Tracing a future-history involving a world in which climate change is taken seriously far too late, all these stories take place in a shared world the students built based on research and experiences rooted in students' present moment, in a fictional future context where students sought to imagine what it might look like to rebuild, repair, and preserve what is left of the natural world. While a problematic narrative thread of technology rescuing society from the worst of climate change ran throughout much of the world-building project, as part of the planning process students also had to envision what, based on research, they believed would happen and, accordingly, how both daily life and the structures governing society would have to change to make life—for any living being—possible at all. Using science fictional storytelling as a form of present critique (Sullivan III, 1999; Thomas, 2013) and as a way of engaging with difference, the world-building project gave students imaginative space to play with possible radical changes in their future-oriented storywork while simultaneously basing their fictional projections on issues concerning them in the present. Through this work, some students explored broader issues (e.g., climate change, political systems), while others used extrapolative storytelling to consider personal challenges, working through both preferred and probable futures rooted in their own perspectives and experiences. Balancing preferred and probable futures was a significant element of the project, even as they brushed up against each other's differing views on what preferred, probable, and possible futures might look like. Having the freedom to explore different possibilities while using extrapolative storytelling to ensure their contributions to the world-building project could *actually* happen also empowered students as they imagined *viable* alternative futures to dominant discourses. At the end of the study, students expressed how important it was to them that the future they envisioned was feasible and rooted in reality, even more so because the futures they imagined were not the grim, dystopic futures they thought they would create.

Given the collaborative process through which students navigated the liminal space between fact and fiction in their future world construction, the *Toronto 2049* project represented an articulation of critical hope. Working together to research, discuss, share, revise, and create, students' imagined future did not abandon the present but rather was, importantly, informed by renewed understandings about what is at stake and how the facts that underpin our present reality can lead to many different outcomes depending on *present* action. As much as possible, this project was facilitated to allow space for open inquiry, wherein participants directed which elements of society to focus on as they imagined the future; a sharp turn away from the predetermined narratives of future possibility largely advocated for in conventional schooling practices. As one student noted:

> [The project] was a new way for me to view the future, especially in an academic setting, where the future is often a very self-centered thing. My experience in talking about the future in school was always nerve-wracking, all about choosing a lifelong career at age 18 and going into post-secondary education, burdened with debt. (Zad W.)

Through a project that engaged openly with myriad possibilities still rooted in the "real," students were able to carve out space within which they could combat hopelessness and helplessness and envision alternatives even as present reality crashed down upon them. As the student quoted below articulated, closed narratives of the future foreclose on the potential for youth to see the point of any action at all.

> I think people often have a lot of hopelessness for the future and I don't blame them. I used to think like that all the time. I used to think like, 'oh my god, what is the point? Where are we going?' And I kind of got sick of that. I got so sick of just hearing myself talk down on everything, thinking everything was going to hell. And I just like to think that maybe, just maybe, something good is going to happen. It [this project] was a breath of fresh air because I used to hate thinking about it [the future]. (Ivy B.)

The kind of narrative, speculative storywork at the heart of a speculative pedagogy, and exemplified in the *Toronto 2049* project, is critical in times like these. Through collective imagining like this, we might once again able to see the future as truly undecided.

CONSTRUCTING COUNTERNARRATIVES: WHAT IF/IF, THEN

In mapping a cyclical structural grammar of SF, Mendlesohn centers the question "what if?" as a distinct speculative move at the heart of the genre: "Identification of novum and cognitive dissonance usually leads to the idea of causality and consequence; that 'what if?' needs to be followed by the concept of 'if, then' " (2009, p. 13). The interplay between "what if?" and "if, then," as exemplified in the previous section, invites endless speculative possibilities. This is an especially important realization now, when we are inundated by news of our impending doom and events that feel routinely unprecedented and always "out of our control." In this context, SF and speculative narratives offer us a path forward through which we can collectively imagine messy, uncertain, but nevertheless radically different futures in community with others. A speculative pedagogy, posing the question "what if?" in a ceaseless pursuit of myriad possibilities, centers this imagining in all aspects of learning; every pedagogical act becomes a potential moment in which the present can be reanimated and the future reimagined. In contrast with approaches to teaching that only make space for teaching toward what *is*, speculative pedagogy involves privileging "what if?" questions—a mode of inquiry at the core of SF and speculative fiction—within dynamic relations of teaching and learning.

Of course, as Haraway warns, it is paramount that we "[stay] with the trouble" in the present (2016, p. 4), to not allow dreams and fears of the future inspire us to abandon the present. But what speculative pedagogy offers, particularly through science fictional and speculative storytelling, is a means through which "what if?" questions can be answered with many, plural "if, then" responses. As is illustrated by the brief example provided, if we can ask "what if?" questions, and we can map out myriad "if, then" possibilities, we might be able to act within the present toward those possibilities—even when the outcome cannot be guaranteed. As Moylan (2021) asserts, "those who consciously desire that better world have to find ways to tease out the tendencies and latencies that will enable all of humanity to build it, here and now, in the shell of the old" (p. 15). A speculative pedagogical orientation offers a redefined relationship toward the future, involving not just teaching how things *are*, but also making space for how things *could be*, in partnership with students whose plural perspectives, fears, hopes, and ideas about possibility offer many, inevitably uncertain, paths forward toward that better world.

While the *Toronto 2049* project occurred in an English Language Arts context, this restorying is no less needed and no less possible within the context of science and environmental education. Addressing climate education specifically, Maria Ojala (2016) calls for this kind of work and its importance:

> climate change educators should allow time and space to consider probable, preferable, and possible futures. For instance, when imagining the personal, the local, and the global futures X years from now, what are the probable scenarios in relation to climate change? It is also important to work with visions of preferable futures […] to promote constructive hope there is also a need to compare the "probable" with the "preferable" and come up with materially grounded and realistic "possible" futures. (pp. 51–52)

Echoing Gough's (1993) call to dismantle the false disciplinary divisions between literary work and scientific learning and inquiry through the use of SF as a window into innovation and science "in the world," a speculative pedagogical orientation fulfills Ojala's call to explicitly engage with future possibility in climate education by not just reading SF but using SF as a way of thinking about change. The tools of literary SF are not incommensurate with fact-based learning but, rather, help encourage new ways of processing what is. Following Mendlesohn's formula—what if?/if, then—speculative pedagogy in science and environmental education can help teachers explore alongside their students a pair of critically important questions: *What now? What next?*

PRACTICAL IMPLICATIONS: TOWARD A SPECULATIVE PEDAGOGY

These are dark times, seemingly darker with each passing day. With recent reports outlining the urgent nature of climate change while simultaneously affirming once again the profound, devastating impact humanity has had on all living creatures and on our own potential capacity for future continuance (IPCC, 2022), it is clear how susceptible to hopelessness we might become; how prone we might be to believing that overwhelming anxiety and grief might be the only path forward as we watch the truths of our complicit violence unfold. However, as Sarah S. Amsler (2015) argues, this is also exactly the kind of context in which we require "radically new readings of the present and the future and new methods for learning to read the world differently; readings that not only encourage 're-thinking' or 're-imagining' but re-doing the world" (p. 21). Accordingly, by way of conclusion, we end with central features of a speculative pedagogy that might help us address the feelings of helplessness and hopelessness in science and environmental education within the Anthropocene/Capitalocene that can help us process alongside our students' feelings of anxiety and grief without being immobilized by those fixtures of contemporary life. With a speculative orientation in hand, bolstered by science fictional storytelling, we might illuminate alternative paths and ways of being beyond narratives of the inevitable death and destruction that lay before us.

First, a speculative pedagogical orientation supports an exploration of the future that is inherently collective and socially constructed in community with myriad others. As seen in the brief example shared in this chapter, imagined futures are more powerful when imbued with the complexity of shared life. This collectivity is imperative to building paths forward that do not recreate the violence of the anthropocentric, capitalist paradigm laid out before us. Echoing Moylan (2021),

> this utopian project must necessarily be collective; for it involves the totalizing transformation of social reality, by all of us, for all of us. Settling for utopia in one person results in nothing but a tantalizing indulgence that is all too easily available for the capitalist disciplinary imagination. (p. 5)

As we envision future possibility in and beyond our classrooms, it is paramount that we embrace the subversive potential of SF to imagine "a future that opens out, rather than forecloses, possibilities for becoming real, for mattering in the world" (Pearson et al., 2008, p. 5) alongside those who have not been included in the dominant (and dominating) construction of the world. Building off of William Gibson's assertion that the "future is already here, it just isn't evenly distributed" (Gibson in Lothian, 2018, p. 4), Alex Lothian in *Old Futures: Speculative Fiction and Queer Possibility* emphasizes attending to "what imagined futures mean for those *away from whom* futurity is distributed: oppressed populations and deviant individuals, who are denied access to the future by dominant imaginaries, but who work against oppression by dreaming of new possibility" (pp. 4–5). Similarly, Harding (2009) notes, "feminism and postcolonialism both argue in effect that how we live together both enables and limits what we can know, and vice versa" (p. 403). Taking all of these perspectives into account, how might a speculative pedagogy open up space for new, collective, messy stories of the future to be told? The kinds of stories we need to build new paths toward those futures in the present? Accordingly, a speculative pedagogy is also one driven by students' own present experiences and their hopes for the future—visions of possibility that are too often denied space to be heard and taken seriously. In "an overdetermined world" where individuals are left feeling powerless to change the trajectory of society (Levinson, 1997, p. 439), a pedagogical orientation rooted in openness to possible futures makes space for student perspectives to disrupt learning what "is" with explorations of what "could be." Future-oriented, speculative pedagogy is necessarily less prescriptive and more community oriented than pedagogies rooted in pragmatic skills and mastery. The necessary work of dismantling the broader narratives of singular, controllable futures hinges on making space for diverse voices to be brought to the fore.

Speculative pedagogy also involves acknowledging the interdisciplinary nature of what it means to know and be in the world and embraces the complexity of the present and future as something that cannot be captured by siloed subjects but, rather, the dynamic interplay of myriad elements of human existence. In the *Toronto 2049* project, a deep understanding of literary SF form brushed up against information students learned in their science, civics, and history classes, an intermingling of disciplinary knowledges that helped capture the nuances of envisioned lives. Personal perspectives and experiences, and discussions of contemporary events and lived concerns, coexisted alongside research and an examination of climate reports, urban planning for increasing water levels, and potential technological innovations. In this world-building project, science did not exist in a vacuum but, rather, was woven into the narrative of our possible futures. Mirroring Lisa Tuttle's (2005) view of world building as an ecology of infinitely interrelated features and phenomena wherein even "in an imaginary world, actions ripple out and have an impact on everything else" (pp. 38), envisioning possible futures breaks down the barriers between discrete scientific strands and disciplines. It is within this openness to complexity that we might find opportunities to move beyond totalizing narratives of mastery and control toward difference, to imagine beyond the Anthropocene as a form of critical hope wherein we might not only learn "to live (and die) on a damaged planet" but work toward "imagining and creating spaces of refuge for a future we cannot predict" (Lakind & Adsit-Morris, 2018, p. 32).

Finally, a speculative pedagogy is one that is open to uncertainty and to the ultimately unknowable nature of the future. This is not a chapter promoting SF prototyping (Johnson, 2011) or overt prediction, wherein we predict and thereby insulate ourselves from future possibility. Rather, we echo Sardar and Sweeneys' (2016) assertion that "our command-and-control impulse will only serve to heighten our ignorance and entrench uncertainty" and that our effort should not be to "manage risk but rather our perceptions of risk" (p. 10). Accordingly, a speculative pedagogical orientation toward environmental and science education, and education more broadly, encapsulates a "move away from attempts to reduce uncertainty, and instead embrace it through diverse, contrasting futures: and the need to approach not only the future but also the present in a constructivist and pluralistic fashion" (Vervoort et al., 2015, p. 63). As Michael Pinsky (2003) asserts, "*The gift is the very possibility of a future that can be anticipated, but will always contain the unexpected*" (Pinsky, 2003, p. 189, italics in original). Given that so much of our current moment is encapsulated by efforts to contain, predict, and control, this chapter contributes to the call for "a new kind of thinking coupled with creativity and imagination," which requires that "we must be able to deal with complexity and incomplete knowledge, link what is compartmentalized, and tackle interconnections and interdependence" in order to adapt to our new position in relation to the future (Sardar & Sweeney, 2016, p. 12). In privileging SF and speculative storytelling as narrative pedagogical spaces that can expand the limits of what we see as *possible*, we advocate for abandoning static processes of knowing in favor of dynamic, unknown alternatives that lay before us and spaces for those narratives in schools and elsewhere: how can classrooms become spaces where uncertainty can be grappled with?

What we propose here is the use of SF and accompanying pedagogical orientations toward storytelling and possible futures as a means of expanding our capacity to imagine better futures and clear paths toward those futures, and as a way to uncouple ourselves from the dominant narratives that leave us resigned to the belief that things must always remain as they are. Amsler (2015) outlines myriad barriers to change and critical hope, noting that: "Within existing horizons of possibility, it is difficult to first conceptualize and then imagine being able or willing to do the kinds of material, intellectual, social and affective work that are needed to give counter-capitalist and democratic ways of life a fighting chance of becoming realities" (p. 19). Science fictional and speculative storytelling, intermediaries between fact and fiction, open up space to conceptualize and subsequently navigate and catalyze change, and to see and work toward possible futures. In times like these, it is critical that we tell stories of the future alongside our students as we explore the horizons of possibility together.

REFERENCES

Amsler, S. S. (2015). *The education of radical democracy*. Routledge.

Bell, W. (1998). Making people responsible: The possible, the probable, and the preferable. *American Behavioral Scientist, 42*(3), 323–339. https://doi.org/10.1177/0002764298042003004

Benford, G., Delany, S., Scholes, R., & Friedman, A. J. (1979). Teaching science fiction: Unique challenges. (J. Woodcock, Ed.). *Science Fiction Studies, 6*(3). http://www.depauw.edu/sfs/backissues/19/teaching19forum.htm

Campbell, J. W. (2019). *The order and the other: Young adult dystopian literature and science fiction*. University Press of Mississippi. https://doi.org/10.14325/mississippi/9781496824721.001.0001

Csicsery-Ronay, I., Jr. (2008). *The seven beauties of science fiction*. Wesleyan UP.

Cunsolo, A., & Ellis, N. R. (2018). Ecological grief as a mental health response to climate change-related loss. *Nature Climate Change, 8*(4), 275–281. https://doi.org/10.1038/s41558-018-0092-2

Dunne, A., & Raby, F. (2013). *Speculative everything: Design, fiction, and social dreaming*. The MIT Press.

Freedman, C. (2000). *Critical theory and science fiction*. Wesleyan UP.

Gannon, K. M. (2020). *Radical hope: A teaching manifesto*. West Virginia University Press.

Gough, N. (1993). Environmental education, narrative complexity and postmodern science/fiction. *International Journal of Science Education, 15*(5), 607–625. https://doi.org/10.1080/0950069930150512

Haraway, D. J. (2016). *Staying with the trouble: Making kin in the Chthulucene*. Duke UP. https://doi.org/10.1515/9780822373780

Harding, S. (2009). Postcolonial and feminist philosophies of science and technology: Convergences and dissonances. *Postcolonial Studies, 12*(4), 401–421.

Hasse, C. (2015). The material co-construction of hard science fiction and physics. *Cultural Studies of Science Education, 10*(4), 921–940. https://doi.org/10.1007/s11422-013-9547-y

Hergenrader, T. (2019). *Collaborative worldbuilding for writers and gamers*. Bloomsbury Academic. https://doi.org/10.5040/9781350016705

IPCC. (2022). Summary for policymakers. In H.-O. Pörtner, D. C. Roberts, E. S. Poloczanska, K. Mintenbeck, M. Tignor, A. Alegría, M. Craig, S. Langsdorf, S. Löschke, V. Möller, A. Okem (Eds.), *Climate Change 2022:*

Impacts, adaptation, and vulnerability. Contribution of Working Group II to the Sixth Assessment Report of the Intergovernmental Panel on Climate Change [H.-O. Pörtner, D. C. Roberts, M. Tignor, E. S. Poloczanska, K. Mintenbeck, A. Alegría, M. Craig, S. Langsdorf, S. Löschke, V. Möller, A. Okem, B. Rama (Eds.)]. Cambridge University Press. In Press.

Johnson, B. D. (2011). *Science fiction prototyping: Designing the future with science fiction.* Morgan & Claypool.

Kelsey, E. (2016). Propagating collective hope in the midst of environmental doom and gloom. *Canadian Journal of Environmental Education, 21,* 23–40.

Lakind, A., & Adsit-Morris, C. (2018). Future child: Pedagogy and the post-Anthropocene. *Journal of Childhood Studies, 43*(1), 30–43. https://doi.org/10.18357/jcs.v43i1.18263

Latour, B., & Woolgar, S. (1986). *Laboratory life: The construction of scientific facts.* Princeton University Press.

Levinson, N. (1997). Teaching in the midst of belatedness: The paradox of natality in Hannah Arendt's educational thought. *Educational Theory, 47*(4), 435–451.

Lothian, A. (2018). *Old futures: Speculative fiction and queer possibility.* NYU Press. https://doi.org/10.18574/nyu/9781479811748.001.0001

Mendlesohn, F. (2009). *The inter-galactic playground: A critical study of children's and teen's science fiction.* Critical Explorations in Science Fiction and Fantasy, Vol. 14 (D. E. Palumbo & C. W. Sullivan III, Eds.). McFarland.

Mirra, N., & Garcia, A. (2020). "I hesitate but I do have hope": Youth speculative civic literacies for troubled times. *Harvard Educational Review, 90*(2), 295–321.

Montfort, N. (2017). *The future.* MIT Press.

Moore, J. W. (2016). *Anthropocene or Capitalocene? Nature, history, and the crisis of capitalism.* Kairos Books.

Moylan, T. (2021). *Becoming utopian.* Bloomsbury Academic.

Ojala, M. (2016). Facing anxiety in climate change education: From therapeutic practice to hopeful transgressive learning. *Canadian Journal of Environmental Education, 21,* 41–56.

Olin Wright, E. (2010). *Envisioning real utopias.* Verso.

Oravetz, D. (2005). Science and science fiction. *Science Scope, 28*(6), 20–22.

Pearson, W., Hollinger, V., & Gordon, J. (Eds.). (2008). *Queer universes: Sexualities in science fiction.* Liverpool UP.

Pihkala, P. (2020). Eco-anxiety and environmental education. *Sustainability, 12*(23). https://doi.org/10.3390/su122310149

Pinsky, M. (2003). *Future present: Ethics and/as science fiction.* Rosemont Publishing & Printing Corp.

Priyadharshini, E. (2019). Anticipating the apocalypse: Monstrous educational futures. *Futures, 113*(1), 1–8. https://doi.org/10.1016/j.futures.2019.102453

Reis, P., & Galvão, C. (2007). Reflecting on scientists' activity based on science fiction stories written by secondary students. *International Journal of Science Education, 29*(10), 1245–1260. https://doi.org/10.1080/09500690600975340

Rubin, A. (2013). Hidden, inconsistent, and influential: Images of the future in changing times. *Futures, 45,* S38–S44. https://doi.org/10.1016/j.futures.2012.11.011

Sardar, Z., & Sweeney, J. A. (2016). The three tomorrows of postnormal times. *Futures, 75,* 1–13. https://doi.org/10.1016/j.futures.2015.10.004

Singh, V. (2014). More than "cool science": Science fiction in the classroom. *Physics Teacher, 52*(2), 106–108. https://doi.org/10.1119/1.4862117

Smith, D. A. (2009). Reaching nonscience students through science fiction. *Physics Teacher, 47*(5), 302–305. https://doi.org/10.1119/1.3116843

Subramaniam, M., Ahn, J., Waugh, A., & Druin, A. (2012). Sci-fi, storytelling, and new-media literacy. *Knowledge Quest, 41*(1), 22–27.

Sullivan, C. W., III (Ed.). (1999). *Young adult science fiction.* Greenwood Press.

Suvin, D. (1979). *Metamorphoses of science fiction: On the poetics and history of a literary genre.* Yale University Press.

Thomas, P. L. (Ed.). (2013). *Science fiction and speculative fiction: Challenging genres.* Sense Publishers. https://doi.org/10.1007/978-94-6209-380-5

Toliver, S. R. (2021). Freedom dreaming in a broken world: The Black radical imagination in Black girls' science fiction stories. *Research in the Teaching of English, 56*(23), 85–106.

Tomin, B. (2020). Worlds in the making: World building, hope, and collaborative uncertainty. *Journal of the American Association for the Advancement of Curriculum Studies, 14*(1) https://doi.org/10.14288/jaaacs.v14i1

Truman, S. (2019). SF! Haraway's situated feminisms and speculative fabulations in English class. *Studies in Philosophy and Education, 38,* 31–42. https://doi.org/10.1007/s11217-018-9632-5

Tuttle, L. (2005). *Writing fantasy and science fiction* (2nd ed.). A & C Black Publishers Ltd..

Vervoort, J. M., Bendor, R., Kelliher, A., Strik, O., & Helfgott, A. E. R. (2015). Scenarios and the art of worldmaking. *Futures, 74,* 62–70. https://doi.org/10.1016/j.futures.2015.08.009

Westfahl, G., & Slusser, G. (Eds.). (2002). *Science fiction, canonization, marginalization, and the academy.* Greenwood Press.

Wolf-Meyer, M. J. (2019). *Theory for the world to come: Speculative fiction and apocalyptic anthropology.* University of Minnesota Press. https://doi.org/10.5749/j.ctvdtphr3

CHAPTER 15

Curriculum Beyond Apocalypse

Matthew Weinstein

INTRODUCTION

There is a chain of associations I want to play with and tug at in this consideration of the Anthropocene. At the start of this chain is the signification of science in the imaginary. I have argued elsewhere when considering the discourse of germ-free organisms (aka gnotobiology) that science as a profession and a discourse is about signifying futures (Weinstein & Makki, 2009). We do not call fiction about the future technical fiction, but science fiction. Science relies on this signification for its funding, because the technical details of this ecological survey or that obscure epigenetic pathway for cancer do not inspire, except as a promise of better (*eu*) worlds (*topias*). Often the public discourse of science plays upon religious motifs and registers; as Mary Midgely has analyzed, science becomes salvation (1992). Consider this dialogue from the big budget movie *Interstellar* (Nolan, 2014), which is set against the background of environmental catastrophe:

M. Weinstein (✉)
University of Washington-Tacoma, Tacoma, WA, USA
e-mail: mattheww@uw.edu

© The Author(s) 2024
S. Tolbert et al. (eds.), *Reimagining Science Education in the Anthropocene, Volume 2*, Palgrave Studies in Education and the Environment, https://doi.org/10.1007/978-3-031-35430-4_15

Cooper: How far have you got [on solving the formula]?
Brand: Almost there.
Cooper: You're asking me to hang everything on an almost.
Brand: I'm asking you to trust me. (time code: 35:00)

Such a moment of faith is the keystone to the entire narrative edifice. On the long odds of solving a seemingly insolvable equation is deliverance tied, and as the story unfolds, its solution rests ultimately on the metaphysical. Against this, consider this quote from Kevin Esvelt, an actual scientist who tried to sell the residents of Nantucket, Massachusetts, CRISPR (genetically modified) mice as the solution to Lyme disease (Quimby, 2019, Episode 7):

> We are biased. You should never trust an inventor to evaluate whether their technology is safe and effective. Because we're still human, no matter how hard we might try, we will fail.

Here the attachment to one's own technology is put in question. It is not just that there are unintended effects but that the scientist is trapped in their humanity: their passions, biases, and attachments in ways that they know they don't know. Salvation or hubris are the futures offered here.

I am interested in pondering and contesting the ways that science and its framing sociotechnical imaginaries (Jasinoff & Kim, 2015), that is, the fantasies of futures which mobilize projects in the present, traffic in the pseudo- or perhaps crypto-religious discourse, especially in the evocation of crisis. Crisis stands in here for one of a family of terms: catastrophe, disaster, and apocalypse. These terms configure tragedy, nature, human agency, and the divine in complex and contested constellations. At stake in dire sociotechnical imaginaries are fault, possibility, human valuation, and our collective bonds. The Anthropocene is such an evocation. It is barely more than a euphemism for collapse. First, it does so in its reference to nuclear contamination and more recently in its association with climate change. I am old enough to have been subject to both of those fears. I am weary of fear, and want to think beyond that state of living with immanent horror. The word "beyond" in my title does double work: what conviviality is possible temporally past the disaster, and how can we think of this moment in terms other than disaster. Certainly part of my inspiration and guide here is Lilley et al.'s inspiring volume *Catastrophism: The Apocalyptic Politics of Collapse and Rebirth* (2012). The authors unanimously reject the use of catastrophe as an organizing tool for the left and note that both left and right embrace forms of collapse as hopes for new orders. They note how capitalism feeds on disaster, and thus it is not auspicious as a tool (an imaginary) for producing more equitable worlds. In particular Yuen's (2012) essay on catastrophism and environmentalism is instructive here:

> Catastrophism is rampant among self-identified environmentalists, and not without good reason—after all, the best evidence points to cascading environmental disaster. Warranted as it may be, though the catastrophism espoused by many left-leaning greens remains Malthusian at its core, and is often shockingly deficient in its understanding of history, capitalism, and global inequality. (loc 842)

Much of his essay is focused on the failure of Malthusian predictions and the ways such scenarios provide fuel for reactionary projects—such as the emergence of eco-fascism in the present moment (e.g., loc 862).

Part of the work of moving from crisis to action, from nature to responsibility, as it were, is to note that the Anthropocene, as many have noted, but my way into this critique is through Sylvia Wynter's analysis of the dehumanizing logic of colonialism, is a way of displacing responsibility for the material effects of the moment: nuclear contamination, ocean level rise, ocean acidification, and massive species loss, which is not the responsibility of "man" (anthropos, so gender intentional) but of some: of the global capitalist class and their rapacious need for power and wealth. It is empires centered in the global North who have generated this crisis because it externalized the human and non-human misery that is generated by its desires and wants. To achieve those desires the Global North (the ethnoclass, in Wynter's framing (2003, p. 260)) has yet to see the humans in the others they dominate and thus has no capacity to imagine the material other of the planet that must serve as a resource to a consuming metropole. In this framing, the apprehension of the Anthropocene is in reality a projection of the self onto the world. Having suddenly glimpsed, that despite its ideology of no limits, the material finitude of the world is manifest.

My medium for exploring these issues is public pedagogy. Public pedagogy is theorized by Henry Giroux (2003) as the way media and popular texts educate their consumers about common-sense arrangements, social imaginaries, and identities. I am interested in contrasting fabulations of ecological futures. In particular I want to cross-read public pedagogies of the Anthropocene (largely reduced to global warming scenarios) as lenses to understand our socio-science-political imaginaries. Here I will focus on the catastrophism of the movie *Interstellar* and Kim Stanley Robinson's (2017) *New York 2140*. However, moving beyond the popular I want to read these futures against a more scholarly text which is as much a science fabulation as the above: Joel Wainwright and Geoff Mann's *Climate Leviathan* (Wainwright & Mann, 2013, 2018). They look forward to multiple configurations of governmentality and its resistance. What are the likely and possible forms of governance we face as we look forward to the climate crisis, and what are the forces arraigned within and against such forms? They posit two dimensions: capitalist-non-capitalist (in the broadest sense of the term, i.e., a social order that is driven by something other than profit) and planetary-anti-planetary, defining the scope of the power of such governmentality. My reading is not so much a meta-discourse as a useful but problematic slit through which to diffract their narratives.

After considering these contrasting Anthropocene futures, I conclude by comparing them to a fictional account of life under the proposed Green New Deal (GND) (Aronoff, 2018). How is our vision of the GND limited as a sociotechnical imaginary? I end by speaking to the limits of extant imaginaries and the necessity of better ones.

APOCALYPSE

Apocalypse references particular types of fictions, futures, and sociotechnical imaginaries. For instance, it is worth considering the differences between apocalyptic science fictions and dystopian science fictions. While certainly there is an overlap in the genres, there are important differences. Dystopian fictions are inevitably fictions of life under states, usually totalitarian in some aspect; for example, the patriarchal tyranny in Margaret Atwood's (1986) *The Handmaid's Tale*, in which women's autonomy is crushed, or the socialist tyranny in Kurt Vonnegut's (1968) short story "Harrison Bergeron," in which all inequalities are rendered to their lowest form (smart people have to have their thinking interrupted constantly, graceful people have to manage cumbersome weights). They are almost always, therefore, cautionary tales about challenging the extant political order. Apocalyptic tales function differently. There is a giddy sense of freedom that is often part of the tale. The state or family relations or other social bonds are dissolved. A slate is wiped clean, and we can begin again. This is made clear by the writer Sarah Vowell (1999) in her memories of the function of apocalypse in various religious, political, and technological guises throughout her life. She explores the giddy dimension of apocalypse in a visit to a Y2K prepper meeting in California. It is in reflecting on this meeting that she comes to realize that apocalyptic talk is cover for utopian talk:

> Just like my old church and my old anti-nuke group, they're using the end of the world as a means to meet and greet, planning block parties so they can come up with Y2K contingency plans in their neighborhoods. They were also very idealistic. This is the thing you might not realize about end of the worlders. They might seem like they're all about fetishizing doom and destruction, but stick around long enough for them to finish their spiel— few people do, I know—and before long, they get to a straight-up Utopian vision of the world. After all, after the biblical tribulation comes the new Jerusalem and 1,000 years of peace on Earth.

Vowell is identifying that the rhetorical function of apocalypse is to prepare for a new and better world order. As Sasha Lilley (2012) notes:

> The collapse is frequently, but not always regarded as a great cleansing, out of which a new society will be born. Catastrophists tend to believe that an ever-intensified rhetoric of disaster will awaken the masses from their long slumber—if the mechanical failure of the system does not make such struggle superfluous. (loc 207)

Lilley's greater point is that the rhetoric of the apocalypse will fail the left as a strategy, whether or not it is warranted; ultimately apocalypse is too central to reactionary movements for whom "worsening conditions are welcomed, with the hope they will trigger divine intervention or allow the settling of scores for any modicum of social advance over the last century" (loc 213). I endorse Lilley's analysis, but my interest is in working with social imaginaries that might better function than apocalyptic event horizons.

At stake in the Anthropocene apocalypse is nature, its forces, responsibilities, victimhood, status, and so on, all of which act as struggles for the boundary of the conditions of politics (Haraway, 1994, p. 59). The germinating idea of the Anthropocene is that we humans have (unnaturally) been writing nature, leaving our marks on her strata. We are, in essence, embarrassingly visible in nature—of which we imagine ourselves apart. In climate crisis talk nature is both the actor and the object of human action, as many have argued, though hardly the same humans. Is the weather written upon—analogous to the geological strata now containing higher levels of radioactive elements—by our industrial misbehaviors? Are the losses, displacement, and dispossession due to changes in climate our capitalist burden? Can we coordinate, salvage, rebalance, and move toward conviviality? Or as Lilley suggests (see last paragraph), that ecological crisis, like all crises, better serves the forces of reaction, in this case an emergent eco-fascism. Eco-fascism embraces the crisis but responds to it by dehumanizing and abandoning those who try for sanctuary in the better sheltered landscapes of the world/west (Darby, 2019). The formulation of nature as a reified set of relations, forces, and powers beyond or maybe including the human is at stake in the sociotechnical imaginaries that drive policies from walls to new green deals, and those imaginaries are crafted in narratives: narratives of the second coming, narratives of technological fixes, narratives of radical democracy, and narratives of bordered nations, hoping to survive the climate crisis akin to images of families in fallout shelters in the 1950s and 1960s (Rafferty, Loader, Rafferty, Archives Project., & Thorn EMI Video (Firm), 1983).

Again, to explore these sociotechnical imaginaries I look to popular and academic media and texts; public and private pedagogies of the future, looking backward to look forward, and the ways these are likely to play out in any policy or collective response. I start with *Interstellar*, a big-budget film with high-salary stars about the response to an un-named but always implied climate crisis. In a radically different view of the future, Kim Stanley Robinson imagines New York surviving after several major flooding catastrophes as the new Venice, and against and within both of these, I examine the four futures foretold by *Climate Leviathan* and consider the narratives needed to produce a more democratic, less fascist future.

INTERSTELLAR: MAGIC AND SCIENCE

Christopher Nolan's *Interstellar* is considered here as a representation of a kind of common-sense consideration of climate change and the Anthropocene. (Note: I will spoil the plot entirely.) Despite its spectacular futuristic narrative, it is the path forward in the face of the global climate crisis for many of Silicon Valley's plutocratic elites. The main character, Joseph Cooper, is an ex-astronaut, who in a post-industrial age, is farming in Wyoming. His daughter, Murphy, is being haunted by "ghosts," ambiguous visions that push books from her shelf. The world is in deep crisis; crops are failing, dust bowl weather is sweeping the land, and the government has taken an anti-technological stance, promoting the idea that the moon landing was a hoax. And yet Cooper soon finds out that NASA is still active and is planning an expedition which will either 1) get humans to a habitable world on the other side of a wormhole (through a complex data collection operation) or 2) seed that world with humans in the form of embryos carried on his spacecraft (so-called plans A and B). Fights happen, people are betrayed, but ultimately Cooper is able to get the data for plan A and communicate back to Murphy what to do (Cooper is himself the ghost enabled by evolved humans who are trying to save the planet through their ability to exist in five dimensions). In short, the story takes a metaphysical turn at the end in positing some higher dimension that will save us.

This is a narrative of the technical fix. Of salvation through technology, but a technology that also evokes the metaphysical. It is not through the collective science labor that the solution is reached but by the *magic* of five-dimensional post-humans who, to save themselves, allow Cooper to communicate with Murphy at multiple time points. In this sense we are saved by some future and more perfect version of our selves. The logic here is that we will go elsewhere (scientificomagically) to save ourselves from the mess we are in, not by fixing the mess but by abandoning it. We don't have to do anything.

This logic builds on a kind of frontier imaginary. At times the logic is made painfully explicit, as when Cooper explains to his father-in-law Donald, "It's like we've forgotten who we are, Donald. Explorers, pioneers, not caretakers" (time code 15:46). Here a sharp distinction is drawn between a narrative of the frontier joined with masculinity and one of domesticity and care. According to Greg Grandin (2019), it is an ideology that stabilized political dissent and materially managed inequality and difference (p. 2). Frontier myths are, as he titles his first chapter, "Fleeing forward." *Interstellar* is building on that history of frontier thinking (as much of science fiction and NASA have). But it strangely makes clear what is at stake: not care for the existing world; instead, burning through it to another world. Our world, in the name of this masculine ideal, is disposable. "We" (the subject of Cooper's sentence) do not care (meaning we have no feeling of care and we do not provide care). The frontier is about escape not solutions; it is flight rather than fight, even if it is simultaneously a site of spectacular violence as one locus of primitive accumulation, that is, the site of fencing the commons.

Interstellar articulates the mythic structure of feeling of the frontier, involving heroicism, novelty, escape, and masculine violence (whatever the gender of the actors provided) with our contemporary ecological state. It offers the frontier of space as the solution to our Earthly problems. The mad money behind such endeavors as Space X and Blue Origin (Musk and Bezos respectively) envision such escape. Musk has repeatedly fantasized of travel to Mars; Bezos of colonies in space. In his words, "We have to go to space to save Earth" (Foer, 2019). This uncaring vision of the Anthropocene is the world made narrative in *Interstellar*. The funding of NASA while the Midwest goes up in dust is the solution that is implied in their emphasis on space as the way out of Earth, the telos of the kind of dual world described by Naomi Klein (2007) of green zones and red zones proliferating around the world (green zones of order and peace for the rich; red zones of chaos and violence for the rest), but taken one step further, a green zone for a time when all the world is ravaged.

But the irony is that any thought about the plot offered here reveals it as no more than magical thinking. In the end, leaving Earth does not solve the problems except if we can run into fifth dimensional, time skirting, evolved versions of ourselves who can collect data in black holes. While emotionally we suspend disbelief—and the film's exquisite special effects enable that—the absurdity of the dream of the billionaire class to escape rather than care for Earth reveals itself in the *deus ex machine* solution the film offers. This is what Noah Gittell (2014) means in his *The Atlantic* review when he opines on Hollywood's lousy environmental politics: they can't get the environment or solutions to environmental problems right, largely because the solutions are not spectacular, and cinema is about spectacle.

But what such a review misses is the way that this is irrelevant. It gets Bezos's politics of the environment largely right: escape nature, do not care for it. It wraps such a stance in compelling narratives of heroism, masculinity, exploration/frontierism, and so on—and I grant that these terms are imbricated. It works in the way that reactionary movements succeed in giving a narrative of masculinity in opposition to femininity and thus feminism. Critically, in considering this as a public pedagogy of science, that this articulation of masculinity valorizes science along the way, but not real institutional science, that is, it heroicizes science as salvation and magic rather human, fallible, social labor.

The Bezos solution with its concomitant frontier narratives represents one larger context for popular narratives of what has come to be called CliFi (i.e., Climate Fiction—see, for instance, Svodboda, 2014). But it should be clear that such apocalypse and salvation (by science's second coming in this case) are only one narrative, a narrative that treats the climate crisis as a frontier to be escaped and the Anthropocene as uninhabitable. But that is not the only narrative climate change affords, and *New York 2140* is a very different kind of CliFi. In most respects it inverts the frontier narrative that *Interstellar* embraces.

New York 2140: Quotidian Anthropocenes

If *Interstellar* is a curriculum of heroic masculinity and miraculous salvation, Kim Stanley Robinson's *New York 2140* is about life on the other side of the Anthropocene's event horizon. It's not that there isn't crisis but that life has a weird ordinariness in the afterlife of the flooding of the planet despite crisis. The plot (and, again, I will spoil) involves a housing cooperative in the intertidal zone of Manhattan; certainly one of the framing presumptions of the book is that the ambiguous state of property law in intertidal zones creates an opening for social democratic possibility. Different chapters focus on different members of the elite and downtrodden in that dwelling. Manhattan has become a kind of high-rise Venice, in which one boats up 5th Avenue. Despite multiple cataclysmic moments, in the 100 plus years' time from our present, life is weirdly ordinary. Capitalism is doing very well, and capitalism continues in its financialized form. One of the main denizens of the Coop and subjects of the book is Franklin Garr, the developer of Intertidal Property Pricing Index (IPPI); which is explained as "a kind of Case-Shiller Index for intertidal assets" (loc 1893)—Case-Shiller being a contemporary measure of the housing market.

In many ways the pedagogy of the text is that life in the Anthropocene will have a kind of shocking familiarity filled with small pleasures and adventures as buildings collapse now and then. In its own way, this level of lack of drama is as unbelievable as *Interstellar*'s miracles. Granted, this level of ordinary living reflects life in the global metropolis rather than its vulnerable peripheries, that is, in a city that is still needed for capitalism—Washington, DC, by contrast is gone; the capital of power has moved to Denver, Colorado, for dryness's sake. Certainly things would be different in Samoa or another locale. As the author notes, "[T]he people in Denver didn't really care. Nor the people in Beijing, who could look around at Hong Kong and London and Washington, D.C., and Sao Paolo and Tokyo and so on, all around the globe, and say, Oh, dear! What a bummer for you, good luck to you!" (loc 2295). Cool alienation is the structure of feeling here. The overall feel of the story, far from action-packed SF of *Interstellar*, is one of daily life, of continuity, of people doing familiar things in solidarity with each other and the planet, often thwarted. But it is a future that other than all the water and a heavy use of zeppelins feels utterly contemporary.

In fact, in so many ways *New York 2140* is as much about the global financial crisis of 2008 as about 2140. Through various events and coincidences in the lives of those living at "The Met" (the housing cooperative) the characters get to reset the neoliberal solutions put in place in 2008/2009. The book is a grand fantasy of undoing—or at least starting to—the financialization put in place as a solution to that crisis. It is, in U.S. terms, a Democratic Socialists of America fantasy of repeating, redoing that moment of possibility when one capitalist order seemed ready to collapse only to be saved by the champion of Hope and Change, President Obama. Thus, the text includes as much talk of Federal Reserve Chairs (the heads of the governmental bodies that control money supply) as species loss. And the crisis that animates the book is less that of the Anthropocene's climate rewriting than a financial one brought by the protagonists of the tale.

And here is the shared organizing moment of both these public pedagogies of the Anthropocene: that crisis as a literary trope (financial, environmental, agricultural, etc.) serves as the opening for a new order. Crisis resolves in better worlds. Both *Interstellar* and *New York 2140* are about transcending crisis. It is about restarting and shifting global orders, as Lilley et al. so critically observe. To make sense of the shifts in these two fictions, it is helpful to consider the matrix of possible worlds they are drawing from.

CLIMATE LEVIATHAN: A MATRIX OF ANTHROPOCENE FUTURES

Against these two texts I want to examine Joel Wainwright and Geoff Mann's (2018) *Climate Leviathan: A Political Theory of Our Planet Future* (hereafter CL). Like the previous fictions, CL imagines life in/beyond/

under conditions of climate change. This is near-future fictioning (in the sense of imaginative construction) of governmental forms under climate change. CL first appeared as an article in *Antipodes* in 2013 and then was expanded into a book in 2018. Driving the analysis in CL is a two by two matrix (p. 30, all citations will reference the book, rather than the article). The top cells represent capitalist futures and the bottom non-capitalist futures; the first column represents "planetary sovereignty," the second "anti-planetary sovereignty." Four social formations form the cells in the matrix. Climate Leviathan is the extant hegemonic global social formation of neoliberal governmentality—the upper left corner of the matrix. Here, the goal is to preserve the wealth of the ruling class through managing risk and harm at a global level through financial, political, and social organizations such as the WTO, the United Nations, and the International Courts. Rejecting this within the capitalist economy is Climate Behemoth (the name comes from Locke), which rejects the international order, and privileges racial/ethnic states as the site of sovereignty; it is the upper right corner of the matrix. This is the capitalism of Donald Trump in the U.S., Prime Minister Narendra Modi in India, and Andrzej Duda in Poland. Ethnocentric embattled states based on racialized—in Foucault's sense in which "racism [is] understood as a "basic mechanism of power" by which the state becomes able to exercise the sovereign right to kill against some so that others may live (Erlenbusch, 2017, p. 139)—hierarchies of belongingness.

But Climate Behemoth is only one challenge to the current global hegemon. As CL points out, the region most impacted by global warming is Asia, and within that region there remains a strong Maoist ideology. Climate Mao is what Wainwright and Mann call that social formation that is global in ambition but anti-capitalist in orientation. "Climate Mao expresses the necessity of a just terror in the interests of the future of the collective, which is to say that it represents the necessity of a planetary sovereign but wields the power against capital" (p. 38). While the current Chinese administration remains committed to Climate Leviathan as a project, there are seeds in place in China (and elsewhere) that are pushing for a more totalitarian solution that rejects (necessarily if one thinks about it) capitalism which has been a large engine of anthropogenic climate change.

Opposed to all of these, in CL's lower right quadrant, are the small-scale, anti-capitalist, anti-global hegemon movements. These movements are not coherent in any sense and include a wide variety of first nations, peripheral (to the centers of Climate Leviathan's governance), indigenous, and bioregional responses. In the mainstream it might take the form of the proposed New Green Deal, but it also takes the forms of resistance movements everywhere: climate strikes, boycotts of banks, and so on. Because of the amorphous form of the activities in this quadrant, these social formations are called Climate X. While the authors are openly ambiguous about the nature of Climate X they do propose three "principles" that ought to be fundamental to such a social formation to distinguish it from the three others: equality of all humans, inclusion, and dignity of all, and "solidarity in composing a world of many worlds" (pp. 175–176). They realize that X is almost an impossible formation to establish and maintain; it will be eviscerated by forces of Leviathan, Mao, and Behemoth the minute it obtains a toehold, yet it keeps emerging: in Chiapas under the Zapatistas, and then again in Syria in Rojava (which was based partially on eco-anarchist Murry Bookchin's political theories). It is also signaled by the front-line role of indigenous and colonized peoples: "While these groups have, of course been subject to capital and state power, to generalize, their present strategies do not emphasize forging internationalist solidarity for a revolutionary communist or socialist future. Their point, rather, is to ensure that the full multiplicity of those lifeways has a vital and dignified future—and in some cases, to communicate to those willing to listen what they might learn from it" (p. 189).

The Wainwright and Mann 4-square gives us a tool to rethink the fictional texts above, though I would argue that one needs to be cautious because those fictions can also be used to problematize CL. CL can't simply be read as a meta-discourse on these fictions. From the point of view of public pedagogy, CL helps analyze the ways these fictions construct an other who is to be resisted, that is, these fictions educate us to respond to certain social formations as the enemy, as the problem, as to the order that must be transcended. In *Interstellar* the dominant social formation appears to be some variant of Climate Behemoth, squashing previous history, condemning science (yet still funding NASA? It makes less sense the more you think on it), and the hope is to reinstate the Leviathan, now covering the solar system. *Interstellar* is a kind of nostalgia for the present—one shared by the liberal elites in response to Trump—though the movie was released in 2014, years before Trump's election. In *New York 2140*, the extant order is Climate Leviathan, which seems to have survived ecological disaster after disaster, fulfilling Frederic Jameson's quip "it is easier to imagine the end of the world than to imagine the end of capitalism" (cited in

Wainwright & Mann, 2018, p. 47). The dream is sort of Climate X, but the book is not utopian but merely optimistic at the end that small gestures could be accomplished, and the fight has just begun in the closing chapters. We do not, as in more utopian novels, get a structure of feeling for life in a different order, for example, in Piercy's *Woman on the Edge of Time* (1976), LeGuin's *Dispossessed* (1974), or Gumbs's "Evidence" (2015), just that such an order might be possible, that Leviathan is not the end of history, to quote Francis Fukuyama's thesis (1992). This is a pedagogy of small optimisms. But the smallness of the vision is in line with the ambiguity and deep democratic, intercommunal nature of Planet X.

It should be clear that the Climate Leviathan 4-square template is not all inclusive, and other futures, not anticipated by the authors, are possible and can be read in the interstices (between the lines) of these fictional futures. Alyssa Battistoni's (2018) review ends with exactly this sort of reading of other possibilities in and outside of the 4-square.

> Must movements really be opposed to all forms of sovereignty, on all scales, in order to oppose a capitalism-reproducing world state or achieve any measure of justice? Is there truly no left-populist Climate X that could act as a counter to Behemoth at the level of the nation, no way to channel planetary solidarity through international—not necessarily global—institutions? The difference between, say, Jeremy Corbyn's pledge to nationalize and decarbonize the British energy industry and Justin Trudeau's sign-off on private pipeline projects in Canada may not be enough to save the planet, but it would seem to deserve at least the status of an opening. Instead, the ways that actually existing states have acted in relation to their subjects as well as in relation to capital are collapsed by the authors into an argument about sovereignty—for or against.

Another possibility, albeit a darker one, is signaled by the Bezos-*Interstellar* articulation: as likely as any future is a feudal but non-ethnic, plutocratic, fragmented, anti-Leviathan social formation. A neo-feudalist vision or corporate future that mixes Leviathan's acceptance of climate change and preference for technocratic solutions with Behemoth's eco-fascism—that is, rejecting the global nature of climate change or even its existence in the name of resource hording for hegemonic groups however imagined or defined. This could look like Neal Stephenson's world of *Snow Crash* (2003), in which the U.S. is divvied up between competing food chains. This is a cyberpunk reimagining of the organization of politics in which corporations are the state in a way very distinct from neoliberal capitalism or fascism. One can sense this in the quasi libertarian utterances of Silicon Valley (writ large) players like Bezos, Musk, and others but also in their political ambitions. My point here is that while *Climate Leviathan* helps us imagine much more than the CliFi offerings of popular (and less popular) fictions, in more analytic and strategic ways, it is not a final word, not even on the fictions I have cross-read against it here.

Conclusion: Formalizing the Curricula of Eco-futures

To think of texts as educative, it is necessary to state that they operate at the level of imagination and habitus. They shape what we can consider and what how we feel within the matrix world of the story extending outward to our affect in the worlds in which we live. The stories I consider here, fictive and academic, help consider and condition the horizon line of potential action. To consider Climate Mao or Climate X as a possibility is to see beyond the Leviathan embraced by *Interstellar* and assumed to be the extant governmentality of *2140*. These texts thus interact with, that is, are intertextual with, current discourse on strategies moving forward, especially after the failure of the U.N. Climate Conference COP25, in which actors really associated with the Climate Behemoth, for example, Donald Trump, scuttled the agreement. The eco-fascist state lightly portrayed in *Interstellar* is suddenly palpable. The question becomes, can we envision a way forward toward 1) actual effective responses to the looming climate crisis and 2) something participatory and democratic within those responses?

Take Kate Aronoff's short fictional opening to "With a Green New Deal, Here's What the World Could Look Like for the Next Generation" (2018). Aronoff tries to capture daily life under Green New Deal (GND) in 2043. The story confounds in a number of ways. First, while presenting a vision of social democracy: rent controlled housing, free wifi and broadband, and free water; it does so under the most U.S. terms, hybrids of public and private, that ultimately seem more like neoliberalism with a safety net than a revolution. This is most clear in her portrayals of American Job Centers, in which people are connected to work and training through these centers, but basically the market and entrepreneurship are the defining qualities—in essence this is ObamaCare for labor.

But the history of ObamaCare should make anyone wary about such a future. ObamaCare has been shredded and defunded by courts and Congress, that is, by people opposed to any moral economy that does not let the poor starve and die in the manner that their God of the Wealthy does not support. In short, Aronoff's story is told without the presence of an active and hostile (and even fascist) opposition, without actors whose identities are tied in deep ways to extractive industries, without climate deniers and their industrial backers, that is, in some other world than this one. It is not the story of struggle and opportunity, of strategy and tactics needed to make the vision a credible tale. This is where *Climate Leviathan* succeeds more than the other fictions considered here: it spins futures that one can taste, battles one can imagine winning and seeds of hope that are embraceable because the opposition, in both the forms of Leviathan and Behemoth, is factored in.

The world needs fictions: great novels that move between our fragile present and better futures. We need both to be educated to exist within a more modest material world, for those of us in the power centers, and a world shared across all types of social and geographic borders, and stories that get us there in compelling ways (plural intentional). What can we say about such stories: they must heroicize care against the narrative of the frontier of *Interstellar*, they must be driven by something different than the Darwinian/Classical Economic logics of survival of the individual against the other (as Wynter has so elegantly argued for (e.g. McKittrick, 2015, pp. 16–18) wherein morality is equated with that survival). Finally, it must be a story that is filled with pleasures small and large, that is, that provides a reason for living in such a world. The texts here hint at such elements as the collective action and small heroicisms of *New York 2142*, the love and connection that drives Cooper in *Interstellar*, and the dream of democracy and life not yoked by a sovereign in *Climate Leviathan*, but the stories I am asking for are still being drafted.

References

Aronoff, K. (2018, December 5). With a Green New Deal, here's what the world could look like for the next generation. *The Intercept*.

Atwood, M. (1986). *The handmaid's tale*. Houghton Mifflin.

Battistoni, A. (2018, July 16–23). States of emergency. *The Nation*.

Darby, L. (2019, August 7). What is eco-fascism, the ideology behind attacks in El Paso and Christchurch? *GQ*.

Erlenbusch, V. (2017). From race war to socialist racism: Foucault's second transcription. *Foucault Studies,* 22, 134–152. https://doi.org/10.22439/fs.v0i0.5239

Foer, F. (2019, November). Jeff Bezos's master plan: What the Amazon founder and CEO wants for his empire and himself, and what that means for the rest of us. *The Atlantic.*

Fukuyama, F. (1992). *The end of history and the last man.* Free Press.

Giroux, H. A. (2003). Public pedagogy and the politics of resistance: Notes on a critical theory of educational struggle. *Educational Philosophy and Theory,* 35(1), 5–16. https://doi.org/10.1111/1469-5812.00002

Gittell, N. (2014, November). *Interstellar:* Good space film, bad climate-change parable. *The Atlantic.*

Grandin, G. (2019). *The end of the myth: From the frontier to the border wall in the mind of America* (1st ed.). Metropolitan Books, Henry Holt and Company.

Gumbs, A. P. (2015). Evidence. In W. Imarisha, A. M. Brown, S. R. Thomas, & Institute for Anarchist Studies (Eds.), *Octavia's brood: Science fiction stories from social justice movements* (p. 296). AK Press.

Haraway, D. J. (1994). A game of cat's cradle: Science studies, feminist theory, and cultural studies. *Configurations,* 2(1), 59–71.

I. Glass (Producer). (1999, April 2). *Again.* https://www.thisamericanlife.org/125/transcript

Jasinoff, S., & Kim, S.-H. (2015). *Dreams of modernity: Sociotechnical imaginaries and the fabrication of power.* University of Chicago Press.

Klein, N. (2007, October). Disaster capitalism: The new economy of catastrophe. *Harper's Magazine,* 47–58.

Le Guin, U. K. (1974). *The dispossessed An ambiguous utopia.* Harper & Row.

Lilley, S. (2012). The apocalyptic politics of collapse and rebirth. In S. Lilley, D. McNally, E. Yen, & J. Davis (Eds.), *Catastrophism: The apocalyptic politics of collapse and rebirth* (pp. Kindle Location 415–894). Oakland, CA: PM Press.

Lilley, S., McNally, D., Yuen, E., & Davis, J. (2012). *Catastrophism: The apocalyptic politics of collapse and rebirth.* PM Press.

McKittrick, K. (2015). *Sylvia Wynter: On being human as praxis.* Duke University Press.

Midgley, M. (1992). *Science as salvation: A modern myth and its meaning.* Routledge.

Nolan, C. (Writer). (2014). Interstellar [Film]. In E. Thomas, C. Nolan, & L. Obst (Producer): Paramount Pictures.

Piercy, M. (1976). *Woman on the edge of time* (1st ed.). Alfred A. Knopf.

Rafferty, K., Loader, J., Rafferty, P., Archives Project., & Thorn EMI Video (Firm). (1983). The atomic cafe [two-dimensional moving image]. London?: Thorn EMI Video,.

Robinson, K. S., & Martiniere, S. (2017). *New York 2140* (1st ed.). Orbit.

S. Evans-Brown, & E. Janik (Producer). (2019). *Patient Zero*. https://www.patientzeropodcast.com/podcast-episodes/episode-7

Stephenson, N. (2003). *Snow crash*. Bantam Books.

Svodboda, M. (2014). (What) do we learn from cli-fi films? Hollywood still stuck in holocene. https://www.yaleclimateconnections.org/2014/11/what-do-we-learn-from-cli-fi-films-hollywood-still-stuck-in-holocene/

Vonnegut, K., Jr. (1968). Harrison Bergeron. In *Welcome to the monkey house* (pp. 7–14). Dell.

Wainwright, J., & Mann, G. (2013). Climate leviathan. *Antipode, 45*(1), 1–22. https://doi.org/10.1111/j.1467-8330.2012.01018.x

Wainwright, J., & Mann, G. (2018). *Climate leviathan: A political theory of our planetary future* (US and Canada. ed.). Verso.

Weinstein, M., & Makki, N. (2009). *Bodies out of control*. Peter Lang.

Wynter, S. (2003). Unsettling the coloniality of being/power/truth/freedom: Towards the human, after man, its overrepresentation--An Argument. *CR: The New Centennial Review, 3*(3), 257–337. https://doi.org/10.1353/ncr.2004.0015

Yuen, E. (2012). The politics of failure have failed: The environmental movement and catastrophism. In S. Lilley, D. McNally, E. Yen, & J. Davis (Eds.), *Catastrophism: The apocalyptic politics of collapse and rebirth* (pp. Kindle Location 415–894). Oakland, CA: PM Press.

CHAPTER 16

Perturbing Current Boundary Conditions in Discipline-Based and Science Education Research in the Anthropocene: Implications for Research and Teaching Communities

Michelle M. Wooten and Katherine Ryker

INTRODUCTION

If we are to develop political vision, if we are to develop some sense of living and dying with each other responsibly … I think the practice of joy is critical. And play is part of it. I think that engaging and living with each other in these attentive ways that elaborate capacities in each other produces joy. (Haraway & Wolfe, 2016, pp. 252–3)

M. M. Wooten (✉)
University of Alabama at Birmingham, Birmingham, AL, USA
e-mail: michellewooten@uab.edu

K. Ryker
University of South Carolina, Columbia, SC, USA
e-mail: kryker@seoe.sc.edu

© The Author(s) 2024
S. Tolbert et al. (eds.), *Reimagining Science Education in the Anthropocene, Volume 2*, Palgrave Studies in Education and the Environment, https://doi.org/10.1007/978-3-031-35430-4_16

285

Hello, reader. *We* are Michelle and Katherine. Michelle is an educational research methodologist who teaches undergraduate astronomy and qualitative research methods. Katherine is a geoscience education researcher who teaches undergraduate geology and graduate geoscience education research courses. In this chapter, our intention is to detail communal problems we experience in our crisscrossing treks throughout the research landscape to which we both contribute—research on science teaching and learning—in addition to how our differing perspectives push each other to think differently about that landscape. We do this with a nod toward Haraway and Wolfe's encouragement to play, showing how engagements across boundaries evoke attentiveness toward one another, elaborate our capacities as educators and researchers, and produce joy. We will also detail some ways we have attempted to conceptualize and work through these problems—what we call *perturbing* the features in the landscape to enable perceived freedom to move across boundaries (see also, Wu et al., 2018).

This study emerged from Michelle's doctoral dissertation, in which Michelle explored material (affectual, intellectual, and physical) dynamics involved in the development and maintenance of communal divides between and among researchers of science teaching and learning from 2015 through 2018. Michelle interviewed twenty-seven researchers contributing to what she conceptualized as a landscape of research on science teaching and learning to which she and the study's participants contributed. Katherine was one of the participants in that study and was invited because of her unique (aforementioned) expertise.

Michelle learned that different groups of researchers who study science teaching and learning used theory or scientific practices to render their research recognizable and acceptable to some education research communities and not to others (Wooten, 2018). For example, Katherine,[1] as a member of the geoscience research community (GER), felt that the GER community had experienced a lack of recognition in the broader community of discipline-based education researchers (DBER) and science education researchers.[2] She described that GER was not (at the time) often recognized among other discipline-based education research (DBER)

---

[1] For this study, Katherine has consented to deanonymize her previously anonymized interviews from Michelle's study.

[2] We differentiate DBERs from science education researchers because during her study, Michelle learned that one researcher who studied science teaching and learning did not identify as a "science education researcher" because they did not have formal training in education.

communities (e.g., chemistry education research, biology education research) as a newer field, which could be seen in rates of tenure and promotion (e.g., Libarkin, 2015, as cited in Dolan et al., 2018; Singer et al., 2012). Consequently, her community was focused on developing standards of quality research for the purpose of developing recognizability in the landscape of research of science teaching and learning, the hope being that "other communities start to cite our research because they believe it's high quality, not just because it's something they have to cover: 'Oh, a geologist did this. We can ignore that' " (Katherine, interview excerpt from March 2017). And yet Katherine also expressed tensions her community experienced in gaining recognizability: the practices they invoked to incur status, such as gatekeeping, also had the effect of potentially eliminating particular forms of research that did not appear as recognizable.

We consider Katherine's experiences of feeling differentiated from other communities—with whom she would like to sense belonging—reflective of normative relations in the Anthropocene. Relations in the Anthropocene have been critiqued for effectuating identities of *things* (humans, nonhumans, nature) separated from the dynamic, fluid, earthen phenomena in which they are embedded (Normand, 2015). The implications of these differentiations are that care for their multiple possible *connected* and *joyful* figurations and relational potentials may no longer be conceptualized or invested in.

> Katherine: We need those changes to happen in order to have a healthy ecosystem. If you don't have *perturbations*, then eventually you hit stasis: nothing is happening. Nothing is growing. Nothing is blooming. (February 2018)[3]

We longed for perturbations in our research and teaching communities to enable their diversification, growth, and blooming. In this chapter, we use the differentiated sensing produced in our shared landscape of research on science teaching and learning as a starting point to motivate and provoke our own (and incite others-in-our-communities') mobility across our

---

[3] This quote was in response to Michelle's advocating for *disrupting* normative methodology. Katherine was referring to the creation of a phytoplankton bloom: "When you talk about disruption one example that comes to mind regards perturbations of a system: adding a particular nutrient to a nutrient-limited ocean environment and getting a phytoplankton bloom. There is an arguably *disruptive* change in something about the system, but it's a natural part of its ebb and flow."

perceived disciplinary boundaries. To this end, we think with Deleuze and Guattari, who write that "to attain the multiple, one must have a method that effectively constructs it" (1987, p. 22). We assert that practices that enable multiplicities (multiple, connected identities) rather than uniformities (singular, differentiated identities) in our shared landscape are supported by playful rather than a prescriptive method. Using a geologic framing of our connected research and teaching practices as a landscape, we considered our method of study toward our multiplicative becoming as "nomadic." This becoming, as we show, was resistant toward a method of "stable identities and fracture[d] temporal linearity" and leaning toward, "affirmative alternatives which rest on a non-linear vision of memory as imagination, creation as becoming" (Braidotti, 2013b; p. 165). Adopting Braidotti's theory of the nomad to inform our method enabled our journeying into uncharted territories in our landscape and enabled our sensing about our multiple possibilities in and across our research and teaching endeavors.

In our presentation of this study below, we begin by describing our entries into the concept of the Anthropocene, rendered somewhat differently in our disciplines of astronomy education (Michelle) x geoscience education (Katherine) x education research (Michelle and Katherine).[4] We then begin traipsing through transcript excerpts from previous interviews-meetings (in February 2018, November 2019, and December 2019) to figure-together the shape (and possible re-shaping) of our shared landscape of research on science teaching and learning in the Anthropocene. As we do, we use nuances in our disciplinary perspectives to, as Haraway puts it, "elaborate our capacities" and become aware of how our practices both provoke and resist communal differentiations (Haraway & Wolfe, 2016). Through continual figuring of our shared landscape, we describe a method of cultivating disciplinary perturbations for the purpose of multiplying our and others' sensing of identities within our communities.

---

[4]We use the symbol "x" in response to the Editors' description of this book's focus on science x education x Anthropocene. We think of this as a playful entry into our interest in disrupting our identification with only one community or another. At the moment that we wrote this text, Katherine described herself as "geoscience education researcher" who is also interested in identifying with the "education research" community, that is, I (Katherine) am using the writing of the chapter to transform my understanding of her own identification.

Disciplinary Perspectives on the Anthropocene

Katherine's Entry into the Anthropocene

Through my lens as a geologist, I think of the Anthropocene as a proposed new geologic epoch. The name "refers to the present, when human impact on Earth's surface, atmosphere, and hydrosphere has been deemed to be global" (Finney & Edwards, 2016, p. 6). There are arguments within the scientific community about whether we can identify the change-over in geologic epochs while we're in that change, as epochs are much longer than our human lives. In order to be fully recognized as an epoch, the Anthropocene needs to have concrete identifiers such that it can be distinguished from earlier and future epochs.

The lower boundary (beginning) of a unit of geologic time, like the Anthropocene, is defined by a golden spike. The International Commission on Stratigraphy (ICS) drives a literal golden spike into the type of location with the clearest indicators of the transition to mark the lower boundary. The location is called a Global Boundary Stratotype Section and Point (GSSP). In order to point your finger and say "There! That's the transition," a location has to meet a number of criteria (Remane et al., 1996). Normally, the boundary is marked by the first appearance of a fossil species with secondary markers (other fossils, chemical signals, or evidence of geomagnetic reversals).

A helpful way to visualize how geologists think about time can be seen in the stratigraphic column, or cross-section showing rock units, of the Grand Canyon in Fig. 16.1. The rock units on the bottom are older than those on top. In the same way, I think of "up" as being younger and "down" as being older. If you were to explore the rocks within the Grand Canyon, you would see distinct fossil species or markers in each layer, like a beautiful brachiopod in the Kaibab Limestone. If we were to imagine a layer of rock being deposited today as part of the Anthropocene, its secondary markers might include signals of notably elevated atmospheric carbon levels (as recorded in, say, a limestone) or the many plastics that will outlive us all. Below, as we envision the future of our disciplines together, I imagine this as "above our heads" because of this bottom-to-top view of time.

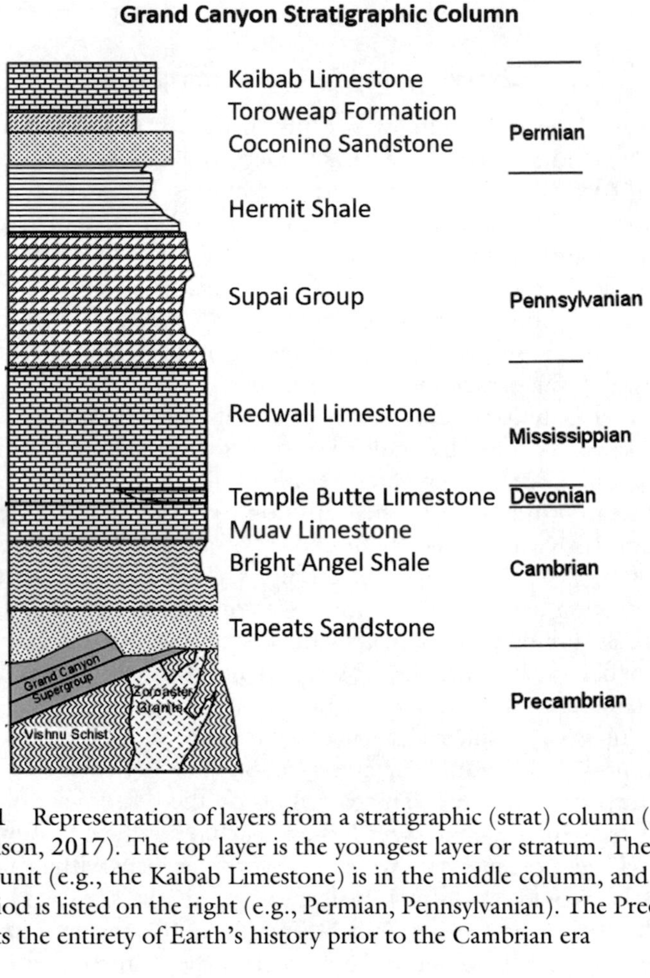

Grand Canyon Stratigraphic Column

Kaibab Limestone		
Toroweap Formation		Permian
Coconino Sandstone		
Hermit Shale		
Supai Group		Pennsylvanian
Redwall Limestone		Mississippian
Temple Butte Limestone	Devonian	
Muav Limestone		
Bright Angel Shale		Cambrian
Tapeats Sandstone		
Grand Canyon Supergroup Zoroaster Granite Vishnu Schist		Precambrian

Fig. 16.1 Representation of layers from a stratigraphic (strat) column (modified from Nelson, 2017). The top layer is the youngest layer or stratum. The name of the rock unit (e.g., the Kaibab Limestone) is in the middle column, and the geologic period is listed on the right (e.g., Permian, Pennsylvanian). The Precambrian represents the entirety of Earth's history prior to the Cambrian era

Michelle's Entry into the Anthropocene

My (Michelle's) introduction to the Anthropocene stemmed from my readings in philosophy that suggested social and environmental relations cannot possibly be jump started to a desirable state (such as a utopia, or "back to the way it was") that will be desirable for everyone for all time. Braidotti (2013b) helped me think about how in the present age of the

Anthropocene, we might continuously reconfigure the relationship to our "complex habitat, which we used to call 'nature' " (p. 81).

Braidotti (2013b) encourages a both/and perspective: that while phenomena like deforestation and climate change have significantly shaped humans' present relations with their complex habitat, it is still possible to perturb[5] these relations, for example, by:

(i) developing one's awareness of the dynamic capacity they have in concert with others—human and nonhuman, living and nonliving—in organizing their dynamic, material world.
(ii) enlarging the frame and scope of identities that disavow traditional identities, visualizing "the subject as a transversal entity encompassing the human, our genetic neighbours the animals, and the earth as a whole, and to do so within an understandable language" (p. 82).

I responded to Braidotti's recommended perturbative practices through my research. Firstly, while I began my study as one in which I studied participants' practices, as if separate from the study's participants, I began to study *with* my participants, for example, by asking them to respond to my representation of landscape features and mapping on large sheets of paper logics enabling and disabling connections in our research practices. Secondly, I began looking for ways to adopt language and practices that seemed mutually generative to the study's participants. For example, Katherine's use of the term "perturbation" had significance for us both, and by using it to frame and represent our shared research-intentions in *this* study, we blurred sensing about differences in our identities. Finally, I thought about how traditional identities within the landscape of DBER and science education research could be rendered perturbable. Haraway suggests that such "permeability of boundaries" involves constructing a "*network* ideological image" (Haraway & Wolfe, 2016, pp. 45–56, italics added). The project of *networking an image* contrasts with the project of creating an ideology of the purpose of things, the latter of which may lead to heightened sensing about differences. In the next section, we begin describing the process by which we networked an ideological image of research on science teaching and learning—in particular, one

---

[5] Braidotti does not refer to perturbations in *The Posthuman* (2013b). Rather this is my own (Michelle's) capitulation of Braidotti's points while thinking with Katherine's metaphor of perturbations.

that did not invoke the categorical differences that provoked our individual and communities' feelings of isolation.

Constructing a Network Ideological Image of Research on Science Teaching and Learning

Constructing a network ideological image to me, Michelle, was necessarily a relational process, learning from others about how they experienced and practiced boundary-marking. I reflected on my own boundary-marking practices. To support constructing a network ideological image, Haraway, like Braidotti, suggested that in the Anthropocene, "if you can't use a different rhetorical toolbox with different audiences...then you're never going to get anywhere" (Haraway & Wolfe, 2016, p. 289). To investigate how my representations were experienced by participants in my study, during one of my interviews with Katherine, I asked about a figure that I had included in a manuscript under revision (Wooten, 2018). It represented a philosophical concept that I used to think about the landscape of research on science teaching and learning. The figure's caption read, "Each of the black dots is a possible practice. Those practices that are similar to one another are closer together" (Wooten, 2018, p. 214). I anticipated that the figure could be helpful to researchers of science teaching and learning to consider how "accumulation of like practices" is not without effect, that is, normalizing practices in our landscape makes some practices (dots) appear deviant. Katherine's response to my figure took me by surprise:

Katherine: I was so fascinated by that [figure] because I can't tell from reading this if this is a representation or if this is mapped out from the data.

Michelle: That's a good point. Because in the visualizations *you* make, everything is data with a referential X and Y meeting place.

Katherine: It could be something like the output of a social network analysis or similar analytical technique that I just don't know...?

Michelle: My audience is possibly used to seeing points on a grid that they refer to as having X and Y coordinates, so (based on your response) I am interpreting that this figure could actually be confusing.

Katherine's response made me think about how scientists' interpretation of dots on a grid would typically involve looking for correlations or standard deviation ellipses. Katherine suggested that "When you [Michelle] are talking about the density of ideas, it seems very rare that any practice would pop up in isolation." While my interest in adopting a landscape metaphor used an assumption that all landscape contributors' practices were connected, the representation I had made detracted from that impression.

Perhaps not surprisingly, as a geoscientist, Katherine had her own ideas about landscape representation: When I, Katherine, think about landscapes, I think about what their shape tells us about the processes that made them, which makes me wonder what caused the dots in Fig. 16.2. For example, there is a feature known as a linear island chain—simply, a line of islands in the ocean that vary in elevation from the highest at one end of the chain to the lowest at the other. If you've ever looked at a map of Hawaii, you've seen this shape. I picked up my pencil and began drawing (Fig. 16.3).

In this figure, the Pacific Place, a piece of lithosphere made up of oceanic crust, is moving to the right over a hot spot. At a hot spot, relatively warmer mantle rock is rising and melting, creating magma that can push through the lithosphere to create an active volcano. As the plate continues to move, the original volcano is moved with it, no longer actively being "fed" by the hot spot, so it becomes dormant and then extinct. It's experiencing erosional forces, though, so it's also getting smaller and smaller

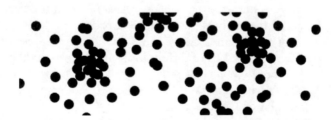

Fig. 16.2 A reprint of a figure in Wooten's (2018) article representing how individual researchers' practices (dots) produce recognizability, or norms, in the landscape when they are practicing similarly to one another. Research practices are deemed deviant, or unrecognizable, when appearing distant from normative practice

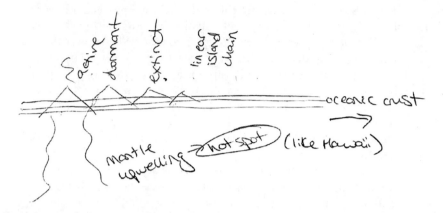

Fig. 16.3 A figure of Katherine's representation of the landscape using a linear island chain (February 2018)

while a new active volcano is forming. The result is a line of progressively smaller volcanoes, connected by pieces of higher-elevation oceanic crust.

Katherine: I wonder if there's an analogy to be made here—we've got the constantly shifting state of the [landscape of research on science teaching and learning]. The [landscape] is moving along with the plate. Practices such as generalizing feed the topographic highs in the landscape like the hot spot. Over time, the specific practices may change. And so how those play out in the literature and in conversations with each other may become "lower" or less relevant. Then the "active volcano" is the hot topic in the field today.

Michelle: That's really interesting. I'm drawn to that representation.

Katherine: See, we have lots of different mountain features that we can build on!

Michelle: That's true. It doesn't have to be one landscape or one [representation], yeah.

Katherine: Pretty much any shape that you come up with regarding the relationship between different practices or logics—you could probably find a geologic process to match it.

By inviting Katherine to literally *network* an ideological image, I (Michelle) was learning what it meant to de-center my practice and disavow my identity as *the* inquirer. I also realized that my initial landscape representation—read by another landscape generator (Katherine)—had the effect of boundary-marking: its form was too far off the map to be recognizable by another community member. An effect of embracing Katherine's logics toward representation was that I was learning what it meant to have my "capacities elaborated" through shared inquiry (Haraway & Wolfe, 2016).

I (Katherine) was also being pushed and humbled (and excited!) by this new way of thinking about topography outside of its traditional meaning within the disciplines of geology and geography. Two years later, I still think about the creation of Fig. 16.3 when I draw a similar figure of a hot spot for my introductory geology students. As I'm drawing, I think back to our conversation and how the joy inspired Michelle's manner and willingness to explore new concepts. I also think about how students' learning can be supported by different manners of knowledge production in the classroom, from drawing to oral and written expression, and thus, how the different representations can be useful in researchers' knowledge generation as well.

Between and Among Education Research Communities

Another way that landscapes can be perturbed and grow or shrink is through processes at a convergent plate boundary. Here, plates move toward each other, causing stress that produces mountain belts. I (Katherine) think back to a stress I expressed in Michelle's article (Wooten, 2018) about going to a conference put on by the National Association of Research on Science Teaching (NARST): I perceived that I would be seen as "less than" with my limited knowledge of theory or even that I would not know what was going on at the conference.

> Katherine: That stress has driven me to claim 2019 as "The Year of the Theoretical Framework" (laughs). I'm just diving in—that's all I'm doing is just reading and reading. So it could be stress that perturbs the topography. (November 2019)

This claiming of 2019 caused me (Katherine) to do so much reading, writing, and reflecting, resulting in growth (a raised elevation in the landscape) in the form of new projects. For example, I have been applying

Ajzen's (1985) theory of planned behavior to develop a better understanding of why students enroll in introductory geology classes and how their thoughts shape their intent to take (or not take!) a second geology class. This has meant playing with and immersing myself in new bodies of literature. It has brought a new sense of joy in my understanding of knowledge production. Reflecting on this growth process again invoked a geologic parallel for me.

> Katherine: When the bottom of the crust breaks off, it's called delamination or decoupling. When it breaks away, it's replaced by something that's less dense—part of the asthenosphere. And that means the whole topography rises. So removal of something could also allow the topography or the peaks to increase. For example, losing confusion over what constitutes a theory or losing the fear of being accepted as a scientist: those could be different forces sort of driving that change. When we're able to let these things go, the topography is elevated—at least for the individual. (November, 2019)

Constructing a network ideological image seemed to afford us our own (re)constructions of perceived communal boundaries as mobile—flexible, bendable, and consequently, capable of permitting a multiplicity of identities. Each image we conceptualized in relation to our landscape generation led us to a broadened social imaginarium about our landscape's possibilities.

Between and Among Education Research and Science Communities

One of my (Katherine's) favorite papers is a play on the idea of misfit toys. It's a qualitative study that explores feelings of isolation and lack of representation or recognition faced by GER scholars among the geoscience community (Feig, 2013). The way that paper resonated with many members of the GER community indicated those *feelings* had widespread existence in our landscape's layers, like an index fossil that marks a rock layer as belonging to a particular time. As the landscape continues to expand, and there's growing recognition of the importance of science education research to science disciplinarians, I would suggest that these feelings of isolation would become *less* prevalent when moving up toward the more recently formed surface layers. This might be one way to identify an idealized Anthropocene for research on science teaching and learning: the *absence* of the fossil of "feeling like a misfit toy!"

Michelle: It sounds like you're speaking with hope.
Katherine: It's that ideal landscape that eventually we're building (refer-
             ring to Fig. 16.1): 10 feet above where we are right now,
             like, what does that look like? In an ideal world all those feel-
             ings of isolation and lack of representation, those are just
             gone. (November 2019)

Using a Network Ideological Image to Conceptualize and Permeate Boundary Conditions

Constructing a network ideological image—a shared landscape at ten feet above where we are now, in the "Anthropocene"—encouraged us to think about ourselves as in the middle of the past and the future. We wanted to consider how we could reconfigure the topography—assert stress to *flatten* the peaks (or hierarchies) that positioned us in seemingly bounded communities.

Through Our Research Practices

Within the Anthropocene, I (Katherine) conceptualized that Michelle's interviews with researchers of science teaching and learning could be represented in a single stratigraphic layer, like the Supai Group in the Grand Canyon strat column (Fig. 16.1).

> Katherine: What would we consider the fossil or other evidence to suggest that there's been a large enough "shift" to justify the new epoch label? In your [Wooten's] 2018 article (Fig. 16.2), you identified two landscape peaks of scientific and theoretical practice. Geologically, these topographic highs could decrease (e.g., through weathering and erosion; a volcano moving off a hot spot, cooling and contracting) or increase (e.g., at convergent tectonic plate boundaries), such as in our 2018 sketching exercise (Fig. 16.3). (November 2019)

Using the strat column network ideological image Katherine had invoked (Fig. 16.1), I (Michelle) asserted:

> Michelle: My wish for that top layer would be a lack of Othering [between science and education research communities]. But I don't know how to get rid of Othering unless firstly, education adopts a really strong goal of sustainability: reducing climate change or hierarchies between economic statuses. These goals seem like they would require disciplinary perturbations. But that means not focusing on discipline anymore, which means restructuring all of higher education!
> [Katherine and Michelle laugh]
> Katherine: That might be 12 feet above.
> Michelle: And it's not that you disagree or I don't recognize your value [for identification in the broader landscape of science and education research]. However when you express an interest in dispelling the negative affect associated with lack of representation, I'm on the other side saying "Yeah, we need lack of representation: we need to eliminate the notion of identity altogether."[6]

Katherine's response reminded me that being identifiable in the landscape had professional implications:

> Katherine: I feel stress to identify as a scientist for professional reasons. When I was a graduate student, I asked myself, if I want to be a geoscience education researcher, "What is my degree going to be *in* [for geoscience education research]?" I wanted my degree to come from a science discipline like Earth Science or Geology because I feared my science colleagues wouldn't listen to The Other from the College of Education. And so I thought that I could be more effective at producing discipline-level systems change by embedding myself in this geoscience community. If I have the same training as my colleagues I can say [uses comical, persuasive voice], "You will accept me as one of your own and I *will* have some kind of magical team change power within this community!"
> [Michelle laughs]

Katherine was using science disciplinary identity to position herself among science colleagues in such a way as to invite an appreciation for education research. She was perturbing boundary conditions through systems of relations in which she was embedded. And thus I was reminded of

---

[6] Thinking with Braidotti (2013a) who writes about "disidentification."

Braidotti's assertion that it is perhaps less generative to strive for utopias, for example, one in which identity is eliminated, because we cannot possibly disconnect from our systems of relations and the possibilities these afford us. Even though these systems of relations had some power in determining our recognizability in the landscape (identities) and our potentials, we could generate (even joyfully together) oppositional stress toward fixed, unitary identities. For example, Katherine's investment in theoretical frameworks and Michelle's de-centering of her research practices produced oppositional stress toward having clear identities in our own disciplines.

Through Our Teaching Practices

As academics, we're perhaps most keenly aware of our communal boundaries through our departmental disciplinary assignments. Harkening back to our network ideological image of the strata (Fig. 16.1), in December 2019, I questioned Katherine, "What do we need to do as a community in our local contexts, or as a faculty member in our disciplinary departments, to start assigning value to disciplinary-perturbations?"

Katherine: I think that if the transdisciplinary gradient is part of our holistic landscape, where the top strata is where we're getting to (Fig. 16.1), what is the evidence that we would see that the state has been reached? Are there valuations that would themselves be evidence? And what would that look like? Would it be the offering of more interdisciplinary or transdisciplinary courses? Would it be valuing that my graduate students in geological sciences take classes over in Education, and vice-versa, because there's value in that exposure or (inter)disciplinary thinking? I don't know what all that would look like, but I can see that as being sort of a hypothetical, hopeful future state that we could get to.

Michelle: And also, what do you think about natural phenomenon markers for the Anthropocene? Would you expect there to be a decrease in the effects of climate change? We bring students through courses to learn stuff that's already been learned when they're actually really capable people to start addressing socioscientific problems.[7] That we're wasting the potential of education is a new idea for me because I haven't thought about it *that* outside of the box: why wouldn't we use education—all the people sitting in our classes—to address socio-environmental injustices instead of teaching only accumulated knowledge?

Katherine: That's a really fascinating idea. I have five to seven people in my [geoscience education research] methods class. They're incredibly talented. Why am I not utilizing them as a knowledge producing body in a better way?

Michelle: So then if you wanted to also collect evidence on a global level, you would hope that you could measure education's coincident effects on socioeconomic hierarchization or climate change. It seems like there would be measurable social and natural phenomenal changes if undergraduate education is no longer just about teaching accumulated knowledge. So maybe disciplinary perturbations are about applying learning in our local communities. I'm wondering if that's how you would also envision it?

Katherine: I think that's a fair claim.

Michelle: Do you have any hesitation?

Katherine: No, my only hesitation is more like, what are the other indicators that might suggest there is change one way or another? There's another simultaneous line of thought going through my head of like, "How do I stress this in my intro class next semester?"

Katherine and I were once again elaborating each other's capacities, this time regarding our personal plans of action for practicing disciplinary perturbations in our teaching in the Anthropocene. Through the discussion of disciplinary perturbations in our teaching, our hopes for the Anthropocene were rendered on a larger scale than our research communities alone. Methodologically, our thinking together about disciplinary perturbations was slipping past its singular focus on our research communities—we were beginning to construct a network ideological image for undergraduate education.

---

[7] Thinking with Michelle's post-doctoral adviser, Dr. Scott McDonald.

A Vignette with Michelle's Astronomy Teaching

Katherine and I were both earth educators of a sort. Because our December 2019 dialogue had slipped into disciplinary perturbations within education, I (Michelle) felt eager to vet my ideas for a sustainability project in my astronomy course the following semester (Spring 2020). I explained that part of the course would be devoted toward traditional disciplinary content, and part would be devoted to a project that connected course topics to the local context: mitigating light pollution in students' local community.

> Michelle: One effect of mitigating light pollution is being able to see the Milky Way, another is nonhuman animals' ability to navigate. This semester, out of seventy-five students, only three have seen the Milky Way. One student saw it for the first time last week when completing my trial light pollution assignment!

Despite Katherine's apparent interest in playing with the idea of disciplinary perturbations in undergraduate education, I felt nervous telling Katherine about the light pollution project and nervous about doing the project in general. Among scientist colleagues like Katherine, I felt that I needed to defend doing a project that was only tangentially related to the science of astronomy. Because by studying light pollution on Earth, students would not be studying astronomical phenomena or doing astronomy, I feared judgment that I was not teaching *real* science. So I quickly began talking about the project's more scientific aspects:

> Michelle: Students will submit measurements of stars' visible apparent magnitude to the Globe at Night app, which stores persons' measurements from all over the world. You can download their yearly maps to see how light pollution changes over time in your neighborhood (https://www.globeatnight.org/maps.php).

I was grateful when Katherine responded with encouragement rather than critique:

> Katherine: Oh! That sounds like undergraduate research. It sounds like a wonderful opportunity to apply what they've learned to take something that's out in the universe and bring it in locally. That sounds like an

assignment that would make a long-term difference for students and actually get them involved somewhat in a meaningful capacity moving forward.

Because of her encouragement, I felt safe to share with Katherine the potential risky professional consequences I envisioned, as I perturbed the norm in undergraduate introductory astronomy education:

> Michelle: I'm not asking for permission. And there is a question of—if all of my colleagues who teach introductory astronomy are teaching straight astronomy content and skills, and I'm dedicating a significant chunk of my instruction toward a transdisciplinary project—"Am I allowed to... ?" Or, "Should I be?" Further, "What do students expect? Do they expect to enter this class doing this project?" And so I also get nervous about students' expectations.

Thinking with Braidotti, taking risks such as these was worthwhile because they are a motion toward academic freedom, enabling generations of students, universities, and communities to form transdisciplinary networks informed by their "yearning for sustainable futures, which "can construct a livable present" (Braidotti, 2013b, p. 192).

A Vignette with Katherine's Geoscience Teaching

I (Katherine) had my own teaching experience to share that illuminated uncomfortable affectual effects associated with perturbing science disciplinary boundary conditions.

> Katherine: [When I worked at] Eastern Michigan University, several of my colleagues from different disciplines (chemistry, Earth science, philosophy, and communications) arranged our students to meet in small, interdisciplinary pop-up groups to discuss climate change. We invited other faculty to have their students participate, and one faculty member pushed back: "Yeah, I don't want my students going to that, but I'll come and watch them." He wanted to judge them doing this exercise that he was highly skeptical of from the beginning.
>
> And he was so convinced by the conversations he saw students having: "I didn't realize that our students would get to that level." He thought that students would sit in their

silos. He was willing to be involved in a future semester, but he needed to see evidence that students would be inspired by the idea and would communicate across their disciplinary boundaries, because it's not how he's been thinking about his students for the twenty-plus years he has been teaching.

Michelle: Yeah. Wow, that's really beautiful.

Katherine: But also scary when someone shows up to experience like "This is new for me too. I can't tell you how it's going to go!" (December 2019)

In later renditions of the pop-up course, many of my science colleagues were appreciative of the pop-up integration of another scientific discipline (chemistry) and could see the value of communication as part of a discussion on climate change. Philosophy was a harder initial sell to the same colleagues, many of whom weren't sure what their students would get out of it. This may be because philosophy appears too far off the map in the landscape of climate conversations. And yet, the contributions of the philosophers in small-group discussions were cited as some of the most helpful by their traditional science colleagues for their ability to dissect arguments and lay out their logical underpinnings. I would describe the day of the pop-up learning communities as one of the educational peaks of my time at that institution.

CONCLUSION

The turn of the year offered us a turn in our teaching contexts. I (Katherine) reflected on the journey Michelle and I had invested in together—elaborating one another's capacities toward enacting shared values. In an email to Michelle in January 2020, I wrote,

Katherine: I've had several of our conversations running in the background of my mind as I prep for this semester. I'm teaching ~450 students in our intro geology courses. I've been thinking about how you changed up your course, which is changing how I talk about science and the class with them. The conversation is much more oriented around creativity, curiosity, and what can we accomplish with so many amazing minds and experiences in the room.

Halfway across the United States, I (Michelle) entered my astronomy instruction feeling bolstered rather than frightened, remembering Katherine's encouragement toward enacting the perturbative project I had proposed: "That sounds like undergraduate research." "A wonderful opportunity!" "make a long-term difference!"

The study presented in this chapter was precipitated by Michelle's dissertation study, in which she perturbed traditional research practices by inviting interview participants to think together with her about how to shift the landscape of research on science teaching and learning toward increased connectivity. Katherine and Michelle both desired producing flows in the landscape oppositional to ones that maintained stable identities leading to othering and feelings of isolation. Through our studying together, we constructed numerous networked ideological images to support our interest in perturbing boundary conditions in DBER and science education research communities. In doing so, I (Michelle) felt an increased sense of belonging to Katherine's community of GER and even attended a conference they organized. I (Katherine) understood myself more acutely as an education researcher, beyond the more limited scope of GER. In a sense, we each became landscape contributors "marked by the interdependence of [our] environment through a structure of mutual flows and data transfer that is best configured as complex and intensive interconnectedness" (Braidotti, 2013b, p. 139).

We consider that our sensing about our interdependent, multiple identities came about through nomadic inquiry, a method that enabled us to construct our sensing about them. Our nomadicity was akin to what Katherine described as fault slippage in geoscience. As our dialoguing and writing about practices continuously slipped past a singular focus on either research practices, research communities, or our teaching, we both felt we were disavowing traditional research practices and consequently required a stabilizing remedy, in the same way that terracing or driving nails or bolts could be used to stabilize a slipping surface. However, we found that the non-prescriptive methodology of nomadic inquiry was an affordance, as it enabled networking an ideological image for diverse ways that our present communities construct notions of science, science education research, DBER, and their potentials. Although disciplinary perturbations at times feel scary, and have professional implications, we argue that nomadic, networking research and teaching practices are worth the risk in terms of how they enable connected, permeable entries into engaging with one another's communities, shaping undergraduate science education, and producing joy in human–nonhuman relations.

REFERENCES

Ajzen, I. (1985). From intentions to actions: A theory of planned behavior. In J. Kuhl & J. Beckman (Eds.), *Action-control: From cognition to behavior* (pp. 11–39). Springer.

Braidotti, R. (2013a). In R. Dolphijn & I. van der Tuin (Eds.), *The notion of the univocity of being or single matter positions difference as a verb or process of becoming at the heart of the matter.* Open Humanities Press. https://doi.org/10.3998/ohp.11515701.0001.001

Braidotti, R. (2013b). *The posthuman.* Polity Press.

Deleuze, G., & Guattari, F. (1987). *A thousand plateaus: Capitalism and schizophrenia* (B. Massumi, Trans.) (1st ed.). University of Minnesota Press.

Dolan, E. L., Elliott, S. L., Henderson, C., Curran-Everett, D., St. John, K., & Ortiz, P. A. (2018). Evaluating discipline-based education research for promotion and tenure. *Innovative Higher Education, 43*, 31–39. https://doi.org/10.1007/s10755-017-9406-y

Feig, A. D. (2013). The allochthon of misfit toys. *Journal of Geoscience Education, 61*(3), 306–317.

Finney, S. C., & Edwards, L. E. (2016). The "Anthropocene" epoch: Scientific decision or political statement? *GSA Today, 26*(3), 4–10.

Haraway, D. J., & Wolfe, C. (2016). Companions in conversation. *Manifestly Haraway.* University of Minnesota Press. https://doi.org/10.5749/minnesota/9780816650477.001.0001

Nelson, S. A. (2017). Geologic time. https://www.tulane.edu/~sanelson/eens1110/geotime.htm

Normand, V. (2015). In the planetarium: The modern museum on the Anthropocenic stage. In H. Davis & E. Turpin (Eds.), *Art in the Anthropocene: Encounters among aesthetics, politics, environments and epistemologies* (pp. 63–78). Open Humanities Press.

Remane, J., Basset, M. G., Cowie, J. W., Gohrandt, K. H., Lane, H. R., Michelsen, O., & Naiwen, W. (1996). Revised guidelines for the establishment of global chronostratigraphic standards by the International Commission on Stratigraphy (ICS). *Episodes, 19*(3), 77–81. http://www.stratigraphy.org/upload/Remane1996.pdf

Singer, S. R., Nielsen, N. R., & Schweingruber, H. A. (Eds.) (2012). *Discipline-based education research: Understanding and improving learning in undergraduate science and engineering.* National Academies Press. https://www.nap.edu/catalog/13362/discipline-based-education-research-understanding-and-improving-learning-in-undergraduate

Wooten, M. M. (2018). A cartographic approach toward the study of academics' of science teaching and learning research practices and values. *Canadian Journal of Science, Mathematics and Technology Education, 18*(3), 210–221. https://doi.org/10.1007/s42330-018-0029-9

Wu, J., Eaton, P. W., Robinson-Morris, D. W., Wallace, M. F. G., & Han, S. (2018). Perturbing possibilities in the postqualitative turn: Lessons from Taoism (道) and Ubuntu. *International Journal of Qualitative Studies in Education, 31*(6), 504–519. https://doi.org/10.1080/09518398.2017.1422289

CHAPTER 17

Let's Root for Each Other and Grow: Interconnectedness (with)in Science Education

Rachel Askew

Seeds often sprout when and where they are not expected. In *Arts of Living on a Damaged Planet*, Lesley Stern (2017) paints a vivid picture of growth in the Laureles Canyonic Landscape. Stern describes the scene of tomato plants growing from tires and sprouting from cracks in concrete saying, "I would have never thought that paved roads might be the key to tomatoes. But so long as the boundary between the road and the vegetable patch is permeable, the potential of yet another landscape unfolds, one in which vegetables and people and nonnative trees cohabit in a reshaped ecology" (p. 27). Plants grow after harsh winters; flowers bloom through cracks in concrete; new plant life blooms in previously desolate spaces and provides the ecosystem with stability. In the Anthropocene, we can see new ways of growth. In our last elementary science methods class, growth authentically became a theme among the students to put into words how they saw themselves as science teachers. But what does that growth as a

R. Askew (✉)
Freed Hardeman University, Henderson, TN, USA
e-mail: raskew@fhu.edu

© The Author(s) 2024
S. Tolbert et al. (eds.), *Reimagining Science Education in the Anthropocene, Volume 2*, Palgrave Studies in Education and the Environment, https://doi.org/10.1007/978-3-031-35430-4_17

science teacher mean? For Elizabeth, a student in the course, it meant a flower blooming (see Figs. 17.1 and 17.2); it meant acknowledging the ways in which she thought about science teaching—past, present, and future.

Preparing Elementary Science Teachers

Past/Present

Science methods courses are traditionally the spaces where preservice elementary teachers receive science education training. These courses, if offered, are usually the length of a typical semester course and incorporate science content and pedagogy, often through reform-based practices. The curriculum and activities involved in a science methods course for elementary teachers vary depending on the school, and some, but not all, may incorporate a practicum experience for students to observe science teaching in an elementary school or even give students a chance to teach science content. Some teacher preparation programs merge science methods with

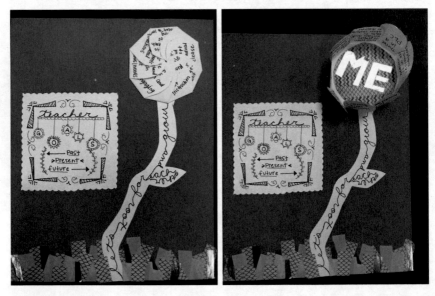

Figs. 17.1 and 17.2   Let's root for each other and grow: Elizabeth's classroom design project

math and social studies, while others offer the course online, and some only for a shorter amount of time. In teacher preparation programs, methods courses are places where students are presented content and pedagogy for teaching, often based on the instructors' ideas of what it means to be a certain kind of teacher. It is in these courses that preservice teachers are confronted with ways of being and doing science which may be vastly different from their own prior experiences.

These courses, along with the ways in which science is or is not taught in elementary schools, reinforce discourses about what it means to do science and/or be a science person. In schools, science often reinforces stereotypical views of what it means to do science as well as what it means to be a scientist. White, middle class, male ways of knowing and being are often centered within science, alienating those who do not identify in such ways (Brown, 2005; Elmesky, 2005; Emdin, 2010; Lemke, 1990). As Foucault (1982) stated, "we have to know the historical conditions which motivate our conceptualization. We need a historical awareness of our present circumstance" (p. 778). When considering elementary teacher preparation, the ways in which science is presented should challenge and call into question these stereotypical views of science and scientists.

Present/Future

Knowing how elementary teachers have and are being prepared to teach science, we can begin to imagine different possibilities. Considering the lack of instructional time often allotted for science, the common trend of low self-efficacy in science among elementary educators, and the current state of science education and our world, it is a vital time to commit to training elementary teachers to teach science in a way that promotes equity and opportunities for all. Science educators have the opportunity to discuss the ways in which humans have interacted with and viewed nature (current issues surrounding climate change, sustainability, disregard for scientific news, etc.)—and the ways in which we can use science to explore new possibilities (Guyotte, 2019).

We need the ways that science has been perceived and taught to be called into question with those pursuing careers as teachers. To do this, individual experiences of and with science need to be examined before new ways can be explored. When I, as a teacher educator, present a curriculum regarding how to teach science, I am doing so from my own experiences. The choice of wording, the set-up of curriculum—every aspect—will serve in part of how this group of teachers construct their subjectivities as science teachers. The ways in which I teach may reinforce the discourses associated with science. Opening up a course to co-construction with students can be seen as a line of flight—or a way to resist the instructor-mandated curriculum—and position students as co-creators, educators, and instructors.

RETHINKING ELEMENTARY SCIENCE TEACHER PREPARATION

Relationality and Slowing

The Anthropocene, our current geologic moment in which human impact on the earth is undeniable, opens possibilities for rethinking the ways in which things have traditionally been done. In elementary science teacher preparation, this means looking at the ways in which we have been preparing preservice teachers (instructor chosen/driven) and the ways in which we might try different things (co-teaching, co-designing, co-analyzing, etc.).

Pushing back, questioning, acknowledging interconnectedness, and growing despite conditions are all areas which the Anthropocene embraces and teaches us about. Guyotte (2019) suggests a theory of STEAM (science, technology, engineering, art, and math) in the Anthropocene to consider slowing and relationality. Taking concepts of slowing from Ulmer (2017) and Stengers (2018), Guyotte (2019) employs slowing to explore how we might do differently in STEAM education in the Anthropocene through relationality. She explains, "Relationality as partial, as incomplete both in our being and in our perceptions requires that we think about our constant connectedness. In other words, if we see the "others" from which we are learning as more-than-human bodies, if, like Ulmer (2017) suggests, we Slow ourselves to take notice of the ways in which we are already materially implicated in our world, our ecological awareness inspires a different practice of relationality with what surrounds us" (p. 8). Relationality through slowing could allow us to explore the connections between and within a science education context.

Approaching science methods with these two aspects gives space and time to rethink what science education is and who science teachers are. Slowing and relationality allow us to ask: What if students, in this case preservice teachers, had more choice and say in how (specific assignments) and what they learn (specific content), and in turn, what kinds of teachers they want to be? Deleuze and Guattari (1988) draw attention to the power of the state to produce conformity, "Doubtless the State apparatus tends to bring uniformity to the regimes, by disciplining its armies, by making work a fundamental unit, in other words, by imposing its own traits" (p. 424). It is this uniformity of regimes that began to unsettle my thinking; as I read and thought through the idea(s) of the State apparatus, I began to think about how these practices are part of a structure set in place to keep us all the same. Teacher preparation programs, standards, instructor-driven curriculum, evaluations, standardized testing—each of these areas presume a certain way of being and doing.

While simultaneously disheartening and enlightening, there are opportunities to act in opposition to the State apparatus. In these spaces that seem to continue to produce the same structures, we can always search for lines of flight—ways to escape and do differently (Deleuze & Guattari, 1988). The previously mentioned issues in science specific to elementary teachers and students, and the connection between experiences and beliefs, are the result of repetitive structures seeking to keep things in place. However, where there are structures there are lines of flight. In teacher preparation, one line of flight can be found in the ways in which we co-construct courses with students and how we position students' analysis of their own identities as teachers. As we think of preparing elementary teachers to teach science, and as we embrace the pushing, questioning, and slowing in the Anthropocene, we can see ways to do differently. In the study I will describe in the rest of this chapter, approaching our science methods course together focused us on student concerns and questions regarding teaching science and moved us into a space where experiences and questions were thought over, discussed, and used to move forward in our learning—a slower approach.

Subjectivity

In this approach, a theoretical background of subjectivities laid the foundation for thinking about what it means to be a science instructor/teacher/student/future educator. Subjectivity can be defined as the beliefs, thoughts, and views—conscious and subconscious—of an individual and how they see themselves relating to the world, all of which are of the produced by positions as subjects (Bazzul, 2016; Jackson & Mazzei, 2012; St. Pierre, 2001). It is an "ongoing process of 'becoming'—rather than merely 'being'—in the world" (Jackson & Mazzei, 2012, p. 53).

Subjectivities are never completely constituted but are always changing and evolving based on relationships and context. Contexts and those involved play a role in how the subject understands and perceives her/himself (Mansfield, 2000). The concern in science education is that which includes teachers' beliefs, thoughts, and views—all of which are included in their subjectivities and all of which can change based on various aspects. As mentioned above, a theory of subjectivity allows for a questioning of traditional teaching methods by interrogating and (re)negotiating the relations between teacher and student.

Overview of the Project

The study took place at a large university in an urban area in the mid-southern United States. As a graduate teaching assistant at this university convenience sampling was used. Participants of this study included six students, all women, enrolled in an elementary science methods course as students, and myself as the instructor. Participants had been accepted into the teacher preparation program and were enrolled in the elementary science methods course section. Students enrolling in this specific section were a cohort, having had previous courses together. Students could refuse to participate in the research with no adverse effects on their grade or standing in the course; however, all consented to participate. Rather than prescribing standard curriculum as an instructor, the curriculum was created based on their perceived needs as future science educators. While many things were discovered from this co-creation, a focus on growth solidified within our final class and prompted the question for me as a science educator: What does it mean to grow, as an elementary science teacher, in the Anthropocene?

BEGINNING OUR JOURNEY TOGETHER: OUR FIRST CLASS

Before creation of the syllabus, we began with a period of self-reflective inquiry about a "high and low moment" in science (Birmingham et al., 2017). This prompt was left open on purpose—to allow each student to reflect based on her conceptualization of science and what is meant by high and low. After completing this initial reflection, students discussed in pairs, noting any similarities and differences. I presented students with the idea of our class journeys' timeline and how we would be adding to this blank sheet periodically throughout the course, using safety pins to allow us to change and rearrange our ideas as needed (Adriansen, 2012). Throughout the course, we revisited our timeline to assess where we are and where we want to be as individuals and as a class. The timeline remained fluid throughout the course and was a place that could be changed and moved to show how we were changing.

In the first class, we each wrote down our high and low moments in science on a neon-colored notecard. (It is important to note that we could choose whatever color card we wanted, with no tie to what was written on the card or a certain color mentioned.) Then, each student shared as they wanted about their high and low moments in science. Low moments varied but were all related to in-school science. After discussing these experiences, students discussed how to place them on the timeline by deciding to order the events chronologically. They created "waves," placing the high moment wave at the top and the low moment wave at the bottom. Since elementary school experiences were left out of initial response, I intentionally asked students to think of a high and low moment from their elementary experience (if possible) and to also think about what they have seen in elementary schools since being in the teacher education program. After adding these experiences, students decided to change the format of the "waves" timeline into individual "circles." On this new timeline, each person had her own circle with her high and low moments in science, as shown in Fig. 17.3. The timeline remained in this format, adding new notecards to our individual circles, until the last class.

"LET'S ROOT FOR EACH OTHER AND GROW": OUR LAST CLASS

Presentations of Science Classrooms

At the beginning of the last class, students presented classroom design projects. These projects consisted of self-reflection of themselves as science teachers currently and imagining what this might be in the future. During these presentations, a student, Elizabeth, shared scrapbook pages created to capture her responses. To reflect on herself currently as a science teacher, she displayed a flower with closed petals. When discussing her future science teacher self, she opened the petals to reveal the kind of science teacher she wanted to be while still acknowledging her current perceptions. Along the stem was written, "Let's root for each other and grow." Explaining her

Fig. 17.3 Timeline iterations: waves of high and low moments and with overlay of individual circles

choices, she pointed to the quote and shared that this class, as a community, had helped her and that she wanted to continue to support each other. Elizabeth was not the only student to suggest growth as an aspect of herself as a science teacher. All six students used some description of still "-ing" verbs in their presentations: Still learning, growing, becoming, and so on.

Returning to Our Class Timeline

At the end of class, we returned to our class journeys' timeline. I asked the students how they thought we could best show our class through this timeline and reminded them that it did not have to look like an actual timeline. It was quiet for a while and so I brought up something I had noticed in the presentations. In each was an idea of still becoming, still learning, still growing, still developing, and so on. Along with this idea,

Elizabeth had written a quotation specifically for our class, signifying a focus on our class community supporting each other. Taking this idea of "-ing" (becoming, learning, growing) I asked them if that could be some way to represent their journeys, since it was something they had all mentioned. They immediately agreed, and the conversation began to flow.

> *What if we...*
> Me*: Used the idea of growing from your presentations?*
> Student*: Like a flower?*
> Me*: One flower or different flowers?*
> Student*: What if the petals were our classwork?*
> Student*: What if we made a tree and used the leaves to show our experiences?*
> Student*: What if we had a lot of flowers or trees?*
> Student*: The roots could be our past experiences.*
> Student*: And had a sun and raindrops being what helped us grow?*

During this discussion, McKenzie got her notebook to begin drawing an idea. After my first two comments, I stepped back and listened. Students continued to throw ideas around such as how to divide our experience notecards. Should the flower grow chronologically, with the roots beginning with the start of class and the petals being the most recent events? Should negative experiences be the root and positive experiences the stem? What about the experiences from class—are they raindrops that allowed the growth? Or are they part of the flower? Or maybe the sun?

McKenzie chimed in showing a drawing from her notebook, "It doesn't matter. We can't separate these by parts. Our past experiences—positive and negative—are important. It's all connected. And all of our experiences are connected because we are in this class together." The final agreement: Let's just separate the cards by color—it doesn't matter what they say. The students then took the cards off and began sorting them by color. They began working on creating a flower, deciding the roots would be yellow cards, the stem would be green cards, petals would be pink cards, and the center of the flower would be the orange cards.

McKenzie decided to draw roots to show more detail of this area of the flower, and they came up with the idea of placing each of their names on a root signifying their connection not only to the experiences but to each other. Since everyone felt moved by Elizabeth's quotation for the class, they chose to write it along the stem. At the end of class, I took a picture of the flower (shown in Fig. 17.4). However, students asked if I could still

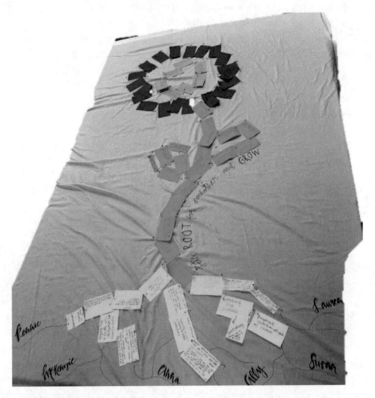

Fig. 17.4 Class journeys' timeline flower completed in last class

add pieces of their work to the flower, saying that it wasn't complete with just their notecards.

After Course Completion: Designing Our Final Representation

Since our class was no longer meeting, I designed two different ideas for how to add student work to the flower. The first option had pictures from class surrounding the flower's outline, with student work being displayed around the edges of the sheet. Student work included students' final classroom design presentations, journal entries, questions, data charts, and various materials from class (e.g., maple seed mail carriers, moon phases model, puddle diagram). Students were sent both pictures to vote on a design, which became our final representation, shown in Fig. 17.5.

(RE)CONCEPTUALIZING GROWTH IN THE ANTHROPOCENE

The resounding idea of -ing in the last class caught me off guard. When (un)planning for this course (the course was co-constructed after the first class meeting), I did so with the idea that I wanted all students to be given the opportunity to see themselves as science teachers. So, during the last class, I was surprised to hear each connection to growth and continuation. It was as if the end of the class was not the end at all but rather a meeting place for us where our stories intersected and learning continued, creating

Fig. 17.5 Final class journeys' timeline representation

offshoots from middles, not a linear progression. To discuss this study, rhizoanalysis was chosen to allow for a contextualization of the process and the data learned and interpreted throughout. Rhizoanalysis is an analysis that seeks "to disrupt, to think, and to do qualitative research differently" (Masny, 2013, p. 339). The concept of rhizoanalysis stems from the discussion of rhizomes by Deleuze and Guattari (1988). Rhizoanalysis uses rhizomes, pieces with "tubes" or "shoots" that connect to other rhizomes, with a focus on the middle as opposed to the beginning or end (Deleuze & Guattari, 1988; Masny, 2013). Rhizoanalysis is "difference that allows for creation and invention to occur continuously" (Masny, 2013).

While interpreting the events of the course (journaling, syllabus, timeline, student work) our class journeys' map was created together to show the overlap of activities and understandings. While traditional analysis allows for a tracing of theory and data, rhizoanalysis suggests map making. As Alvermann (2000) states, "Maps, unlike tracings, are always becoming; they have no beginnings and endings, just middles. It is by looking at middles that we begin to see how, in perspective, everything else changes" (p. 116). While not looking for a single truth or traced way of interpreting, rhizoanalysis is exploring and attending to the multiple connections and relations that occur.

Each student contributed pieces to the map throughout the course and participated in constant (re)connecting of the pieces we each brought in. Each instance of connection, or assemblage, is specific to that moment within this certain group of participants, which ties to rhizoanalysis as an assemblage (Masny, 2013). Mapping their experiences visually led to different choices in the analysis. At the start of the class, students mapped onto each other's experiences, showing their conceptualizations of themselves as science teachers in relation to chronological order. These connections showed that most students did not recall high or low "science experiences" before middle school. This common connection led to a discussion of what science is—and why we (1) tie science to school and (2) did not tend to think of science in elementary school. As a fluid analysis, when we returned to the timeline a few weeks later, students then shifted from mapping on to each other, to creating individual circles where they connected their own experiences. And finally, when asked how to create a timeline that best represented our class, students returned to connecting all experiences together (as done in the first class).

Interestingly, students chose to represent a large piece of our map with a flower and roots. As Deleuze and Guattari (1988) discuss the rhizome, which is the basis of rhizoanalysis, they often discuss the distinction between tracings and maps in relation to trees, roots, and rhizome. When probing into the idea of maps and tracings, they say, "Thus, there are very diverse map-tracing, rhizomeroot assemblages, with variable coefficients of deterritorialization. There exist tree or root structures in rhizomes; conversely, a tree branch or root division may begin to burgeon into a rhizome" (p. 15).

What can we question, push on, and deterritorialize from our representation? A flower that was created by students to show our collective growth? It may be that the choice of a flower signifies some sense that students had to grow in a certain way or into a certain teacher; however, the fluid nature of our map along with the multiple names along the roots pushes us in thinking that the map-tracing is never final, and the rhizome-root assemblages of each individual meet, connect, and split within different parts of the flower. In other words, we became a flower for the end representation but never with the assumption that this would stay but rather that we would meet in such a way for this time and continue to shift, move, and meet in different ways and places throughout the remainder of our journeys.

Our class journeys' timeline was constructed in a fluid manner to allow for constant analysis of our experiences—individually and collectively—and was revisited throughout the course. During the last meeting, the final activity was analyzing our class journeys' timeline through this question: What have we learned about ourselves as science teachers through this course? And how can we show who we are as science teachers at this moment? Through a class conversation about what our interpretations of the course meant for preservice elementary teachers, the students decided how to arrange an assemblage of our experiences to depict our time together in this class and specifically ourselves as science teachers.

We discussed the commonalities and differences in our stories as presented through various responses to prompts from the duration of the course. The commonalities among experiences and conceptualizations lead us to "middles" and showed us the connections and disconnections made in our journeys to becoming science teachers together (Alvermann, 2000). These middles, connections, and lines of flight appeared throughout the course, and we attempted to capture them in our final representation. One specific connection was made in the acknowledgment that there is some aspect of growth that has occurred, and will occur, in each as she decides what it means for her to be a science teacher. But what does growth mean in becoming science teachers? And more specifically—what does this growth mean in the Anthropocene?

Growth and Interconnections

The concept of growth provided the direction for our final class journeys' timeline. However, it was not focused on individual growth but rather a community growth that included human and more-than-human participants. Students' reflections of themselves as science teachers included interactions with each other in the course as well as interactions with materials: standards, models, videos, and so on. Throughout the course, our map changed only twice. The first timeline showed our experiences together, with time being the determining factor for the placement of cards. However, students decided within a few weeks to create individual circles, separating our experiences out, so that we each had a place for our cards, with no overlapping. Both "waves" and "circles" showed something occurring or changing—in other words, growth. However, they did not seem to capture the essence of our combined experiences and learning throughout our 16 weeks together. The discussion of how to represent us as a class stemmed from presentations that each mentioned a continuation of becoming: science teachers who are but who are still growing. When McKenzie suggested that the cards cannot be separated by experiences or what they represent, because they all represent something different for each person, all agreed. Each experience played an important part in how we saw ourselves as science teachers and the kinds of science teachers we want to become. This acknowledged the connection of experiences, individually and communally.

Students not only represented the interconnection of their individual experiences (i.e., their personal experiences were all connected) but also were all connected to each other—part of one flower with separate roots. Despite being connected, they each are growing and becoming individually, which was shown by writing each person's name on a root of the flower. Students did not feel their reflections on experiences alone represented the pieces of our map. They requested the addition of work—of things produced—to show and give ideas about what we did in the class that supported and challenged them as science teachers. Likewise, students did not just request to show their work as a representation of what we did in the class, as the reflections and our timeline were pieces they wished to include as well. In their analysis of our timeline, they showed and discussed how our class was a community. It was this community-focused view that supported our constant questioning of how we view science and ourselves based on our experiences.

Throughout the course students were able to "ask-after their own subjectivities by questioning how they have come to understand various practices or situations as commonplace" (Bazzul, 2012, p. 3). Students' end-of-course analysis revealed that it was easier to describe the kind of science teacher they want to be as opposed to the kind they currently are, as they were still growing. Each experience, reflection, and piece of work makes up a different part of their views on science and themselves as science teachers. These pieces cannot be classified, sorted, or separated; rather they are all important pieces of how they view themselves.

DISCUSSION: LOOKING FORWARD

Multiple Ways of Becoming

The ideas presented support teacher education programs embracing the idea that becoming a teacher is not a linear process but rather an individualized one based on differences (Jackson, 2001; Gaches & Walli, 2018; Sharma & Muzaffar, 2012). Within our course, each student's differences and experiences were centered as they were able to reflect on who they were and wanted to be as science teachers. As Avraamidou (2016) stated, "to become a science teacher is a distinctly personal and intimate affair influenced by myriads of interactions, events and experiences that cause shifts in beliefs, values, emotions, knowledge and understandings—essentially, on identity development for science teaching" (p. 172).

Specifically, in science education, the idea that there is not one way to do science or be a science teacher, that it is personal and different for everyone, can support preservice elementary teachers in their identity as science teachers. Showing them that science can be different from their previous experiences, regardless of whether their experiences were positive or negative, gives room for them to ask questions and try different ways of teaching and becoming. Teacher education programs can support preservice teachers by showing there is not one set path or one set teacher.

Acknowledging multiple ways of becoming a science teacher gave students an opportunity to envision the kinds of teachers they want to become or their future science teacher selves. As a piece of reflection, students were consistently asked about the kind of science teacher they were and want to be, creating space for viewing themselves in different ways as science teachers.

Future Science Teacher Selves

Although I constantly strove to let students' experiences and ideas guide this class, I found myself still expecting a certain outcome. Beginning the course, my assumption was that by allowing them to create the course, they would be able to see themselves as science teachers. (And, to be honest, I was still sorting through the ideas of what kinds of science teachers they would/should want to be.) However, this came to be a point of stuckness—of not currently "being" a science teacher due to their position as students while simultaneously acting as preservice teachers, who hope to one day "be" a teacher. Each could give an answer to how they currently see themselves as science teachers, but most ended with a concept of—but I am not yet. These shifts and tensions were occurring while they were also rethinking what science is, furthering the questioning of themselves as science teachers.

Throughout our final class presentations, there was a consistent acknowledgment from each student regarding what kind of science teacher they want to become. This course allowed them to think about how they want to teach science; however, they were still focused on multiple aspects of their subjectivities as students in their pre-residency phase of teacher education. This led to a divide for most students: the kinds of science teachers we are versus the kinds of science teachers we want to be. Most students had difficulty discussing the kinds of science teachers they are currently. Not in that they could not come up with responses, but their responses included an aspect of incompleteness, of still becoming, still learning, still developing, still -ing, still -ing. However, when discussing the kind of science teachers they want to be, they listed multiple, specific aspects and things they envisioned. Although this class was co-constructed within our time together, the experiences students had in other contexts (e.g., courses, K–3 school setting) positioned them as learners and future educators—a tension within the structures which did not yet consider them as teachers.

While each student presented a very different view of herself as a science teacher, showing the ties to personal experiences and reflections to conceptualizations, the comments focusing on "This was difficult for me"; "I'm not sure about this"; "Here's what I think, but" to introduce the current science teacher section caught my attention. Was this an issue with the ways in which I framed questions? Or could it be something else? How does this speak to ways in which we expect preservice teachers to become teachers? Subjectivity is an "ongoing process of 'becoming'—rather than merely 'being'—in the world" (Jackson & Mazzei, 2012, p. 53). These ideas, being vs. becoming, can be discussed in how students were experiencing the conversation of themselves as science teachers. Their responses support the idea that *being* a teacher is not actually so, rather, it is a constant process of *becoming*. The assumed linear progression of becoming a teacher is not an area new to preservice research. Jackson (2001) specifically discussed this struggle when exploring "the making of a teacher" and concluded by saying,

> When we see how certain structures and discourses get produced and regulated (and others silenced), then we might contest them, reconfigure them, and make space for new ways of learning to teach that reward difference rather than identity. It is then that we can give up the idea of expecting a predetermined teacher "self" to emerge from a linear path of the student teaching experience and instead open new possibilities of multiple and contingent knowledges, experiences, and subjectivities that are productive in the making of a teacher. (p. 396)

We must work to disrupt this notion that one magically, or legally, "becomes" a teacher when certain work has been completed (Gaches & Walli, 2018; Sharma & Muzaffar, 2012). The process, as shown through our experiences in this course, is unpredictable for everyone and requires a messiness that does not flow in a line.

Reconceptualizing Growth in Elementary Science Teacher Education in the Anthropocene

*Growth as...*disruption
*Growth through...*tension
*Growth into...*unknown

Each year, K–12 students receive a growth score on standardized tests showing the difference between where they were and where they are, comparative to other students. It is said to tell of learning, change, and accomplishment. Like a seed sprouting, a student's growth can be traced, bubbled in, and predicted. But what about growth in the Anthropocene? What about the seed, in a world full of poison, concrete, and a lack of what we thought the seed needed to grow?

In this class, students self-identified growth as a result of 16 weeks together. What was this concept of growth in the course? It was not to master an objective or even to grow toward a prescribed something, but rather it was a growth through tension. A growth that was pushing back against the normative expectations, cultivated in a place of slowness, that allowed each to explore what teaching science meant to them. It was individualized and deeply personal, with each growing where they previously had not seen a space to be. And while growth was individualized and deeply personal, it was simultaneously collective and not final. Growth was not the reaching of a goal but rather an acknowledgment of a collective process in which materials, experiences, and others were interconnected and not easily untangled.

So, we ask "what if…?" What if growth were non-linear, a multiplicity of possibilities? In elementary science teacher preparation, this conceptualization of growth into multiple ideas of science teachers allowed each to embrace their experiences in deciding what it meant for them to be a science teacher. Specifically, in science education, the idea that there is not only one way to do science or be a science teacher, that it is personal and different for everyone, can support preservice elementary teachers and push back on traditional, neoliberal ideals of what it means to be scientific. The continuous un/making (Wallace, 2018) of an elementary science teacher in multiple ways with various points of entry and exit: growing toward many instead of one.

What if, as opposed to tracing the difference between two points, growth were in the unknown. Each student found it difficult to discuss the science teacher they currently were but embraced the idea of an unknown future self, describing a desire to continue, to dream. Acknowledging their becoming, the process, as shown through our experiences in this course, is unpredictable for everyone and requires a messiness that does not flow in a line. Rather, it flows in and out, between and within, to a point in the future which cannot be known. What if, instead of growth as only one, it were the growth of a collective—the interconnections of human and non-human entangled so tightly that they cannot be taken apart. Growth together, growth because of, growth with. What if …

Wallace et al. (2018) conceptualize another way of analysis through Thinking with Nature (TwN). TwN could create new questions when used to analyze work in science education in the Anthropocene. Specifically, using TwN to analyze the concept of growth and ways of becoming science teachers could uncover the interconnections between nature and science teacher education. However, as discussed, these concepts should not be explored alone, as we grow through connections and in relation to others. Using TwN with participants would invite them into spaces to think through their experiences and conceptualizations in relation to nature (i.e., growth), providing more ways to acknowledge and allowing more questions to arise.

This (re)conceptualization challenges us to think of growth though differently: growth in the Anthropocene. While this work is a point, a rhizome, in the assemblage of our course, we will not stop growing. The way that our course was conducted allowed reimagining ways of being, learning, and growing in a science methods course that push against neoliberal control and status quo. However, the rethinking and reimagining should not stop here. We must grow through tensions—pushing ourselves to continue asking how we can do differently in elementary science teacher preparation.

References

Adriansen, H. K. (2012). Timeline interviews: A tool for conducting life history research. *Qualitative Studies, 3*(1), 40–55.

Alvermann, D. (2000). Researching libraries, literacies, and lives: A rhizoanalysis. In E. St. Pierre & W. Pillow (Eds.), *Working the ruins: Feminist poststructural theory and methods in education* (pp. 114–129). Routledge.

Avraamidou, L. (2016). Telling stories: Intersections of life histories and science teaching identities. In *Studying science teacher identity* (pp. 153–175). Brill Sense.

Bazzul, J. (2012). Neoliberal ideology, global capitalism, and science education: Engaging the question of subjectivity. *Cultural Studies of Science Education, 7*, 1001–1020.

Bazzul, J. (2016). *Ethics and science education: How subjectivity matters*. Springer.

Birmingham, D., Calabrese Barton, A., McDaniel, A., Jones, J., Turner, C., & Rogers, A. (2017). "But the science we do here matters": Youth-authored cases of consequential learning. *Science Education, 101*(5), 818–844.

Brown, B. A. (2005). "It isn't no slang that can be said about this stuff": Language, identity, and appropriating science discourse. *Journal of Research in Science Teaching, 43*(1), 96–126.

Deleuze, G., & Guattari, F. (1988). *A thousand plateaus: Capitalism and schizophrenia*. Bloomsbury Publishing.

Elmesky, R. (2005). "I am science and the world is mine": Embodied practices as resources for empowerment. *School Science and Mathematics, 105*(7), 335–342.

Emdin, C. (2010). Affiliation and alienation: Hip-hop, rap, and urban science education. *Journal of Curriculum Studies, 42*(1), 1–25.

Foucault, M. (1982). The subject and power. *Critical Inquiry, 8*(4), 777–795.

Gaches, S., & Walli, S. (2018). 'My mom says you're not really a teacher': Rhizomatic explorations of ever-shifting student teacher identities and experiences. *Contemporary Issues in Early Childhood, 19*(2), 131–141.

Guyotte, K. W. (2019). Toward a philosophy of STEAM in the Anthropocene. *Educational Philosophy and Theory, 52*, 1–11.

Jackson, A. Y. (2001). Multiple Annies: Feminist poststructural theory and the making of a teacher. *Journal of teacher Education, 52*(5), 386–397.

Jackson, A. Y., & Mazzei, L. A. (2012). *Thinking with theory in qualitative research*. Taylor & Francis.

Lemke, J. L. (1990). *Talking science: Language, learning, and values*. Ablex Publishing Corporation.

Mansfield, N. (2000). *Subjectivity: Theories of the self from Freud to Haraway*. NYU Press.

Masny, D. (2013). Rhizoanalytic pathways in qualitative research. *Qualitative Inquiry, 19*(5), 339–348.

Sharma, A., & Muzaffar, I. (2012). The (non)making/becoming of inquiry practicing science teachers. *Cultural Studies of Science Education, 7*(1), 175–191.

St. Pierre, E. A. (2001). Coming to theory: Finding Foucault and Deleuze. In K. Weiler (Ed.), *Feminist engagements: Reading, resisting, and revisioning male theorists in education and cultural studies* (pp. 141–163). Routledge.

Stengers, I. (2018). *Another science is possible: A manifesto for slow science*. John Wiley & Sons.

Stern, L. (2017). A garden or a grave? The canyonic landscape of the Tijuana-San Diego region. In A. L. Tsing, N. Bubandt, E. Gan, & H. A. Swanson (Eds.), *Arts of living on a damaged planet: Ghosts and monsters of the Anthropocene* (pp. 17–29). U of Minnesota Press.

Ulmer, J. B. (2017). Writing slow ontology. *Qualitative Inquiry, 23*(3), 201–211.

Wallace, M. F. (2018). The paradox of un/making science people: Practicing ethico-political hesitations in science education. *Cultural Studies of Science Education, 13*(4), 1049–1060.

Wallace, M. F., Higgins, M., & Bazzul, J. (2018). Thinking with nature: Following the contour of minor concepts for ethico-political response-ability in science education. *Canadian Journal of Science, Mathematics and Technology Education, 18*(3), 199–209.

PART V

Complicated Conversations

CHAPTER 18

In Conversation with Sharon Todd: Rethinking the Future in a Time of Sorrow

Sharon Todd and Jesse Bazzul

Jesse: Thank you for doing this interview Sharon! One of the things we're trying to do with these *Science Education in the Anthropocene* volumes is stretch science education towards new ethical, political, and transdisciplinary horizons. For a long time science education has been an insular sub-field of education for a variety of disciplinary and geopolitical reasons. But we hope that's changing.

You've been a very prominent thinker in education, and I'd like to pull on some of the threads of your thinking in order to get at the affective–ethical dimension of the climate crisis and environmental destruction. I found the way you frame climate sorrow in your work especially interesting. As science and environmental educators, oftentimes we feel stuck in our teaching. Crippled by the immensity of the problems faced, whether

S. Todd (✉)
Maynooth University, Maynooth, Ireland
e-mail: sharon.todd@mu.ie

J. Bazzul
University of Regina, Regina, SK, Canada
e-mail: jesse.bazzul@uregina.ca

© The Author(s) 2024
S. Tolbert et al. (eds.), *Reimagining Science Education in the Anthropocene, Volume 2*, Palgrave Studies in Education and the Environment, https://doi.org/10.1007/978-3-031-35430-4_18

it's the political elements of an issue or relaying the simple facts themselves. I wondered if you could begin by talking about how you think about climate sorrow and how educators might work with this sorrow, process it, etc.

Sharon: So there are a number of things going on that more or less coalesce together around something like climate sorrow. Along with sorrow there's anxiety and anger. Especially when you consider how youth are facing a precarious future. These elements or emotions come hand in hand, with one of them often taking over or subsuming the others at certain points. What really struck me about the problem of climate sorrow was the existential piece. How do these elements of sorrow, anxiety, and anger open up our teaching? We can no longer rely on the same kind of flow from past to "now" and then to the future, as some of our ancestors did for centuries. So how do we think about that disruptive sense of the future? As Franco Berardi's work makes clear, the idea of futurity is being put into question when you have ecological collapse on the horizon.

There's a couple of things for educators to note here. One is the increasing anxiety youth are feeling in relation to the ecological emergency. This is clearly what I saw during the climate marches in Dublin and Montreal, as well as my recent time at Concordia University. Youth don't see a future for themselves—and of course, we can all see this with immanent facts and statistics about tipping points, etc. A caring person would naturally feel overwhelmed.

As adults, I think, we have a much wider and longer set of resources to draw upon when we feel anything overwhelming. But youth don't necessarily have the same set of resources to immediately draw from just by virtue of their age and experience. So the existential piece, for me, is not just for our psychology colleagues to come up with a new disorder for the diagnostic manual. Instead, it involves engaging with sorrow, anxiety, and anger pedagogically and thinking through these really crucial existential issues with youth in ways that don't just inscribe a discourse of collapse but actually give them some language, experience, and tools to draw on. In a nutshell, establishing relationships with the enormity of what people are facing. Scientists like Kimberly Nicholas are arguing precisely for more feeling in all that we do. So for me feelings need to have a prime place in the human condition. This also means not beating ourselves up because we can't fix everything overnight. Kids are stuck in the middle of responsibility and these feelings of helplessness.

Jesse: The idea of futurity comes up a lot in environmental education, through a connection to those who have yet to be born, which features in many Indigenous ways of thinking. How does futurity come to play on the emotions we are speaking about? Why is it necessary to consider temporality?

Sharon: Yeah, I mean I'm currently writing about time, and I'm a Zen Buddhist and follow a different view of the present and time itself. So, that whole notion of time as something that's sequential doesn't quite capture what I mean by futurity shutting down for youth. There's no capacity any longer for youth to open up the present to the future. So, the present is always seen to be something that's quite instrumentalized or functional. It's become something youth "need to do." I'm speaking here from an educational point of view, obviously. And the present is functional in terms of getting somewhere else, but the effect is that there is little imagining possible. And at the same time, there's an inability to appreciate this present. The ecological collapse is essentially making the future impossible to imagine in a positive way. Educationally speaking, it's important I think to reorient to time itself more broadly.

Jesse: In your work, you speak about the aesthetic encounters with things like nonhumans or the more-than-human. In a field like science education, the aesthetic dimension is almost never considered. I wondered if you can speak to the importance of aesthetic encounters and how educators might remain open to these encounters in a general way.

Sharon: I think there are two sides to that, I suppose. So, one for me is to recognize that aesthetics are everywhere in some sense. Not just in the sense of Art and Design and its relationship to visual art or performing arts. It really has to do with, at least for me, basic sensory perception—whatever is encountered corporeally. Not just through the five senses in a very straightforward way but through contact with what I usually call elements of the environment, which can be living things, but they can also be just general material elements in our environment. In this way, teaching science is already very much engaged with the aesthetic. Even though it's not talked about as an aesthetic, teachers are always staging encounters with the world for students. And as teachers, we're always presuming what those encounters ought to look like and what kinds of relationships students ought to be making.

What I'm trying to understand in my work, and in relation to the climate crisis, is what kinds of relationships students are making beyond what we've already set out for them. What kind of sensory encounters are made possible through education spaces, classrooms, and outdoor environments: Things that invite what Jacques Rancière calls perpetual shock—things that disrupt our conventional ways of seeing the world. That's the power of aesthetics. So aesthetics is not just something contained only in art, but something that sensorily makes us rethink who we are, the conditions under which we live, and who we might become. So it is something much broader. In this sense, the arts partake in this kind of activity but are just one small part.

Jesse: Yes, in science education you have big value put on things like wonder—and scientists often recognize a certain aesthetic quality to something that already has some kind of scientific value. And yet, we don't seem to grant that expansive sense of wonder to our students—for them to go beyond the boundaries of our classroom or course but also, to recognize that this boundary is political too. You can't have anything like justice without transgressing or disruption of aesthetics and the aesthetic dimension. So how do educators put that aesthetic sense together with a sense of justice? Because it's so easy to just separate these wondrous or shocking experiences and relegate them to somewhere outside daily life or political life and so on.

Sharon: To me it's hard to strive for justice without understanding something of aesthetics and to me the big elephant in the room—something we haven't mentioned yet—is the body. You can have an attitudinal disposition towards something, but it's also something one feels, and this is where aesthetics slides into affect. It's about the sensations that you're having in relationship to something you're encountering and having contact with. It resonates literally in the body. Your living body that is responding constantly to this kind of constant contact and you're reformulating yourself constantly as well. What justice requires is not just someone who thinks about justice. We all know those people who think about equality and behave in the most dreadful ways. So, justice has to be something one embodies. It's not something one can think about and take part in like it's a legal problem or mathematical proof. Justice is an orientation to the world and is itself a sensibility. So it's something we have a sense of. And this sense of justice needs to be decolonized—and to me these are profound ways of embodying. If you separate justice from the body, you are just left with a bunch of words and abstractions. Justice is something that is performed. It is about things like action, doing, practice, as opposed to some abstract thing.

Jesse: I want to try and get at the ethics at work in aesthetics. What ethically guides you in your connection to aesthetics? You've mentioned Buddhism, and I wonder if we can get at a sort of messy ethics of the aesthetic. Does that make sense?

Sharon: It's a really fascinating question. First, I'm going to answer it in a personal way I suppose. You know 'I had been writing about ethics for pretty long time, and lately I've been moving away from Levinas' work, but one thing I've always held onto is that it's never for the one doing the action or forming the phrase or doing the writing, etc., to say when justice is done. Or to say that this has been a nonviolent ethical act, etc.

What Levinas has taught me is that it's about openness to listening when someone has been harmed despite our best intentions. That's one part of this ethical stance. Rancière's work does a similar thing in saying that it's never for somebody to give another person equality. It's in the act of claiming it. It's the performative act of claiming something. This I see as profoundly aesthetic in some ways because you are moving against something in enacting a kind of equality. Moving against regimes of perception that have boxed a person in. For someone to mobilize something beyond this regime in a stance of equality is a profound act because you're basically tearing down the very pillars of perception that are keeping them, and everything else, in place. So, it's never for me or anyone else to struggle for someone's equality but to set the conditions by which people claim their own equality. To me that's different. As teachers, we need to think about what it is we are setting up for students—what kind of environment we are creating materially, and emotionally, and aesthetically.

Jesse: I think education is a profoundly ethical act in terms of setting the stage for things to happen. I wonder if you could talk about how you've moved beyond the more humanist traditions of ethics. Scientists for one have very strong attachments to nonhuman beings, but perhaps get overly stuck in scientific regimes of perception. How do you open up what would be more humanistic frames of ethics and politics to something more inclusive of the more-than-human or nonhuman?

Sharon: Yeah, it's a difficult thing to pin down; I think there's been a number of things that have kind of led to this. One of them has to do with the idea of nonduality, which doesn't mean unity. It actually means seeing things in their particularities, but in ways that recognize that these particularities penetrate each other at the same time. That also comes from a sort of Zen disposition. But there's also Bruno Latour's work—the whole notion of symbiosis and the ways in which life is manifold. There's no human that is not formed by other life forms . We are composed of a plu-rality of different life forms. As "individuals," we're a consortium. That implies a whole other ethical dimension that occasions a different kind of language for thinking through our relations to the nonhuman. Since, you know, we're intellectually conscious beings who are causing a lot of harm to the planet. Ethics becomes something that is an exchange—it is no longer figuring out what to do correctly and so on. Words like fairness, respect, and justice too all have to be rethought from a relational perspec-tive. It's about seeing how relationality actually becomes the site of ethics. I think that's the difficult thing to put into an everyday common lan-guage, because we don't always have the words for that, because most of us come from the modern system of separation.

Jesse: I wanted to ask you, since we're doing this interview in Ireland, and we're both not exactly Europeans, how much enlightenment thinking and the controlling aspects of modernity are trappings. Since science and philosophy have a colonial history that's hard to shake—how have you seen Europeans try and wrestle with this problem?

Sharon: Well, the first is the way in which we carve up our reality, right? Like it's really as basic as that! The language that we use to carve it up into say, nature and culture, humans and animals. Everything has a very par-ticular place and in that way, our language has a colonial form to it. I don't necessarily mean like empire, but certainly, hierarchies and positions of power are very clear in language. There are varying relations of harm that are enabled by this "divvying up," so I think our languages are, for the most part, languages of separation. When you think about it, we don't even like to use the word cosmology and relationship in the sciences because somehow, you know that just seems way out there! Instead of try-ing to account for a reality that's relational through different ways of sort-ing this reality.

And I'm not trying to say that science is just another way of storytell-ing! I think it's more complicated than that. I don't think anybody would want to greatly privilege one cosmology type over another—but science has such a great influence and power in and over the world. It's also led to great amounts of harm. But more than that—and this is something Vanessa Andreotti has taught me—it's not that science is some external field of endeavour that can be problematic, but it's the way in which our minds, our modern minds, have been sort of co-opted in some ways. We start to only see the world as though a scientific worldview is the only sort of real-ity that matters. That in itself is quite powerful, because then the question becomes "Okay, if we are wanting to actually have a much more expanded view of what reality can be then we need to change ourselves." Because it's just not about adopting somebody else's cosmology. It's about breathing and working through things together. This is something education can actually deal with: the way we systematically see or categorize the world, etc.

Jesse: It's funny because there's always the reality: "well if you're going to do things differently, you're going to have to risk looking really bizarre." In Ireland especially, the education system seems a bit more regimented or controlled than in other English-speaking countries. What do educators have to do to "break out" of these regimes of perception, so to speak? This is always something educators love to talk about: what we can do that might be radically different.

Sharon: Yes, I agree, and I didn't even touch on Ireland. Of course, there's the whole context of colonization, which is in the backdrop of everything. I think you're right—if you're going to break regimes of perception you're going to be doing things that are unintelligible. Breaking this regime then becomes that "aha" moment you're looking for—which might work for some people and not others. I think it means getting comfortable with experimental pedagogy if you will. Not just doing them, but being aware of why it is you're doing them, and in what context. I get students to have conversations with inanimate objects and ask them how they might respond to what it might be saying to them. Some people might call it play, rather than experimental. And just like young children, there's a purpose behind it. You need to guide the students in how or why they might want to purposely engage in play. You know, how might you want the play to get serious.

The tools we have in education, particularly higher education, are mostly intellectual, right? We seldom bring paint boxes, sandboxes, and crayons. This is where more outdoor education at the third level would be amazing.

Jesse: I wanted to ask you about this concept of hospitality, because it can also relate to the nonhuman world too, and I wondered if you could speak a little bit around your thoughts on ethics, and hospitality, and the climate crisis.

Sharon: I think hospitality is an interesting one. Because you know, it was always double. Hospitality is never innocent. You're in this situation where you're welcoming the stranger into the home *as* a stranger. It always undermines what it's trying to do.

But what it's also related to is the word hospice. Here I'm going back to Vanessa Andreotti again, whose new book is called *Hospicing Modernity*. And so, there is a way in which we might need to "be with modernity in its death" if you will. Really just like how you might be with someone who's hospitalized. Not just as a humane act, but also in some ways feeling a tad grateful or accepting during this death. As in how you might respect another life form on Earth. One that, in some ways, has helped us survive. So, in this sense hospitality is not something I show, it's something given to me as well. It then matters what we do with how we feel welcomed or not. Hospitality is a relationship that needs to be taken care of, and one of the problems is that usually the discourse always comes back to the human as the "Centre of the House" that welcomes other human guests, so to speak.

Jesse: I want to talk a little more about modern Western science, because it carries a legitimacy that governments and people can't seem to do without. What do you think about the salvageable parts of modernity? What parts of modernity might we keep?

Sharon: Hopefully I didn't misrepresent Andreotti's words earlier. I've been teaching with this book (*Hospicing Modernity*) so that's why I keep coming back to it. One of the things she's trying to argue is that we need to approach change in different ways. Some will want change, some will want to keep parts of modernity, some will want none of it. But one of the things she's trying to say is that we're all in the same boat when it comes to the death of modernity. So, she's essentially trying to say that we're all sitting beside these modern entities on their deathbed. And what we need to do is think about what we're going to make of that relationship. What kind of difference is it going to make in our lives going forward?

For Andreotti, modernity and coloniality go together. You can't talk about modernity without coloniality. If this is true, then it makes the act of holding on to some bits a lot more complicated. It may not be possible to just hold on to the parts we desire because the parts have come together. They're actually two sides of the same coin. Of course, you know, we also have a third partner—and this is capitalism. So modernity, capitalism, and colonialism are all woven together. If you take something like science—a modern phenomenon to be sure—you then have to ask how it gets turned into a capitalist phenomenon as well as an instrument of colonialism. And does that mean that we just get rid of it? The question for me is not what to get rid of, but to reformulate the important relationships that are pre-supposed in modernity. For example, the epistemological relationships can be undone in ways such that science is no longer able to support coloniality or that capitalistic drive, and so on. There are many scientists, again like Lynn Margulis, Kimberly Nichols, and even James Lovelock that don't think along these lines—that actually see that interconnections and relationships are essential to the science work they do. So, it's important to not set up science as a kind of monolithic category. Overall, I think there are many ways of doing science that seek different relations, for example, by looking at the relation between modernity and coloniality and capital. Capitalism really confuses things, I think!

Jesse: I find it interesting to take this relational approach and focus on details, because some relations may seem productive, some harmful, some ambiguous, etc. One of the things I'm understanding in what you're saying is that we can position ourselves to let go of some of these relations of modernity. And this leaves an exciting landscape to refigure and recreate things like epistemologies. Instead of educators trying to find the way through like some kind of navigator, we're recreating the relational landscape by which we come to do everything! It makes topics like global health inequality and masculinity much more nuanced!

Sharon: I was just going to say, because you know feminist philosophy of science has been doing this for years. Showing how the endeavours, and the practices of science, have gone against science's own ethos in many ways. The idea here is to do more of the work that shows how science constitutes itself. Often science doesn't take ownership for this or is busy denying that it plays any role in oppressions, or that there are complex relations between subjectivity and science. There's Stenger's work around ecology of practices and then seeing how those practices are actually working in the lives of people in and around science. Which is profoundly relational and about building connections and seeing things that are not based on these rigid separations. The kind of dualistic conception of the world. So I think it is to do more of that work.

Jesse: I wonder if that then constitutes a major ethical task for us as educators. Trying to constantly bring in matters of care and refigure that ethical landscape. That ethical task of creating space for new kinds of relationships is, I think, our first task as educators. We get stuck ourselves in things like government apparatuses and disciplines like the social sciences—where we always have to mimic or echo certain disciplinary markers in our work. I feel that's something that needs to be abandoned.

Sharon: Yeah, I would tend to agree with that. I think there's also the opportunity for hybridity that can be enormously productive in education.

To me education is a process and a practice it's not, you know, just a system. What I would like to see more of in education is not so much what we ought to do, not so much performing decolonial strategies or performing ethics, but really just taking a closer look to what it is we are doing from another vantage to what kind of ethical landscapes and relationships we are actually creating. Then how do we account for those? We want education to do something good, but we also want education to simultaneously be the vehicle through which we're doing that good. But, of course, that's where things become a little screwed up because nobody can live up to some ideal of what we're supposed to be affecting and doing at the same time! And this is in tension with the idea that it's hard to just not do what everybody is telling us to do! How do we think about what we're doing in ways that actually start to shift the ground under which we're doing them? I don't mean a critical analysis, like in the old days of the hidden curriculum, where you know you have the voices of boys speaking more in the class, etc. But rather asking what kinds of encounters are we putting out there for our students? How are they meant to create various relations needed to nurture environments? How do we generate new vocabularies and new kinds of concepts that come out of that practice? As opposed to imposing a whole set of relations and concepts and then just leaving it to teachers to get the job done. What's actually "bubbling up" with what they're already doing?

Jesse: I feel like science is extra-intense. People could leave the arts and social sciences relatively alone, but the sciences are caught up in governmentality and biopower. They're integral to international competition. So, science education is completely integrated with capitalism, war, and colonialism through resource extraction, and so educators are a site of resistance against power and the truth discourses it employs through science. This conversation has been great in terms of giving some shape to that resistance, against those things that prevent us from creating these new relational-scapes. Is there some final thought you'd like to leave us with? Some conviction? What have you been working on lately perhaps?

Sharon: Well, I'm speaking in the context of the environment now, and there's a whole bunch of things we need to do away with, but what we really need to do is learn how to talk and be with each other. Speak with children and youth in ways that are non-punishing. And I feel that education as a system has become very punishing for all kinds of reasons, and I think science education has been part of this, just like every other discipline—so it's not unique in that! But in my heart of hearts, I still have this idea of having small schools like just a regular house where neighbourhood kids would come and people would engage students with things, educate with them and alongside them in more public environments. Not just in homes but in much wider places. So education becomes something that's much more relaxed but also much more caring at the same time. I also think adults would thrive on that as well. Teachers generally have really big hearts and the kind of systems they're forced to work in sometimes position children as mere statistics. Any resistance on the part of students is perfectly understandable. It's perfectly understandable why there's such high teacher burnout, and you know schools haven't traditionally been enriching environments. We started this conversation off with something existential, and I think relationality is the key to the existential transformation we wish to participate in.

Jesse: Do you think schools will exist in 100 years?

Sharon: I would think so. What I fear is if they don't exist, then it becomes a privatized world, so what I would really argue for is public spaces. And that schools would be in public places, whether we called them schools or not, where people of different backgrounds come together in a very nourishing way to actually educate each other. Obviously, there's a generational thing here, where newer teachers would work with more experienced ones, etc. Anyway, that would be my vision. To see alternatives, even things like alternative high schools which I know exist in Canada and other places. And science would be very good to learn in those alternative spaces—because it's something that easily permeates all spaces in terms of inquiry and wonder and all those things we mentioned earlier!

Jesse: Yes! Sharon, it's been so nice to talk to you this spring morning. I think this is a great place to stop for now. Thank you so much for your time and thoughts.

CHAPTER 19

In Conversation with Max Liboiron: Towards an Everyday, Anticolonial Feminist Science (Education) Practice

Max Liboiron, Marc Higgins, and Sara Tolbert

If one of the reasons the natural and social sciences and humanities have turned to the figure of the Anthropocene is because it describes a condition in which current ways of life (human and otherwise) are no longer able to continue their recent path, then concepts of politics and polity based in those ways of life also will change … A permanently polluted world is one that, because of its deep alteration, reclaims the need to incite new forms of response-ability. (Liboiron et al., 2018, p. 343, 323)

M. Liboiron (✉)
Memorial University, St. John's, NL, Canada
e-mail: mliboiron@mun.ca

M. Higgins
University of Alberta, Edmonton, AB, Canada
e-mail: marc1@ualberta.ca

S. Tolbert
University of Canterbury, Christchurch, New Zealand
e-mail: sara.tolbert@canterbury.ac.nz

© The Author(s) 2024
S. Tolbert et al. (eds.), *Reimagining Science Education in the Anthropocene, Volume 2*, Palgrave Studies in Education and the Environment, https://doi.org/10.1007/978-3-031-35430-4_19

343

Sara Tolbert: Welcome, Max. I am speaking to you from the lands of Ngāi Tuahūriri, who are mana whenua of this region. I am a white settler, of Scottish, Irish, and British descent, formerly living in the United States, now Tāngata Tiriti in Aotearoa New Zealand. My pronouns are she/her. I became familiar with your work after seeing you present at the 4S conference in 2015. I've found your work and the resources from CLEAR Lab particularly important in my position as a science and environmental educator. We really appreciate you joining us in conversation today. Thank you for meeting with us here on Zoom.

Marc Higgins: I'm Marc Higgins. I am a fourth-generation white settler of Irish descent currently living and working in Treaty 6 territory, specifically in amiskwaciwâskahikan (i.e., Edmonton, Alberta, Canada). My pronouns are he/him. Thank you for meeting with us here on Zoom.

Max Liboiron: *Taanisi, ki'ya?* I'm Max Liboiron. I'm Michif, or Red River Métis from Treaty 6, from the place where Treaty 6, 7 and 9 come together in a place called Lac La Biche, although my Métis family's originally from Manitoba. I live and work now in St. John's, Newfoundland and Labrador, which are the homelands of the Beothuk, in a province that are the homelands of the M'ikmaq, the Beothuk, the Innu, and the Inuit. My pronouns are they/them. Thanks for inviting me. I think that's all the paperwork stuff.

Marc Higgins: One of the invitations that you make clear from early on in your recent *and* influential *Pollution Is Colonialism* is that the naming of the Anthropocene holds its own sets of problems as a rallying cry. However, rather than dwelling on and further articulating the critique, you move towards a politics of "accountability in practice *through* practice" (Liboiron, 2021, p. 119), which is all about the "obligation to enact good relations as scientists, scholars, readers, and to account for our relations when they are not good" (p. 24). Could you speak to the significance of obligation and relations in their specificity rather than universals as orientations to responding to this contemporary moment often referred to as the Anthropocene?

Max Liboiron: That travels a lot of ground in a short amount of time, so I think I would start by just rehashing a little bit of the critique. Critique is necessary but insufficient—you need critique to be able to define the problem in order to try and address it, and, most importantly, how you define the problem forecloses on certain forms of addressing it versus others. Part of the Anthropocene critique I draw on is from Mary Annaïse Heglar's (2019) "Climate Change Isn't the First Existential Threat," which talks about slavery, and Kyle Whyte (2016a, 2016b), Davis and Todd (2017) and Zoe Todd's (2018) various works that talk about the colonial déjà vu (we've been here before!). This is what we've been dealing with since at least the 1400s: life cannot continue in a genocidal way. Sometimes I want to throw the Anthropocene folks a bone because you do need a way to talk about this hot mess. But, of course, that hot mess has been going on for fundamentally different ways and different amounts of time for different groups. That unevenness is the ground we stand on and where our obligations and relations come out of.

We start from this incommensurability. This means that your obligations and relations don't universalize, they don't even out. A lot of people sometimes ask for prescriptions, "Okay, but how do I do this in a good way?" My answer is, "I don't know. I don't know who you are, where you are, or what your deal is. I can't answer your question. You have to do your basic homework: where are you and what are the relations that are already here that you're already obliged to because you exist there?" Even if you haven't chosen this particular hot mess, it's like a family: you're still obliged to it. You don't just get to skip it, or skipping it has acute ramifications.

I think a lot of scholars and educators reach immediately for "how can I be anticolonial in my work in my teaching, research, administration, and/or activism?" I think, first, you have to ask how you are, *specifically*, already in colonial relations. Being at a land-grab university is different from not being at a land-grab university. Being at a university that is sponsored by oil is different from at a university that is not sponsored by oil. Being at a university that is taught mostly in English is different from being at a university that is not mostly taught in English. Being at a tribal college is different from not being at a tribal college. You can't even begin to answer how to do anticolonial work until you've looked at the very specific forms of colonialism in your place and in your discipline. For example, geography has different hang ups than mathematics. We have different colonial formations through those, and while they have some stuff in common, they're also going to be very specifically different. We call that "doing your homework."

Marc Higgins: It's so important to get to the situatedness of our respective work. I'm reminded of Rauna Kuokkanen's (e.g., 2007, 2008, 2010) work on the homework of responsibility. Even if you have a responsibility, if you're not able to respond, because you don't know how and where you're situated, you're likely to reproduce the same problems, although differently. In your work, you remind that there's no *terra nullius* for any of this work; it's rife with complicities.

Max Liboiron: I think the metaphor that really works for a lot of folks is family. Would you like to start fresh with the relationships in your family? Go ahead and give that a shot. You can never start fresh. There's no fresh. Even if your family is the most awesome, greatest, functional family in the world (I would really like to meet you and try and figure out how that happened), there is still all this fraughtness, these dynamics, heritages, inheritances, and all that. Even in the best-case scenario, it's compromised. And that's where you start from.

Sara Tolbert: I want to circle back to *terra nullius* again because it's a really helpful analogy for people to think about; it's complex in that it is situated within some sort of universal-isms while also being very localized and situated in particular contexts and specific historicities of relation.

Max Liboiron: One of the places where I really discovered the continued strength of the *terra nullius* mindset was when I became an executive administrator at my university. I was Associate Vice President of Research and people were saying, "oh, you're selling out," "you're becoming the man," and "you're getting into the muck." I was like, "where the fuck do you think I was as a professor?" This isn't actually different; I just get to see the machine better, which was very helpful for activism. I understand colonialism much better, so I could change things much better. But there's no purity, there's no pure spot, there's no mountain top from which to start fresh. There's no "start fresh." I learned how to frame a lot of this from la paperson's *A Third University Is Possible* (2017), which provides an anticolonial theory of change for administrators. Brilliant.

One of the best parts of being an administrator is that your obligations are super obvious because they'll yell at you—there are concrete structures that keep you accountable in many ways. While, as a professor, it's often less obvious, it was more useful moving further into the colonial machinery, or what people *imagine* to be further into the colonial machinery. I think you can see it better because it has names and acronyms, and you have to have meetings with it, as opposed to a faculty member where it's a lot foggier because the obligations aren't clear. It *seems* that as a professor, you can *choose* your obligations, which is not true, as opposed to an administrator where you don't get to choose them; those obligations yell at you, and you have to have meetings with them.

Marc Higgins: As we're talking about specificity and relationships, that seems like a great point to speak to laboratory life as its own set of relationships. Speaking of the entirety of *Pollution Is Colonialism* is full of generously pointed reminders to account for and be accountable to the relational obligations that we all differently possess, but most specifically with attentiveness to everyday anticolonial science practices that "do not reproduce settler and colonial entitlement to Land and Indigenous cultures, concepts, knowledges (including Traditional Knowledge), and lifeworlds." (Liboiron, 2021, p. 27). How do you define anticolonial science and can you speak to its significance? Further, as you've stated elsewhere that you've "been working on plastic pollution since before it was cool" (Liboiron, 2019): why is plastics a significant site for doing anticolonial science, and how do these everyday anticolonial practices take shape in the lab?

Max Liboiron: I don't think plastics are a particularly good place to do anticolonial work—I think you can do it literally anywhere. If I was making umbrellas, or if I was a sanitation worker, or if I was a mathematician, I think those are all equally good places to do this work, because the good news about dominant systems is that they're definitely what you're doing right now.

When you're trying to do anticolonial work, and therefore trying to surface all of the very hidden ways that colonialism has crept into all structures, everyday common sense, as well as what seems normal and natural, you must *do* things. When I sat at my desk and theorized, I got maybe two feet in this discussion. But when I was in the lab and had to start making decisions about what to kill and what not to kill (and how to kill!), where I was going to get samples from, how I was going to handle them, who I was going to work with, what we clean the shelves with, where we get the shelves from, these discussions about power dynamics and colonialism in particular went much faster, much deeper, and sometimes to an overwhelming degree. The *doing* was crucially important: through making decisions with things you're holding in your hands, which become your collaborators that, again, you don't really choose.

There was an interesting, fairly colonial and elitist critique of the book, which is that there's no theory in it. At first, I was "Oh, maybe there isn't" and then literally the next day, I won a book award for theory, so I was "No, that was gaslighting. Okay, moving on." But I think that critique really highlights how there's a difference between theorizing at a desk, versus theorizing through chores. I theorize almost exclusively through chores, where the specificity of things just hits you in the face and argues with you. Like when trying to solve a physical problem: How do I get from point A to point B? Who's in point B? Who do I need to ask to get to point B? Should I even go to point B? All these things start to happen in ways that don't necessarily happen when you're at your desk, as your brain can slide right over those questions without solving them or even considering them.

One example is that in my lab, we don't go anywhere we're not invited as scientists. In the Q&A of a lot of my talks people ask, "how do you get invited?" "how do you set up an invitation?" and, I respond, "you don't. You literally show up at work and perform your work in a good way and then people will invite you." If you seek out an invitation or give a presentation with the aim of receiving an invitation, you've assumed and reached out for access. That's a colonial move. Even if it isn't, it looks exactly like existing colonial moves because that's exactly how colonialism has happened via research, for a very, very long time. Colonialism happens through outreach: reaching out and grabbing hold. I don't reach out; people can reach in, and that's fine.

The example I give is from when I was doing work on settler lands around here on cod. I was on the radio and in the newspaper a lot, but I didn't ask to get into any Indigenous land claim areas. People knew my deal, where I was from, and then I went up to Nunatsiavut (Inuit Land Claim areas) fishing with a friend of mine who's from there. I like to think that it's not until people saw how I fished that they invited me as a researcher: when you fish all your baggage is apparent to everyone, everyone knows what you're like. How you deal with long silences. How you bait your hook. How you are with patience. How you deal with the fish once you have it, who you give fish to once you're off the boat. That's when they were like, "we can handle you," and I started getting invitations to come up for research. Of course, I have to keep earning that re-invitation, so I structure things so that there are very clear moments for me to exit and ask, "want to go again?" or, I can say "I'm on my way out" and they can say "no you aren't." It gives that option. It's how consent works: there has to be a robust and easy way to say no. That's not rocket science, except I think the presumed access to land and this idea that you can procure an invitation based on your desires, is baked into academia quite solidly so it gets hard to unbake. It becomes less obvious how consent and invitations should work.

Marc Higgins: This has me thinking about some of your models of dissemination and collaboration in terms of what it means to not only be invited into communities, but also maintain those good relations throughout, be it data analysis, dissemination, or other. When you find information that may or may not be in the best interest of communities, how do you negotiate and navigate that?

Max Liboiron: You probably won't know in advance. You have to set those up so you are told what is good, because you may think the data is great and then find out that the data is harmful, or that has to stay, it can't move, it belongs only right there. If you're not from a place, it's likely impossible to know those things yourself.

That's happened to me. The sorts of moves that I've learned through academia really don't work in community. Because I can't unlearn them all by myself (as unlearning usually takes at least another person, if not more), I have to set up all of these structures that constantly stop me because otherwise I would just plow ahead. It doesn't matter how good or ethical I am, or how many Indigenous godparents I have, I'm going to plow because I'm an academic and that's not good.

Marc Higgins: Unlearning as collective.

Sara Tolbert: You have written, "you can stand with a group without standing in their midst" (Liboiron, 2021, p. 25). Solidarity is so critical for anticolonial science as well as activist work in general. You've also written extensively elsewhere about solidarity (e.g., Liboiron, 2016). Can you say a bit more about solidarity as part of anticolonial science? What are the tensions of working within a feminist and anticolonial lab (p. 41)? What shape do and can anticolonial and feminist solidarities take within science, recognizing that this can be a murky space as well?

Max Liboiron: I've largely stopped using the word solidarity, not because I think it's a bad word, rather because it gets taken up in such a broad range of ways that I don't understand what it means anymore, sort of like "decolonization," where it's gotten haunted with all sorts of baggage and its meanings can run the gamut from "oh yeah that's really good" to "whoa that is sketchy as shit." I don't know what people mean when they use the word solidarity. Sometimes it means creeping on "diverse" people. Sometimes they use the word allyship, but it has the same baggage, so I don't know what word to use. Coalition? Same deal. I think that for a lot of people, especially in academia, where a lot of things come out of our brains instead of our chores, there is an idea that statements and intent count as solidarity. That can't be true. That's one of the reasons I've moved away from the term. People have said they're in solidarity with me before and I'm like "I don't know who you are. That seems impossible."

I think that there are quite a few models of solidarity, allyship, coming-together-ness towards a shared goal to make change, or whatever we want to call that. Part of this knowledge comes out of being in a lab that's mixed Indigenous and non-Indigenous folks and likely always will be for various historical and structural reasons. The first practice that I preach for solidarity, allyship, coming-together, or whatever we want to call it in the lab, because I think it's overlooked all the time and in science, teaching, and academia, is to *stay out of the way, systematically*. Not to rush in and join up, but to get out of our way, and help others get out of our way too. Where possible, get out of our way in an infrastructural way, not just an in-the-moment way. In the book, there's a story of where there was this gathering for Indigenous and queer folk: this one white queer lady wouldn't leave when everyone was asked to leave who wasn't Indigenous and she couldn't bring herself to stand up and leave. That was one of many, many concrete examples of "your job was to leave. It was even explicit for you." Once you're out of our way, maybe there are other things for you to do. Maybe.

With making infrastructures, in the book, there's a method called cock-blocking, where you stop people's desire to rush in, their desire to grab hold, their desire to be invited, as the dominant way of standing-with (standing-with comes from TallBear, 2014). For example, stopping your school from taking trips to the rez in anthropology, stopping your geology classes from taking rocks from Land and even bringing them back (rema-triating them), running the numbers on your awards and admissions to see where ideas of merit might be based on whiteness and settler-ism, so that you can get out of the way of certain applicants coming in for those awards. These sorts of things are extremely important, and they don't involve reaching in, they involve getting out of the way and making structures that keep folks out of our way. The other side of that is getting in the trenches. There's this great piece called *Accomplices Not Allies: Abolishing the Ally Industrial Complex* by a group called Indigenous Action (2014). It's all about "we don't want your nice words, your thumbs up, your stickers, and your flags. You need to get on the front line with us in front of the police, get your face punched the way our faces are being punched." That's accompliceship and if you're not willing to do that, then get out of the way. Another model of getting out of the way is the Two Row Wampum, which is a Haudenosaunee treaty in and around the area currently called Upstate New York. The treaty is like "okay white settler folks, you march forward in that way, and we'll be over here marching in the same direction, but our paths do not cross. We're going in the same direction, but we are separate, and we will stay separate in that pursuit." Differences is the starting point, and difference never gets dissolved.

There's this great piece by Alison Jones and Kuni Jenkins (2008) called *Rethinking Collaboration: Working the Indigene-Colonizer Hyphen* that talks about this colonial ideal of mutuality. Where, in the interest of mutuality and the progress of equality, structural differences and other differences get collapsed in the interest of shared perspective. There's a lot of violence in that collapse. Instead of prioritizing face-to-face conversations, where we come out "on the same side," they suggest taking up a politics of disappointment, ambivalence, and wtf. This maintains this incommensurability while also being on the same Land.

I think those would be three examples of styles of what you might call solidarity, standing with, or whatever that cluster of stuff is called, and you can do all of that in science and education. Staying in your lane, getting out of the lane, building a lane: all of those are things you can do.

Sara Tolbert: Do you think there's a relationship between the coming together of anticolonial and feminist work?

Max Liboiron: Sometimes, yes and no. My first lesson in the lab about incommensurability was when militant vegan feminism met Indigenous food sovereignty. That was quite shitty. On every other instance, those two movements agreed, except for the case of killing animals.

I think in any coalition, if you want to call it that, or whatever, those three techniques will appear. I think a lot of techniques should be getting out of the way, structurally, making things to get out of each other's way, mostly because it's overlooked. Even when accomplices are joining the trenches, you still have to get out of each other's way a lot. Feminisms and different Indigenous movements don't always jive. It sounds like I'm really interested in dividing different groups, but I only bring it up insistently because the dominant framework is to smoosh and smooshing can be quite violent, and always for the same groups.

I have one specific example from the lab. One of the things we do is rematriate (sometimes called repatriating) our samples, our fish and seal guts, by putting them back on the land afterwards, which is what most hunters and fishers do. Because we've got our guts from hunters and fishers, and we've interrupted the way the guts would normally be put back, we take on that responsibility and put them back like you're supposed to. For some, but not all, of the Indigenous folks in the lab, that's ceremony: there's protocol, ceremony, stuff you're supposed to think about and do, and conversations you're supposed to be having when you do it. For other folks, it's not ceremony, but everyone in the lab takes part. Some folks are doing ceremony at the same time as other folks are thinking about nutrient cycles. That is fine: at no point are white folks invited into ceremony and at no point are Indigenous folks expecting white folks to be doing ceremony. The guts get in the ground or in the water, and it is good.

Sara Tolbert: That's a really good example, as there are so many tensions as they relate to a lot of the work that's going on here under the treaty framework of Te Tiriti o Waitangi, as ethical and political obligations. Some of my Māori (and non-Māori) colleagues argue for the importance of Te Tiriti as guiding framework for all that we do, in both science and education. At the same time, other Māori colleagues have said, you know, "I don't give a shit about Te Tiriti," that regardless of Te Tiriti, this is what should be done and what should happen. This concern is also an expression of how the treaty is used as a tool for "reclaiming rights" under some sort of Western legal framework. Yet, it's a really important tool for political and economic justice for Māori in a settler-colonial society. In teacher education, we run into that as well. Being good Tangata Tiriti (treaty partner), as Tina Ngata (2020) discusses, is about active citizenship, rather than passive, but engaging in ways that don't require appropriation or that aren't violent. That's a really important point that you make that there are ways that this can happen, where it might be ceremony for Indigenous people in a particular area but it's also like yeah nutrient cycling in science.

Marc Higgins: Comments and conversations about staying in your lane as well as who is expected to do what and who's invited to do what, in particular spaces, transitions us nicely to the next question. As you state in your book, "colonialism lurks in assumptions and premises, even when we think we're doing good" (Liboiron, 2021, p. 45). In some, but definitely not all settler colonial educational contexts, there is a move from the exclusion of Indigenous ways-of-knowing and ways-of-doing towards their inclusion, a move that is increasingly being mandated by policy. For example, in the province of Alberta in Canada, it is expected that "a teacher develops and applies foundational knowledge about First Nations, Métis and Inuit for the benefit of all students" (Alberta Education, 2018). As this is a policy-mandated criterion for performance review, with the way it is worded (i.e., "foundational knowledge"), and a confluence of settler colonial desires, the statement gets misread or reduced to an understanding of having to perform Indigenous ceremony in classrooms, something that non-Indigenous students are rightfully anxious about.

Sara Tolbert: There are similar initiatives in Aotearoa New Zealand, particularly under mana ōrite of Mātauranga Māori (loosely translated as parity/equity for Māori knowledges, including in the NZ Curriculum and official Assessments). Mana ōrite is part of a larger Māori-led anticolonial epistemic justice efforts to (re)position Māori knowledge as valid—and so that Māori students (the majority of whom attend kura auraki ["English-medium schools"] vs kura kaupapa Māori) can be/learn as Māori, even in settler colonial institutions.

Marc Higgins: However, colonialism lingers and lurks even in, perhaps especially in, efforts to work against settler colonial logics: the ways in which Indigenous knowledge systems are included often differently reproduce the settler colonial norms and practices through which Indigeneity was excluded in the first place. What might an anticolonial science offer to think and work through what we see as, after Tuck and Yang (2012), "settler moves to innocence," practices "that problematically attempt to reconcile settler guilt and complicity, and rescue settler futurity" (p. 1)? Any words of advice on rendering transparent and troubling settler colonial complexities, complicities, and compromises?

Max Liboiron: And in 10 words or less? No problem. What you're talking about is what we call "inclusion into empire." It's a colonial form of inclusion where everyone gets more access to Indigenous stuff, like you said, "for the benefit of all students," but that's just reaching into Indigenous life, knowledge, and land to benefit mostly settler students. We become an enrichment tool, which is what we've always been for settlers. It doesn't have to be that way but that's usually how it plays out, again because it's the dominant model. That's why we critique "add Indigenous and stir": no matter how awesome the content, if it goes into a container based on the dominant system where colonial relations make the most and only sense, then it's going to go sour.

When I give talks about anticolonial methodologies, we first talk about what colonialism is and how it's different from just casually being an asshole: it's very specific. In certain colonial contexts, like Canada, the United States, and other settled territories, it means non-Indigenous access to Indigenous life and land for the benefit of non-Indigenous folks. That's the bumper sticker version. So anticolonial means *not that*: just don't do that. The most straightforward point of anticolonial work is to interrupt access, the presumed access of non-Indigenous folks to Indigenous stuff.

Just don't. Refusal is a core technique for anticolonialism. So instead of teaching "here's what we know about Métis people," which will inevitably be super racist because that's how we've been defined by the settler state in terms of blood quantum, try "let's learn about how the settler state defines Indigenous people and let's look at the Indian Act in Canada so we can identify on our own how fucked up it is." So, studying up to colonialism, instead of down to Indigenous-ness.

I think a lot of white settler and other settler teachers feel uncomfortable because they know something's hella creepy about the mandate to access Indigenous life and knowledge, and they are right: the gut is a very smart place to put your brain. So, going with your gut, knowing something is weird, interrupts that access, that reaching in, that creepy reaching in that's based on desire, and instead look up to the structures, because you can properly teach that. If you're a math teacher, let's look at the ways that colonialism and math go together, and also how Indigenous people use that same math to resist the state which isn't about Indigenous-ness, it's still about the state. Let's look at Diane Nelson's (2015) *Who Counts*, on math systems and genocide or Walter and Andersen's (2013) *Indigenous Statistics*.

I think studying up, punching up, or whatever you want to call it, plus refusal are key moves to make here. This is part of the several reasons why I talk about anticolonial methods rather than decolonial methods: I'm from the Indigenous tradition that says decolonization means Land Back. Full stop. Rematriation of land and life, which are the same thing. That's the full list. But refusing to grease the wheels of colonialism is something else, and is what you might call anticolonialism and that can happen all the time, at many more scales, by many people.

Marc Higgins: That's a very rich invitation to take up. There are a lot of students in science teacher methods that respond well to that, who recognize that a part of the responsibility, especially if they are settlers, and particularly white settlers, is to do that work. To take seriously your work, Kim TallBear's (e.g., 2013), or others who are working against settler colonial systems, in thinking about the day-to-day practice, there's a whole constellation of moments in which, via either the language of policy, how policies are implemented, teaching, teacher education, etc. There are many instances in which those colonial systems continue to operate. The wheels continue to be greased. That's a really rich metaphor to think with and through. Thank you.

Max Liboiron: I want to mention the Jones and Jenkins (2008) piece again: there's a really great moment in there. It's about a white settler and a Māori collaboration. There's one moment where they're asked to teach a class together. It becomes quickly apparent that teaching a unit on Māori history that the settler students want to learn from the Māori students, but the Māori students are not interested in giving that education or learning from their settler peers about their own history for obvious reasons. So, they split the class. That has very good ramifications for the ethics of learning, even though everyone's "oh my God, segregation" based on this idea that inclusion is inherently good; segregation is inherently bad. But it's more like the Two Row Wampum form of being together by staying apart. That really gets to the better ethics. Even though they still had to teach Māori history, they managed not to grease the wheels of colonialism. I thought that was a really smart move on their part as a way to deal with that curriculum requirement. They did great work.

Sara Tolbert: This is a great pivot to think about this move from awareness to infrastructure. On the *Discard Studies* blog, you've described how:

> Despite its ubiquity as a campaign goal, awareness does very little to create change ... The premise of awareness campaigns is that individuals are the best unit for change. The individualization of action is a way to fragment it, slow it down, and redirect it to ineffective routes ... Instead, there are different loci for change that scale better, including but not restricted to infrastructure. (Liboiron, 2014, p. 2)

In your book, you also address relationships between awareness and infrastructure. I've used that *Discard Studies* blog piece with early childhood and primary pre-service teachers. It's interesting for them to reflect on their own neo-liberal subjectivities because they feel a lot of guilt around not doing enough as individuals to engage with these big problems. It's been really helpful for them to take a step back and think about what the best locus is for change. I see this move from awareness to infrastructure as one that has implications for anticolonial science, but also pedagogy and activism—perhaps as part of a pedagogy of making and doing? Do you see these as related as well, or do you see implications for pedagogy in the move from awareness toward infrastructure, and do you talk about that with your own students?

Max Liboiron: I also assign my students that reading! Whenever we have final projects that are about social, cultural, whatever issues and change-making, they always reach for awareness, because that's the well-greased wheel. But I ask them "are you aware that there was a genocide of Indigenous people here? That we're on ground zero of one of the most successful genocides in history, here on the island of Newfoundland? Are you aware that that's still going on with missing and murdered Indigenous women and girls?" and they say "yup." So, they're already fully aware, some know the numbers, some know the dates. Teaching more on that doesn't make change. Also, teaching to awareness almost always involves deficit narratives or damage-centered research (Tuck, 2009). It means rehashing the crappiest crap. Eve Tuck (2009) and others have pointed out that it causes harm to Indigenous people: it makes us inherently kill-able, killed, rapeable, genocideable, suffering, and we internalize that. Especially if you're an Indigenous student sitting in the classroom and the lesson is theorizing the death of you and your people: that's a rough ride. So, what if we skip that, because most people are generally aware? If they need some pointers, great, we can do that on the way, but people are aware. Instead, let's think about how change actually happens.

First of all, I think if my students hear me say "infrastructural theory of change" one more time in class they might riot, and rioting is becoming-infrastructure so that would be fine. I think that working with the class to develop really well-developed ideas of how change actually happens is important. Like, how did something that was normal at point A get uprooted, become abnormal, and a new normal take its place by point B? That is not an easy question to answer. It's never universal and is very context dependent. "How did it happen?" is what we spend our time in my classes researching and then trying to leverage those insights into new spaces. I think teaching about how change actually does and does not happen is way more important than teaching about the things that need to change, because students are aware, down to their bones. If I asked them to pitch a final project about changing the world on the first day of classes, they can pitch it: they know what the problems are. But they don't know how things got that way, or how historically, and in other contexts, change on those same topics has happened. They're always surprised to learn that it's not policy, it's not charismatic leaders, it's not *one* thing, and that defining the problem is maybe the hardest part.

I teach this piece by Rittel and Weber from 1973 called *Dilemmas in a General Theory of Planning*. It's a piece by these two white engineers in the 1970s during the civil rights movement in the United States where they realize, "wait a minute, wait a minute, we're such good engineers and we've been designing these civic things for good social outcomes for ages, but all the Black people are rioting, so clearly we suck at this." Then they come to realize that all of their engineering problems they've been solving are actually social problems and they've been fundamentally using the wrong way to define their problems. They come up with this idea called wicked problems. Then they work really hard at describing ten specific impossibilities about solving those sorts of problems. Every semester, the students say it's the most helpful reading, because it helps them articulate the feelings that already tell them that the silver bullet won't work. It helps them know which direction they might point their nose in given that failure. This is more doing-oriented and more helpful than a lot of these other blue-sky pieces or critiques that we sometimes read, or that they get in their other classes. A pedagogy of impossibility and how change actually happens are the only things I teach, over and over.

I think it relates to the first question: all of my classes don't have any final exams, only final projects because it's in the project-*doing*, the chores, that the specifics of the problems start to become apparent enough that you start to understand its wickedness. That is the only way to deal with it, to do your unlearning. None of their projects ever work and that is the point: that they can precisely describe why their projects won't work, and it's mostly because it was based on awareness, individual action, or something like that that they come to realize doesn't deal with the wickedness of the problem.

Sara Tolbert: We've talked about how there's such an emphasis in our context of teacher education on getting things right. A lot of it's about getting it wrong and coming to terms with that. You write so explicitly and eloquently about that: the complexity and messiness of theorizing through doing chores. Which leads to our next question, something that comes up all the time in this work, particularly as a white settler: the phenomenon of white guilt, coming to terms with it, and the violence of inappropriate tears. You write in your book about one of these events and state that "great, big, charismatic, obvious, event-based, well-witnessed mistakes are gifts to the lab if we work through them … [providing] insights into the ways colonialism orients us to relations, regardless of intent" (Liboiron, 2021, p. 69). Part of anticolonial praxis then is about naming when and how we make mistakes, as well as working through those mistakes—again, regardless of intent—but with attention to our different roles and responsibilities in constructing alter-relations, as you mention in your book. I see this as an ongoing tension in a context where we're asked to, but refuse to, prepare teachers who are "perfectly ready" to go on day one in real classrooms because that's an impossibility. As a teacher educator, I don't do that: I'm not "perfectly ready" to teach pre-service teachers. How, or do you see these big mistakes as part of an "ethic of incommensurability within anticolonial science" (p. 137) (or anticolonial pedagogy), such that we fuck things up because "science is always already fucked up, which means that our work is already always compromised" (p. 20–21)? Education is the same: it's always already fucked up.

Marc Higgins: We have a particular intersection between these two worlds.

Max Liboiron: We could be doing an interview on nursing or social work or something else; it would be an equally fucked up intersection. The reason that I give the example of the great big charismatic, obvious, easy-to-point-to, everyone can agree that was a mistake in the book, is because colonialism is so dominant that it has become so normalized and naturalized that it can be really hard to point to. It is such a gift when someone makes it easy to point to through a big mistake. We make sure that in the lab that it is articulated and demonstrated as such. Everyone in the lab, when they come in for the first time, they nod when we say mistakes are okay. But no one believes us because they've been employed before where that was stated but then they're threatened with being fired or let go when they make their first mistake.

One thing that was really important for the lab book, which is our guiding social and collective document, is to have an apology section. Because I and everyone else in the lab didn't know how to make a proper apology for the mistakes we knew we were going to make, I knew they were going to make, and I knew I was going to make. A lot of apologies are like "sorry, I won't do it again"—which is the worst! Sorry for what? And it recenters intent as a good theory of change. Yikes. That's exactly what I teach against: the foregrounding of intent. Informed by transformative justice, there are some more steps to apology, including identifying the specific type of harm you caused. In the lab, doing a proper apology often leads being told that the type of harm you thought you were engaged in isn't at issue, and some other form of harm is at stake. That is incredibly valuable. Often people will apologize to me when we meet because they're taking up my time. That is not harm, folks. What do you think my job is? My job is to teach you and give you my time. That is not the harm, that is my job, and what I get paid for. Coming to realize that actually "oh, I damaged partner relations, even though I've never met the partners, by not working my hours and not telling anyone I wasn't working them" as opposed to "you lost money or my time."

Now they know about the harm, the next step is to articulate how they're going to deal with that in concrete terms, beyond intention. That almost always requires talking to people as well as figuring out and learning from how different parts of the lab work or don't work in ways that they might have not been aware of before, such as who's in charge of what. You get out of guilt pretty fast because you're now doing chores to finish your apology. Then there's agreement to do different next time. Through that process, there's usually a coming together.

The example in the lab is the one student who came into CLEAR as a "field site," a term that I really dislike because it's always someone's home, including the lab in a way. So, they came in as an anthropologist, and started studying us before they had our formal consent to study us: gathering data and thinking they're so good at spying on us. When it was clear that they needed collective consent first, they had to apologize to us for stealing our data before we consented to it. That process of them apologizing, having feelings come out on the table, how embarrassed they were, realizing it had to do with anthropology as a discipline, and then learning about data sovereignty as a way to deal with that disciplinary colonialism, resulted in everyone in the lab being like, "you're cool." That person got ingrained in the lab much faster by making that mistake than if they had not made the mistake in the first place. Mistakes are really good as it's really hard to see the ground you stand on without other people and without making mistakes. I would say it's impossible without mistakes. Learning how to fail gracefully through a collective process is a huge gift, and one you can craft when you have jurisdiction over the space as a lab manager, as a principal investigator, as a teacher in control of a classroom. It has to be structural—it can't rise out of your students by itself. It's important to structurally support people to fail gracefully because it's not a skill most people have, and it is a life skill, regardless of whether you're trying to do anticolonial lab work or if you're trying to get married. It works in a lot of different contexts.

Marc Higgins: Thank you for your time, your energy, your scholarship, and particularly this conversation around everyday anticolonial science practices.

Sara Tolbert: Many thanks for joining us, Max—very grateful for your time, expertise, and contributions to this chapter.

REFERENCES

Alberta Education. (2018). *Teacher quality standard.* https://open.alberta.ca/dataset/4596e0e5-bcad-4e93-a1fb-dad8e2b800d6/resource/75e96af5-8fad-4807-b99a-f12e26d15d9f/download/edc-alberta-education-teaching-quality-standard-2018-01-17.pdf

Davis, H., & Todd, Z. (2017). On the importance of a date, or, decolonizing the Anthropocene. *ACME: An International Journal for Critical Geographies, 16*(4), 761–780.

Heglar, M. (2019). Climate change isn't the first existential threat. *Medium.* https://zora.medium.com/sorry-yall-but-climate-change-ain-t-the-first-existential-threat-b3c999267aa0

Indigenous Action. (2014). Accomplices not allies: Abolishing the ally industrial complex. *Indigenous Action Media, 2*, 2.

Jones, A., & Jenkins, K. (2008). Rethinking collaboration: Working the indigene-colonizer hyphen. In N. Denzin, Y. Lincoln, & L. T. Smith (Eds.), *Handbook of critical and Indigenous methodologies* (pp. 471–486). Sage.

Kuokkanen, R. (2010). The responsibility of the academy: A call for doing home-work. *Journal of Curriculum Theorizing, 26*(3), 61–74.

Kuokkanen, R. J. (2007). *Reshaping the university: Responsibility, Indigenous epistemes, and the logic of the gift.* UBC Press.

Kuokkanen, R. J. (2008). What is hospitality in the academy? Epistemic ignorance and the (im)possible gift. *Review of Education, Pedagogy, and Cultural Studies, 30*(1), 60–82.

la paperson. (2017). *A third university is possible.* University of Minnesota Press.

Liboiron, M. (2014). Against awareness, for scale: Garbage is infrastructure, not behavior. *Discard Studies.* https://discardstudies.com/2014/01/23/against-awareness-for-scale-garbage-is-infrastructure-not-behavior/

Liboiron, M. (2016). Care and solidarity are conditions for interventionist research. *Engaging Science, Technology, and Society, 2*, 67–72.

Liboiron, M. (2019). Anti-colonial science & the ubiquity of plastic. *Frankinterviews.* http://frank.rngr.org/interviews/326/anti-colonial-science-the-ubiquity-of-plastic

Liboiron, M. (2021). *Pollution is colonialism.* Duke University Press.

Liboiron, M., Tironi, M., & Calvillo, N. (2018). Toxic politics: Acting in a permanently polluted world. *Social Studies of Science, 48*(3), 331–349.

Nelson, D. M. (2015). *Who Counts?: The Mathematics of Death and Life After Genocide.* Duke University Press.

Ngata, T. (2020). What's required from Tangata Tiriti. *Tinangata.com.* https://tinangata.com/2020/12/20/whats-required-from-tangata-tiriti/

Rittel, H. W., & Webber, M. M. (1973). Dilemmas in a general theory of planning. *Policy Sciences, 4*(2), 155–169.

TallBear, K. (2013). *Native American DNA: Tribal belonging and the false promise of genetic belonging.* University of Minnesota Press.

TallBear, K. (2014). Standing with and speaking as faith: A feminist-Indigenous approach to inquiry. *Journal of Research Practice, 10*(2), Article N17.

Todd, Z. (2018). Refracting the state through human-fish relations. *Decolonization: Indigeneity, Education & Society, 7*(1), 60–75.

Tuck, E. (2009). Suspending damage: A letter to communities. *Harvard Educational Review, 79*(3), 409–428.

Tuck, E., & Yang, W. (2012). Decolonization is not a metaphor. *Decolonization: Indigeneity, Education & Society, 1*(1), 1–40.

Walter, M., & Andersen, C. (2013). *Indigenous statistics: a quantitative research methodology.* Left Coast Press.

Whyte, K. (2016a). Indigenous experience, environmental justice and settler colonialism. *SSRN Scholarly Paper.* https://doi.org/10.2139/ssrn.2770058

Whyte, K. P. (2016b). Is it colonial déjà vu? Indigenous peoples and climate injustice. In J. Adamson & M. Davis (Eds.), *Humanities for the environment: Integrating knowledge, forging new constellations of practice* (pp. 102–119). Routledge.

CHAPTER 20

In Conversation with Isabelle Stengers: Ontological Politics in Catastrophic Times

Isabelle Stengers, Marc Higgins, and Maria Wallace

ON NAMING THE ANTHROPOCENE OTHERWISE

To name is not to say what is true but to confer on what is named the power to make us feel and think in the mode that the name calls for. In this instance it is a matter of resisting the temptation to reduce what makes for an event, what calls us into question, to a simple "problem." But it is also to make the difference between the question that is imposed and the response to create exist. (Stengers, 2015, p. 43, emphasis in original)

I. Stengers (✉)
Université Libre de Bruxelles, Brussels, Belgium
e-mail: Stengers.Isabelle@ulb.be

M. Higgins
University of Alberta, Edmonton, AB, Canada
e-mail: marc1@ualberta.ca

M. Wallace
University of Southern Mississippi, Hattiesburg, MS, USA
e-mail: Maria.Wallace@usm.edu

© The Author(s) 2024
S. Tolbert et al. (eds.), *Reimagining Science Education in the Anthropocene, Volume 2*, Palgrave Studies in Education and the Environment, https://doi.org/10.1007/978-3-031-35430-4_20

Marc Higgins: Thank you for agreeing to participate in this conversation. Part of the intent of the two-volume collection that is *Reimagining Science Education in the Anthropocene* is to bring together folks to speak from either within science education, to science education, or from adjacent but highly relevant locations around the question of what it means to teach in response to, within, and to trouble the Anthropocene.

However, not unlike other scholars with whom we have had conversations, you put to work another set of signifiers to engage these types of questions. Can you speak to why you do not engage on these terms and what terms you employ instead (e.g., the intrusion of Gaia)?

Isabelle Stengers: I don't use it for reasons related to both "Anthropos" and "cene," which allude to geology. As Donna Haraway wrote, if we ever want to use the term "geology," it would be more accurate to associate it to the brief transition marking the end of the Cretaceous Era. That may take a couple of centuries in our case, but on a geological scale, a few centuries are only a very, very thin layer. So, there is nothing to justify calling it an era in the geological sense, even if we extend it to the nineteenth century or beyond.

So, I don't see how geology is going to help us unless it is to actually glorify the capacity of Man, Man with a capital M, to become similar to and to have the same power as a geological force. However, there is nothing to glorify here. All traces from Anthropocene we are supposed to leave behind for future geologists that we don't and cannot know anything about—we rather imagine ourselves as geologists of the future saying, "Oh, it's not nature that intervened here"—all those traces are traces of non-sustainability, traces of lack of attention, of imagination. The description the future geologists will give of this Anthropos won't be too flattering. They really made a mess of their world. However, we are told that this same Anthropos, who has now become aware of its power, is going to use it to fabricate a beautiful Earth, a "good" Anthropocene as some say. But what this prospect ignores is that awareness is quite insufficient. There is no symmetry between destroying and repairing. While it is easy to deconstruct and destroy, it is very difficult to reconstruct, rebuild, and regenerate. The logic behind the notion that whoever had the power to destroy will have the power to repair and do better, it's completely ridiculous. It takes time to fix whereas we often destroy without even knowing what it is we are destroying. We need to have deep and discerning knowledge of what was broken in order to try repairing it, and it needs time and care. In a nutshell, the Anthropos of the Anthropocene, those humans—not the

Human—who took possession of the world, who colonized people and defined whatever they could as resources are not fit to repair.

Other names were suggested. I obviously favor Capitalocene since we can say that capitalism was born—and I'll come back to that—together with the beginning of colonization. But the problem is that it marks the historical causes of what happened but is mute about what to do now. That's why I personally chose to name it the *Intrusion of Gaia*. She is intruding like someone we weren't expecting, like someone we, myself included, would have preferred seeing later, maybe when we were better able to confront capitalism, or at least that she gave us more time in order to try and address our issues than the very few decades Gaia seems willing to give us. Gaia is not really an intruder; we, all those who blessed the power to reconfigure the world which is called modernity, did "awaken" her. But now that she is awakened, we cannot ignore her, we cannot send her away, we cannot put her back to sleep. Indeed, the Intrusion of Gaia also indicates that what we are involved in today is not a crisis from which we can escape; from now on we will need to deal with this Gaia, which I qualify as being sensitive [chatouilleuse] and unstable. The Intrusion of Gaia is a name that puts the emphasis on a present which is only just beginning. Dreams of mastery are over for good.

Further, I chose the name Gaia because it is the crossing between two temporalities. First, there is the temporality associated with sciences and technique. The scientist who first used this name was James Lovelock, who recently [in July 2022] passed away. He was the first one to think that the Earth should be treated as a being with a history, not just an evolution like any planet. Any planet's climate may be associated with a system of physio-chemical variables, but Lovelock did show that Earth was anomalous if considered from that point of view, and he related this anomaly with its being inhabited by living beings, the activity of which is part of its definition. Earth has been kept able to sustain life because living beings are ongoingly keeping it so. But since Lovelock, we also know that some of these living beings—Anthropos—have been able unknowingly to destabilize Gaia. The current climate regime we associate with the Holocene could be destroyed. Thus, it is a modern scientific and technical development which enabled us to understand what was looming and threatened us and many many other beings with us. And so, it's our scientific world who enables us to say, "This is what is going on and this is what we can expect." But Gaia is also a goddess from the past, a past in which she was the goddess of people living in rural environments, agricultural environments and strongly depended on her. She was someone who should not be offended. She didn't have "human" emotions like pity or a sense of justice but demanded or required attention and respect to remain propitious. When offended, She was not a retaliator, punishing the culprits. Everyone would pay for the offense. That is exactly what is going to happen now. Gaia makes no difference. Animals and plants will be victims as well as the poor and those who had a very small role to play in the climate upheaval. They are even the first ones to pay a very, very high price, rather than the (rich and offensive) Anthropos.

ANOTHER SCIENCE IS POSSIBLE: TROUBLING *FAST SCIENCE* IN TROUBLING TIMES

> Slow science is not about scientists taking full account of the messy complications of the world, it is about them facing up to the challenge of developing a collective awareness of the particularly selective character of their own thought-style. (Stengers, 2018a, p. 100)

Marc Higgins: Within the scienceeducation world, your manifesto for slow *science* is gaining traction as scholars and practitioners are (re)envisioning what and how they engage with science. For readers unfamiliar with this work, could you offer a quick orientation to how you problematize *fast science* and its "imperative not to slow down" (Stengers, 2018a, p. 115)?

This is particularly relevant given the ways in which science education is caught up in similar modes of being *fast*: "working in a very rarefied environment, and environment divided into allies who matter and those who, whatever their concerns and protests, have to recognize that they ... should not disturb the progress of science" (Stengers, 2018a, p. 116, emphasis mine). Significantly, how might *fast science* be caught up in these *catastrophic times?* (e.g., sleepwalking science).

Isabelle Stengers: First, I am going to tell you that when I problematize fast science, I am problematizing something that started in the nineteenth century. In the present, science has become very fast—and we'll get back to that—but its speed is partly forced on it. However, fast science was already operationalized in the nineteenth century as its normal, desirable way of progressing. In turn, when I talk about science needing to slow down, I don't mean that we have to go back in time before fastness was forced on it—"when science was respected," as scientists complain. Rather, it's about attending to what this fast science invented in the nineteenth century has destroyed.

Fast science is the one which we most often associate with "progress made in science": physics, chemistry, the fields that evolve, that unassailably incessantly teach us something new, became professionalized during the nineteenth century, and actually gained autonomy against those powers, Industry and the State, which they needed for sustenance. I am particularly thinking about chemistry, but also physics. Chemistry provided industry, and soon the state, a scientifically skilled workforce as well as protocols and the means for a very rapid development and received from it instruments and purified, standardized products needed by their protocols. Chemists are the co-inventor of the network of chemical industries born in the nineteenth century, but they had to struggle not to be absorbed and forced to concentrate on the questions imposed on them by the very industries they helped develop.

Indeed, scientists, and rightly so, knew that their fast advance was conditioned by their freedom to choose the "right" questions, those they might be able to answer. Which is to say, it was not an advance that was brought by a method, a rational, general way to address a problem and solve it. It is not a "We know how," rather a "Here we could ..." Progress, the advancement of knowledge, depended on not being burdened by questions which researchers did not feel liable to contribute to the progress of their science, at least given the state of their art. While I understand their position, the price they had to pay was their correlative definition of the advancement of knowledge as their exclusive vocation. They considered all the issues that weren't contributing to the advancement of their science, as "non-scientific." Scientists should not waste time on questions of common interest, such as the use of what they provided. They should instead trust that advancement of knowledge was the royal road for human progress.

What was produced is what I call a mobilization model. A mobilized army must be disciplined, it must advance at all costs, nothing must slow it down; those who doubt are potential betrayers as they might activate insubordination. And even today, when you speak to a scientist and question the concept of advancement of knowledge as relevant in today's situation, it is as if you had said something completely absurd or catastrophic. It is as if progress in science was the real vocation of mankind, of the Anthropos, and that nothing should disturb that sacrosanct undertaking.

So that means that this particular science that we inherited cannot be slowed down by any question that wouldn't contribute to its progress, that is, by any question concerning its meaning or its consequences. It demands and justifies a practice of irresponsibility. That doesn't mean that science is not interested in creating alliances which will bring "non-scientific" consequences to what it offers, but it wants to be able to claim that it's for the benefit of mankind, that everything will be alright and that if everything is not all right, it is none of its concern or fault. Only the principle of precaution attempted to restrict this optimistic blindness. For a lot of scientists, the principle of precaution, and particularly the idea that scientists must waste their time thinking about consequences, is an attack against progress.

So, the fast science born in the nineteenth century is a disciplinary practice that is mobilized such that asking questions other than those which will be interesting to colleagues, which are the truly scientific questions, is a waste of time. And wasting time is a sin, or else it is for retired scientists, etc. But otherwise, if a student asks questions of this kind—I used to be that type of student, that's why I turned to philosophy after completing my studies in chemistry and understanding that the questions I felt in need of asking would be answered by "You shouldn't be in science." Not wasting one's time has become the trademark of the scientific mind. Scientists present themselves as the goose that laid the golden eggs: "Let us lay our eggs in peace, and you'll turn them into gold." Thus, "Do not impose upon us; you would kill us if you force questions onto us."

Returning to the Intrusion of Gaia, it has been what I first thought when confronting: if we have very powerful scientific techniques, but nevertheless we are dramatically ill-equipped to face the problem at hand. Why? Because we face the consequences and do not trust progress to save us. And the situation is all the more dramatic that one can say that neo-liberal capitalism has decided that it did not need the goose with the golden egg after all. Autonomy was a privilege of the past, and scientists should be forced to concentrate on innovations contributing to economic growth. In a knowledge-driven economy, knowledge must produce patentable, marketable results which do not need "good questions." Scientists complain, and some have turned to the theme of "slow" science: "We need time to think instead of competing with each other in a rat-race for who will offer a promising innovation. Give us time to think if you wish 'good science.'" Which is very nice but sounds very much with the plea: "Treat us like the goose that lays the golden eggs."

This is not at all what I mean by slow science. I can understand the plea, I can feel sorry for the disarray of researchers forced to do substandard, sloppy research or even cheat. But "slow science" as I intend it demands something else. It is a science whose glory is not the advancement of knowledge, but rather must be slow because paying attention to and connecting with concerns is often disregarded as "non-scientific." This is a slow, time-consuming, but necessary task for today.

The slowing down of science needs struggling against all the concepts that we are used to teaching to students: for instance, the scientific mind [l'esprit scientifique], which defines itself in opposition to people's common sense, reduced to mere opinion. Gaston Bachelard stated that opinion is always wrong, and if it happens to be right, it will be for bad reasons. He also said that society should be adapted to the school and not the school adapted to society. So, he supported fast science, those sciences that create their questions but who also ask that all else be ignored, that we ignore the world in which they are produced as well as how the answers to those questions will affect this world.

Students and young researchers are inculcated with the idea that if they get interested in questions that don't contribute to progress, they are dead to science. They are asked to mute their imagination of the world they are living in. It takes time to develop imagination. It takes time to learn how to seriously consider matters we don't know anything about and to learn how they are nevertheless connected with what we know. Slow science, as we consider teaching future young scientists and students, is a science of not expecting right answers to questions from science but to be aware that in all fields where common interest matters, the answer will be the right answer only if it made matter the particularities of the situation in which it is going to intervene. In other words, not to dream of a rarefied situation, when the imagination of the scientist is free to ignore or neglect, but "this" situation that comprises all those concerned by it. They have to listen to sad or dark stories when scientists did not manage to talk to, to hear, and to take seriously those who are saying, "Okay. But did you think that it would have such consequences?" And too late they recognize, "Oh. No, I didn't think about that."

In my opinion, the issue of GMOs constituted a very good lesson for slowing down in Europe (perhaps not in Canada where they grow everywhere). In Europe, we witnessed a resolute refusal, for many different reasons which, in their convergence, generated a profoundly clever opposition. It had the specialists and the experts stumbling. The refusal brought in areas officially not considered related to science: for example, the economic and social impact of patents on agriculture. Or it was a science that was too naturalistic: gene transfers can't be carried out in a laboratory, etc. Basically, everything that didn't concern fast science, the science taking place in a laboratory where scientists take charge of the conditions of asking and answering their question, came into the picture and generated an environment where GMOs were stuck with their consequences. This matter of GMOs happened to be the beginning of many changes that took place in Europe at the beginning of the twentieth century. Agriculture has become a matter of public interest. It also taught me a lot, and both the theme of slow science and the necessity to empower "non-scientific" voices originated from this experience. These voices showed me that it was necessary for scientists to nurture an open imagination toward concrete situations where their knowledge pretends to intervene, pretends to make a difference.

Another science, slow science, is possible, that would not define situations but rather irrigate them, bringing them not water, which is essential, but propositions that might be relevant without providing solutions deemed adequate whenever those concerned by "this" situation object. Those who would say it is not possible are those who state that a scientist who takes interest in those issues will no longer be a scientist. No, they will be a different type of scientist, trained and assessed otherwise, living in a different ecology. The current ecological environment in which science functions is one that is desertified: science has a true relationship only with the state and industry; the public is supposed to keep its distance. Slow sciences would be sciences that would require the empowerment of concerned individuals and groups as active interlocutors. Not an anonymous public, but rather people who know something significant about the situation in question, whether they are scientists or not, coming from other fields of science, or without formal training. Really, I would say slow science "is not impossible," but its possibility is of the same order as that of answering in a non-catastrophic manner the intrusive challenge of Gaia.

Today this possibility becomes more concrete. In France, we see agronomists and engineers from Polytechniques who, at the end of their studies, talk publicly to others, announcing that they are going to branch off from, or that they are abandoning their field and their prospects for a career. Which is to say, they know that the studies which they succeeded in and which provided them with a career path, did not equip them to confront what-is-to-come. Those public announcements of divergence and deserting have made some noise in France.

There is a pressure on our institutions, and it will become stronger because the situation we are all facing is generating pressure. It first affects young people who know how much what we are teaching them does not meet the demands of this era. But it will also affect teachers. If for yesterday's scientists, slow science was but an incongruity, a philosopher's dream, a possibility, it is now becoming an insistent stranger.

Marc Higgins: That is very interesting. In my science teacher education classes, sometimes the discussions are as follows: "What do we do about catastrophic times or the issues of injustice that science is linked to?" Those are not issues addressed in science textbooks, who rarely ask, "what is science's responsibility?" when we teach young people scientific concepts and practices, like how to balance equations in chemistry or consider momentum transfers in physics, etc.

Isabelle Stengers: We rather teach them to understand what it means to become fast scientists, through the telling of science as a success story. If we taught science with an element of adventure which comes in locating it within the world which produced it as well as in discussing what it produced in the world, students would be more demanding than those who want fast science training. They would ask to be better equipped for what-is-to-come. That is why I am happy that the field of science education takes interest in this.

Slowing down science is not a concept requiring a better pedagogy about how to balance chemical equations because it primarily requires a change of practice. It is a transformation of the institution which celebrates chemical equations as something everybody needs to be aware of. Obviously, this change will have different consequences for different scientific fields, but one way or the other, the consequences it will have for one scientific field will concern the other fields.

What does it mean to think? This is what Bruno Latour calls "coming back to Earth." What does it mean to ground thinking in Earth's contingencies, rather than dreaming of escaping its eventualities? What does it mean to come back to Earth? I believe that the new generation, who will face things that I cannot imagine, deserves to be equipped. Eventually, there will be a slowdown in innovation, but what threatens us is precisely the idea that there will be innovative, undreamed-of solutions, that Anthropos will find, must find solutions.

ONTOLOGICAL POLITICS: NAMING SCIENCE'S
EPISTEMOLOGICAL PRESUPPOSITIONS

Modern practitioners are those who belong, whether they like it or not, not only to the fossil-burning world that bears responsibility for the trouble but also to a world able to formulate the problem, define what is globally at stake, and conclude that unanimous mobilization is necessary, whatever our divergences ... [with] peoples all over the Earth [who] are already affected in specific ways. (Stengers, 2018b, p. 97)

Marc Higgins: A significant way that you lay out that scientists can "fac[e] up to the challenge of developing a collective awareness of the particularly selective character of their own thought-style" (Stengers, 2018a, p. 100) is to acknowledge, and work within and against, science's specific relationship to ontology: one of *capture by testator* (or Modest Witness). Could you further expand upon the epistemological presuppositions that are habitually made operational within scientific practice? What are some of the consequences of "the specific way in which [scientific] practitioners world and word their world?" (Stengers, 2018b, p. 91).

Isabelle Stengers: Testators are those who defend Science with a big S against this collective awareness. We hear a lot about the ontological turn, and it has many significations. My view is there is no ontology that would be associated with modernity, or Science, if we take "ontology" in the strong sense, as implying that we think of the world. Science does not think of the world at all. It entertains an image of thought as progress, which produces a vision of a world as needing to accept progress. Thus, whatever we may call our ontology does not constrain us.

I further associate what we call ontology for other peoples to their way of thinking, of being in and belonging to the world, of making a world with others. So, ontologies correspond to practical constraints and concerns this world is imposing on them, what it is asking from them. Whereas it seems to me that if there is one dimension of modernity that colonized peoples know well, and also Europe, it is that it is a world-destroying machine. So, it is a machine that destroys any ontology, whatever it is. A reasoning such as, "We defined nature as being exploitable" does not relate to ontology. It is the destruction of all those who, in the past and even today, are preoccupied with what a healthy forest needs, or what animals are asking for, or what ancestors, or future generations demand, etc. Modernity is a machine that destroys such concerns.

So, if you seriously consider worlds rather than ontology, these worlds are populated in different manners, and the question becomes a political one. It is the one raised by the Zapatistas: "is it possible to create a world in which many worlds fit?" [*Un Mundo Donde Quepan Muchos Mundos*]. But our [modern] world won't be able to fit, or to cohabit with others because its only relation to others is either one of exploitation, or of patrimonialism and tourism: we'll go see the Indigenous dance; we'll go sightseeing, etc., so we preserve it for our enjoyment. Just as we try to preserve some rural landscapes and folkloric events. Because the process of devastation, the destruction of our own "native" ontologies started precisely when colonization started its destruction elsewhere.

Effectively, the testator is one figuration of this destructive machine. Why? In fact, the historical testator is actually the ancestor of the chemist. His role was, on behalf of a prince who had paid alchemists to turn lead into gold, to analyze and conclude whether what was produced by the alchemists was real gold or not. What interest me are the modern testators and the way they test scientific propositions in the name of true Science. They reject the plurality of scientific practices—I'll come back to that—the plurality of what those practices address, and what they demand for this address to be relevant. What they reject indeed is what could be a pathway for the activation of ontological concerns. Sciences in the strong plural may be ways to learn how they are situated by what they address, or to explore what is demanded from them by what they address. Plurality as both anticipated and worked up by sciences could be a way for many worlds to co-exist.

The modern testator asks but only one question: has this been proven? The testator requires a proof; whatever is not proven is but noise, an opinion, and is worthless. In the same way that the testator of the past would say that an alchemist was a fraud if he didn't produce real gold, the modern testator would now say of something not proven that it is an opinion, it is subjective, and we don't need to take it into account. Anything and everything that has value has been proven.

That's what makes the testator part of the destructive machine. Of course, some specific sciences provide evidence: those are experimental sciences, in the strong sense of the term. That is, the sciences that can transplant what they are researching into controlled environments such as laboratories without destroying them and find out which question it brings an answer to in a reliable way. When they succeed, and if they succeed, they have indeed turned what they address into reliable witnesses, liable to participate in a proof. Which is to say, they will be able to produce claims which will pass the test of colleagues who will have to accept that they are not able to reduce this claim to a fiction, as they are not able to interpret the evidence otherwise. This is what is called the successful production of an "experimental fact."

The first one who conducted this kind of science was Galileo:. not the one who found out about the revolution of the Earth, but the one who studied marbles rolling down an inclined plane. It says something about this type of science. What they address is not something which others would value, like the testator's gold. Their value is that they have enabled them to provide a reliable answer to the experimenter's question. Everybody knows that a rolling down marble will gain speed. "Yes, but 'how' does it gain speed?," which also means, "how do we define speed?" Only his colleagues really care all the more so that this definition only concerns a move without friction, a very unusual move indeed. When you get out of the lab, you experience it only when you fall on an icy ground. The great values of experimental sciences are their capacity to create situations which provide the possibility of a proof. But testators require proof for everything. Where there is no proof, there is no value. And that's where the destruction comes from, that's how it destroys. They make the very specific success of the experimental sciences the standard of any "real" science. It turns some sciences into caricatures: accumulating data that enable them to pretend they have got proof. Those data would not resist objection, but nobody is interested in testing them the way experimenters do with a colleague's claim because here the colleagues know that without this pretense their science would not be "possible." What is impossible indeed, because of the testators, is to discover for each science the kind of situations that allow them to learn from, or with, what they address. Obviously, Galileo did not learn "with" the marble, but what of a science like sociology?

Collecting data is not a success for sociologists. They may get about everywhere and get such data and then correlate them and "prove" this relation or another. Let's say he can correlate what women choose to buy in a supermarket, what the present as their criterion of choice with their social class or anything else, anything else that can be associated. Why? Because those ladies will be nice enough to answer questions about which they are not too concerned. But if that sociologist ends up talking to a feminist who is figuring out what the purpose of those questions is, he may well get slapped in the face. So, he then will have run into a situation that (maybe) gave him a lesson; that taught him what the wrong questions are, the ones insulting the people to whom they are asked. However, most of the time, sociologists think they can ask their questions anywhere, anytime on behalf of science and they will prove everything and anything. And they methodologically avoid addressing people engaged in the situation they wish to approach as "non-representative."

In my view, sociology makes sense only when it addresses groups who contemplated a topic and from whom, or with whom, it can learn what questions they feel are worthy and interesting. That makes a difference in terms of ontology. The electron will suggest that a question is wrong in a different manner, namely by not showing up as experimentally expected. But the big problem with most human beings is that they are nice, they don't want to insult, they are ready to help science, and open to answer any question. They are "food for proof," the perfect opposite of the experimental reliable witnesses, and they cannot help sociologists crafting the right questions to address a social situation.

Plurality can affect and inhabit the sciences as soon as we realize, in opposition to the testators, that each domain of reality must be interrogated contingently, creatively inventing a mode of inquiry that will allow us to learn in a situated manner. How can I learn in this situation? That is a reasoning that takes place in ethology nowadays: how can I learn means more and more how not to impose a question but how to create an environment that will enable such or such an animal to show me what it may be able of, a situation that makes sense for the animal itself? So that could be a way to reintroduce ontological distinctions mediated by scientific practices affirming the plurality of what is required in order to learn.

There are also the situations where we have to learn *with*: it is another ontological situation where the only way to learn is by participating.

ONTOLOGICAL POLITICS: MOVING BEYOND ONTOLOGICAL BRACKETING

Marc Higgins: Significantly, as you state, ontological politics is "connected with the possibility of resisting our world's ongoing destruction" (Stengers, 2018b, p. 86). Can you speak to some of the challenges of moving beyond the habitual model of capture or bracketing of Nature as something to be epistemologically represented? Particularly, can you speak to the ways in which taking Nature on its own terms (e.g., animism, other-than-human agency) poses challenges to science's mandate of "do not regress" (e.g., a teleology of progress)?

Isabelle Stengers: I strongly criticize the notion of general epistemology. Today, philosophers who deal with sciences, even when they address ontology, they stay with the idea that reality is largely silent, that it is our human significations which give it shape, and that's what constitutes ontology, and so ontology relates first of all to epistemology and belief, and we are allowed to denounce as "regressive" other "representations," instead of learning to pay attention to what matters for the concerned practices.

This is why I personally use the term "practice" rather than "representation." What I said earlier about the difference between sociology and Galilean physics, let's say, in relation to the marbles rolling without causing any concern or the feminist slapping the sociologist relates to a practice not to a representation. Practices prevent us from talking about epistemology, about some general model for a reliable knowledge model. Learning how to address something in its own terms, is obviously not addressing it as it is "in itself" but learning how they can learn from. A sociologist should never address anyone, a homemaker for instance, as a "representative" sample constituted of homemakers, all available to answer his questions. Availability is a trap. He will learn only with recalcitrant people, who will teach him how their questions insult homemakers. He can then interpret what he learned the way he wants, but at least he will have been transformed first. We have to accept to be transformed by what we are addressing so that we can learn from it. Being able to learn amounts to being transformed.

So, ontology is not a synonym of knowing a kind of nature in itself, it is learning what is required when addressing a forest, how it compels us to be careful and not to ask it the wrong questions. And that is why the kind of success of experimental practices are interesting in itself. It does not "represent," it constantly brings new beings into our worlds, creating a proliferating ontology. But the conditions of such successes are highly restrictive. What we call ontology is not "beyond" pragmatic concerns, it positively and actively relates to them. It always comes down to how can I worthily address this being, this manner of existing; that is to say without imposing on it what I think I know. But this is not epistemology. It is rather a concern: are my questions the right questions? That is to say, it is a constant commitment to what we are engaging with, to what we are addressing.

ONTOLOGICAL POLITICS: MORE THAN
A CRITICAL QUESTION

Marc Higgins: Within your work, you do not hesitate to acknowledge the complicities between science and hegemony: "right from the beginning of modern sciences ...[they] have been complicit with imperialist claims and enterprises" (Stengers, 2018b, p. 86). Yet, you call for approaching the question of complicity in ways that move beyond accusation and abolition of scientific hegemony, in the midst of a contemporary moment when calls for abolition are multiple: It is more than "only a question of the long-entrenched life of colonial thought habits" (p. 95), as you state. Can you tell us more about this choice and its politics?

Isabelle Stengers: Scientific hegemony must of course be resisted, but it matters to understand how it came about. Modern sciences were born in a particular environment. Actually, philosophers such as Leibniz were very interested in China, where knowledge was honored but not divisive, not associated with breakthroughs and quarrels between innovators. It was Joseph Needham who asked, in his work about the history of China, a question that is still relevant today: why is that, and for a very long time, Europe was very far behind China from a technical, governance, and scientific point of view, how can we understand the way in which Europe took the lead? Following Needham, it is due to the fact that, since the Renaissance, freedom was allowed to entrepreneurs who were often working for competing princes who benefitted from their innovations. But why then were sciences and not techniques glorified? Scientists are indeed entrepreneurs, but an entrepreneur depends on who is interested in their endeavors.

It is not Galileo rolling marbles down an inclined surface that contributed to colonization or to the devastation of modernity, especially given it already required a lot of empirical adaptations before it worked, even for a cannon ball, because even the trajectory of a cannon ball is affected by friction. Who was interested in the "new science" which Galileo heralds? Who was interested in a science which claimed to make traditional ways of understanding rationality crumble? In Europe, it echoed with the new idea of progress which appeared in the seventeenth century. Progress is no longer spatial. It means going forward to the future, walking away from the past. But again, why did this all start in Europe? It's a chicken-and-egg dilemma.

But here we connect with hegemony, with conquest, that is with the colonizing enterprise directed against "backward peoples." But this enterprise may be related to a destruction happening in Europe. I am talking specifically about the destruction of the way of living of rural communities, which started first in England with what is called "the enclosure." Starhawk, whom I like a lot, writes in an appendix to her work *Rêver l'obscur* [Dreaming the Dark], that the devastation of the social and cultural European landscape started already with the first form of colonial enterprise which was a matter of commercial enterprise rather than a colonization of occupation. This process brought back precious commodities, for which money was required. The rich needed to become richer, but not only through their lands, but rich in "cash" to be able to buy, and this contributed to an accelerated change in the modes of material possession. It meant the destruction of peasant communities, their way of organizing, how they practiced solidarity with one another, the means by which they cared for and maintained their environment—in short, their communal ways-of-being and -doing.

So, it all came together, because progress and destruction became inseparable together with the idea that freedom was first of all entrepreneurial freedom, the freedom of the individual escaping the constraints that attach them to others. Indeed, the sciences benefitted from all of this because, first of all, they provided a new idea of rationality grounded on so-called indisputable facts, and correlatively the idea that progressing in a rational way means doing so without scruples. The notion of scruple: "What are we doing?" wasn't present in this philosophy of progress: there was no notion of scruple when peasants were made destitute in the seventeenth and eighteenth century. Modern scientific entrepreneurs till today have inherited the fact that progress must disregard all scruple: scientists have bought the narrative that science demands irresponsibility in the name of the "advance of knowledge" even if this advance is measured by the number of publications in high-ranking journals.

Indeed, all of this means that the world, the way it exists and lives on, has no value as such for the entrepreneur. What counts are the ways in which we can transform it: what is and could be possible. This notion of possibility is not a bad notion in and of itself. However, when it is paired with a profit-making industry and a governance grounded on the imperative of competitiveness and economic growth, what matters mainly is what those possibilities could yield, it becomes fearsome. The need for a strong plurality of scientific practices should not make it forget that modern scientists are, as such, entrepreneurs, the creatures of possibility, in contrast with most knowledge holders and keepers who cared elsewhere for the ongoingness of their world. But those entrepreneurs got the privilege of embodying rationality because they allowed the promotion of progress as justifying irresponsibility, neglect, and destruction. Modern scientists take pride in dedicating themselves to the advance of knowledge, and time taken to think about consequences is wasted time, something like a sin. A "true" scientist will unthinkingly answer to concerned people, "Okay, there is a price to pay for this progress, but it will be addressed and repaired by another progress." Even as biodiversity is vanishing, scientists are suggesting: "We will keep their genomes, and with them, we will be able to recreate populations one day."

Another element in this situation is the opposition between "science" and (irrational) belief, which justified the price of progress. When people protest "If you destroy this, you'll destroy us," they will be answered: "no we will only destroy your representation and beliefs; you will adapt and say thank you." Those who define themselves as rational can only be tolerant, but they never ask others to tolerate them. For some contemporary anthropologists, however, interpreting others in terms of beliefs, not contradicting them but recording them as interesting features of worlds bound to disappear is a form of professional malpractice. It presents an obstacle against learning, learning to think of themselves as strange and asking strange questions they ask. We should, they say, learn ways to understand our own strangeness in relation to others. That doesn't mean that the others are not also strange; we are all strangely different. That difference should be used to constitute connected worlds as every way of creating worlds is differently strange. This illustrates the plurality of scientific practices as those anthropologists fighting against thinking in the colonial mode are also entrepreneurs, thinking in terms of the possibility we might become able to think otherwise.

In contrast, when we speak, as we do nowadays, of the Anthropocene, we are in full colonial mode of thinking. The Anthropos is utterly not a human being. He, who is the carrier of rationality, is alone in the world. Those who generate and inhabit an animated world, a world of obligations and responsibilities, a world where there are protocols for living in a good way with others, be them trees, wild or tame animals, or spirits, will be described as believers who have but representations: "but we know better, we know that we are alone in the world and that all values are coming from us." I defend the possibility of sciences, which would honor their own, very particular, obligations, but it's complicated. It's complicated because we never find science on its own. The pride to be the ones who have claimed as a fact that the Earth revolves around the Sun has been used in so many arguments claiming the obvious superiority of our so-called rationality. Can we acknowledge this fact but actively remember that it really concerns rather few people? And not be sorry about it but rather ashamed by the way such a fact was used as a weapon against all others.

TOWARD RECLAIMING ANIMISM: LEARNING TO SMELL THE SMOKE

> Reclaiming the past is not a matter of dreaming to resurrect some "true," "authentic" tradition, of healing what cannot be healed, of making whole what has been destroyed. It is rather a matter of reactivating it, and first of all, of feeling the smoke in our nostrils…. Learning to smell the smoke is to acknowledge that we have learned the codes of our respective milieus: derisive remarks, knowing smiles, offhand judgments, often about somebody else, but gifted with the power to pervade and infect—to shape us as those who sneer and not among those who are sneered at. (Stengers, 2018b, p. 103)

Marc Higgins: A refrain that you often return to is that of neo-pagan witch Starhawk: "The smoke of the burned witches still hangs in our nostrils" (Starhawk, 1982, p. 19). As you mention, part of the politics of moving toward an ontological politics may be to recognize that science always already has a relationship to animism: "we forget that we are the heirs of an operation of social and cultural eradication—the forerunner of what was committed elsewhere in the name of civilization and reason" (Stengers, 2018b, p. 103). The *Burning Times* not only had effects on how science differentiated itself from non-science, but also how science was practiced: "behind Newton's cautious declaration—'I frame no hypotheses'—concerning the nature of the forces lurked the passion of an alchemist" (Prigogine & Stengers, 1984, p. 64), which could not and would not be utterable within the moment. This is not a new relation between science and animism.

However, rather than engaging the impossible task of reproducing a past that is already passed, you call for a reclamation of animism in science that is rooted in "reactivating it, and first of all, of feeling the smoke in our nostrils" (Stengers, 2018b, p. 103). Why is such affect, and what it makes possible beyond modes of critical reflexivity, significant? Why is it important to attend to the embodied sense of fright that comes with taking seriously other-than-human agency and the ways in which the modern imperative "do not regress" that we possess and possesses us? Significantly, you invite us to consider that this may well "make a difference between living in the ruins and just surviving" (2018b, p. 109)—could you elaborate upon this?

Isabelle Stengers: I would say that Starhawk was indeed referring to today when she wrote about the smoke. It's an important passage for me because it is also a way in which we could situate ourselves: either we situate ourselves through either pride or guilt, or we consider our position in relation to the disaster from which the modern world emerged and that is ongoing in our so-called post-colonial times.

We do not think any longer of the witches as satanic. But they were not poor, deluded victims either. As Starhawk and others insist, many of them had an important place in their communities. The coming of the witch-hunting inquisitors was bringing terror and mutual distrust—who would denounce whom to save their own life? But remembering the persecution of the witches as part of an operation of destruction of the rural communities is situating ourselves as resisting against the tale of progress which made it possible not to believe they were the servants of Satan. Not so long ago, they were a subject for historians and social sciences specialists who wondered if witches were just old, a bit crazy, women or if they possibly held the satanic representations their executors had of them.

I would qualify such specialists as "new inquisitors." Indeed in a way they continue the operation of destruction. For them the case is closed, it is a thing of the past, and too bad for those who believe it has something to tell us today. To remember the witches, even if we know very, very little about what they "really" did or thought or honored is willing to be situated by their destruction, refusing to forget and confuse modernity with what did free us from murderous superstition. Presenting and thinking themselves as witches, contemporary witches know very well that they will not be burned, but they know the new inquisitors will deride them, take them as examples of "regression" or "escapism," escaping into an imaginary world. The same is true for "magic" as witches' craft. We have to accept both that magic does not exist and that to use means to bewitch people is bad. Hitler did it, marketing people are doing it, politicians learn to spin a tale. Magic as addressing the others' weakness is potent but bad. Magic as empowering is a silly belief.

That is why I was so interested by neo-pagan witches; they demonstrate to us that even if we know nothing about witches, the hateful fear of inquisitors is still present. Presenting oneself as a witch brings them alive "How dare you? You are crazy!" and one feels the smoke indeed. Contemporary witches have created the possibility to situate ourselves in relation with the past. Are we heir to the inquisitors or to the burned witches? They allow us to situate ourselves in the world in a way that makes us understand why we may indeed be afraid and recognize that facing this fear requires us to come together, not on one's own but defiantly together. Because I believe that this can only be a collective matter.

Remembering that we live in a dangerous world, in an unhealthy world is also important for people like ethologists. They are also watched by the inquisitors: "You are an animist; you accredit abilities to animals whereas we know they only are but animals; you accredit them with something that belongs to humans." Here it is crucial not to speak about a conversion to animism because conversion means converting to a belief. It rather demands to not "recognize" animism, which is a general category created by anthropologists in order to classify others, but to open our imagination and feel that, "yes indeed it makes sense to say that." For instance, you know that in Argentina, genetically modified soybeans have spelled disaster for numerous small farmers. One of them told an inquirer, "those beans are evil." It resembles "animism"—we know that they are only plants, devoid of bad intentions. But we can also take it seriously and think. Indeed, those genetically modified plants owe their very existence to something which the farmer is quite justified to call evil. They cannot be separated from what they were invented for: to turn a fat profit protected by intellectual property rights, to support intensive monoculture farming on innumerable hectares, to allow for the killing of all other plants which bother us, to poison the environment, and of course to put small farmers into poverty, bring them to their ruins. Yes, genetically modified soybeans were made to do all of these.

The point is that feeling the full weight of the farmer's argument is also feeling that we are living in an environment that will state "Ah, this is what they believe." If someone tells you, "this is what you believe in" they are an inquisitor. We need to be able to recognize them and eventually, effectively, tell them what they are. If the testators may be compared to theologians writing the books about how to recognize the witches, inquisitors are those who propagate fear, silencing any attempt to try and craft the questions our sciences would need in order to become relevant. Bruno

Latour, in *Facing Gaia*, has written that Science functions as a vast enterprise to "de-animate" the world, to negate the agency proper to what it describes. Chemists speak of their molecules as "chemical agents." They very well know that they have an "agentic power" since that is what they work with. But the physical interpretation of this power succeeds in getting around the question, and chemists know it. So, if asked about it, they will say that it was just an expedient manner of speech.

Agency has been kept for humans only, and thus equated with intentionality, consciousness, and even free decision. Thus, animism has become the "belief" that plants have all that, while we know that we, "who are gifted with agency," are alone in the world. How to characterize the agency of terrestrial beings is at the center of a strong pluralization of sciences, but it demands that scientists feel the smoke and openly challenge the inquisitors.

Challenging them is also disconcerting them. For instance, many are disconcerted by the fact that I am very interested in tarot reading. They ask me why. My answer is that I am not very interested in fortune telling but rather in the possibility, psychologists might recognize it as a tool, instead of referring to scientific theories. A theory will never help a psychologist if their practice aims at activating possibilities of change. Using tarot, giving the cards the power to constraint them to think means putting their judgment aside—"that is normal, that is not"—and to work with the consultant about the relevance of the draw. I am part of a group who actually created a tarot game, with cards we collectively invented. We thus do not deal with clairvoyance, but the game has an agentic power [efficace][1] which really surprised me. Nothing prevents me from stating that other tools associated with other rites elsewhere are also agentic power: having a tool which forces you to think transforms you. To be transformed by this world is also a way of being; it is not to become animist, as animists don't exist in the general or non-specific sense, but rather accepting that the world did not become de-animated without practitioners being silenced or forced to accept endorsing the cause of the inquisitors. Obviously to fight against this de-animation will bring accusations of animism, or worse. So be it. Learning to feel the smoke in our nostrils is vital today.

---

[1] Translator's note: Stengers qualifies that she does not use *efficace* in the sense of effectiveness [efficacité] but rather in the sense that it is transformative.

What I am greatly interested in are all those practices that we can create, that are created in relation with First Peoples or within activist struggles, etc., that teach us how to accept being transformed by others, affected by others, and that nurture the way we connect with others. That is something that—and we can also say that this is what witches suggest—could be designated as magic.

If I insinuated that successful psychology is equivalent to unrecognized magic, the inquisitors would come after me. But I would also say that the success of an educator in getting their students interested and creating a learning, cooperating group with its own strength, in turning the classroom into a living place instead of an environment where learners chase grades, is a magic work. I use the term "magic" in the very practical sense of the art of transforming something. Transforming together because an educator dealing with a lively classroom makes them a different teacher than the one who has to answer stupid questions or face stupid uprisings. Treating people as if they were ignorant doesn't generate interesting questions or challenges. For activists it was a vital question. The invented facilitators, while in France we are still speaking of "animateurs," as if a group was to be given its "animacy." The art of facilitators is to take care of a process which needs being freed from what thwarts it, habits produced by our insalubrious environment.

Marc Higgins: If we talk of metaphor, in terms of representation, we can also talk about the re-animation of the substantive content. A content already dead that we must revive each time.

Isabelle Stengers: Yes, but there is a way. Because I believe that if we are not ourselves interested in what we are teaching, we won't have any success in interesting the students. There are no "dead contents" if the point is to share their agentic power.

Marc Higgins: This brings us to the last part of the question: being open-minded toward an animated world, to those who already know about its animacy and whose being in relation with an animated world is part of their methods to build a world, it makes the difference between survival mode and a bearable world.

Isabelle Stengers: Anna Tsing's book *Arts of Living on a Damaged Planet* made a lot of people think, myself included, because it is about how to live our lives in ruins, but not the ruins we will face if we keep carrying on our present ways. Not the ruins that would be an environment similar to the one in *Mad Max*. That is to say an environment marked by survivalist violence with no concern for others. When I wrote *Au temps des* *catastrophes* [*In Catastrophic Times: Resisting the Coming Barbarism*], I started with Hurricane Katrina that had recently happened in New Orleans. The rich leave, and the poor stay there to die. So that is a scenario we can imagine living in the ruins. However, what ruins may also be considered as places where nothing is guaranteed. That means we have to pay attention and renew our ways of paying attention at each step. We thus have to reclaim the art of paying attention. We cannot any longer rely on infrastructures like supply chains to provide us with all that comes from the world anymore. Supply chains are already shaky; one day they will disappear.

Somehow, we are going to have to think, and if we are alone, we won't succeed. So, we have to reclaim and revitalize collectives in order to enable them to nurture both and inseparably the culture of interdependent, mutualist relations and the art of paying attention. Something I liked a lot in Anna Tsing's book was her art of *noticing*. It is crucial in the ruins, because it means living fully, without believing that things will stay the way they are. But it also means learning to notice the silent disagreements, growing gaps and distrust because of which so many collectives fail to carry on. Making an art out of such challenges is the best way to answer them. Art in a non-modern sense, of course, which included technical practices and artifices, such as rituals and transformative roles. Such arts have a magic of their own as long as we nurture them and experiment with them together.

What has no magic, in contrast, are abstract discussions about principles. Principles are to be obeyed or else you are excluded. A prevalent concern in Europe where commons are resurging and struggling to exist in spite of property rights that cover most of the land, is what should be expected of "true" commons which are truly different from profit-making enterprises, which present themselves as commons in order to capture sympathy. Everybody agrees that commons should not be only about the collective exploitation of a resource because then the Earth would again be treated as a resource. But, to turn this into a principle is a trap. New kinds of inquisitors may appear. Becoming commoners, learning to trust each other not in spite of but through their difference, making the heterogeneity of the group a strength is engaging its transformative dynamics; that is generative in itself. Indeed, if those who belong to a commons are heterogeneous, different modes of paying attention will emerge, and if they trust each other it will become concerns to be shared. Defining in the abstract terms of resources and profit a place you live in and with demands an anesthesia of the senses, a disciplined differentiation between what matters and what should not. Learning to become a commoner may be hard, but there is joy in the generative awakening of the experience that we are not alone in the world.

It seems to me that there is a sort of generative creativity in all areas where trust in others is both required and cultivated. My optimistic view is that in order to enable the next generation to face the challenges that are awaiting them, the highest priority is to learn and nurture practices everywhere of effective collaboration, of trust, of mutual support, which also means a determined refusal of individual evaluation. I like the idea of resurgence because it pertains to something that was meant to be eradicated but can reappear, even differently. Because, indeed, the eradication was carried out violently but was never a totalizing one: it could not wholly suppress what it attacked. And it had to be carried out again and again for each new generation, notably starting in schools where children learn to compete, to not cheat, to not help each other, and experience the shame of being slower or clumsy. A culture that makes us feel that we are not alone in the world, that interdependency is vital and will become crucially more important as this world will become increasingly unpredictable and shifting.

We have to prepare ourselves for what is coming, that's what it is all about.

Marc Higgins: Collectively, preparing ourselves for arts of listening, of paying attention.

Isabelle Stengers: Yes, and collectively equipping ourselves with all these practices they have gotten us used to not practicing: resisting, defiantly laughing at the face of the inquisitors who, about everywhere, try to make us forget what is happening. Recently, Emmanuel Macron was saying to those who viewed 5G as anti-ecological, etc., "You want to go back to the Amish way of life?" He was an inquisitor. He didn't bring us back to the caves as others do, by the way, but to the Amish community a rather remarkable example of a culture of interdependency. But for him it made no difference, it was just a tagline addressed to a supposedly docile flock. We might make a song of it, make it memorable in our own way. We need not to engage inquisitors on their terms: arguing. Because they are good at arguing in a way that drowns the fish; they drown us with evidence that they are leading us in a responsible manner toward the only future they deem realistic.

References

Prigogine, I., & Stengers, I. (1984). *Order out of chaos*. Bantam Books.
Starhawk. (1982). *Dreaming the dark. Magic, sex & politics*. Beacon Press.
Stengers, I. (2015). *In catastrophic times: Resisting the coming barbarism*. Open Humanities Press and Meson Press.
Stengers, I. (2018a). *Another science is possible: A manifesto for slow science*. Polity.
Stengers, I. (2018b). The challenge of ontological politics. In M. de la Caladena & M. Blaser (Eds.), *A world of many worlds* (pp. 83–111). Duke University Press.

CHAPTER 21

In Conversation with Steven Khan: Sensible and Sense-able Qualitative Literacies for Multi-species Flourishing

Steven Khan and Marc Higgins

As our planet undergoes the equivalent of a violent and rapid (in geological terms) phase-state transition—multiple extinctions looming—brought on by the activities of a single species of thinking and way of Being—Man— what are our responsibilities to … other non-human nations? (Khan, 2020, p. 239)

Marc Higgins: Thank you for agreeing to take part in this interview: I continue to be grateful for your scholarship, practice, collegiality, and friendship. As I have mentioned to you, part of the intent of this collection is to bring together folks to speak from either within science education, to science education, or from adjacent but highly relevant locations around

S. Khan (✉)
Brock University, St. Catharines, ON, Canada
e-mail: skhan6@brocku.ca

M. Higgins
University of Alberta, Edmonton, AB, Canada
e-mail: marc1@ualberta.ca

© The Author(s) 2024
S. Tolbert et al. (eds.), *Reimagining Science Education in the Anthropocene, Volume 2*, Palgrave Studies in Education and the Environment, https://doi.org/10.1007/978-3-031-35430-4_21

the question of what it means to teach in response to, within, and to trouble the Anthropocene.

Within your recent work on multi-species flourishing (e.g., Khan, 2020; Khan & Bowen, 2022), there is a felt sense that perhaps the naming of the Anthropocene serves to mask particular power dynamics that are front and center in your work. Could you speak more about the ways in which the Anthropocene at once poses an important challenge for STEM education and simultaneously needs to be challenged (e.g., attended to otherwise as Plantationocene)? Specifically, what does it mean to do this work in the space of mathematics education?

Steven Khan: I won't try to define the Anthropocene[1] but acknowledge that what falls under its umbrella, or parts of it, definitely falls within the gambit of STEM education or in terms of what it imagines it's about. Also, there are things that are outside of both of those as well. When I think about the Anthropocene, I want to start with the fact that it is a concept that comes out of a particular disciplinary discourse and has moved into public consciousness in a fairly rapid and meaningful way for particular purposes. Similarly, we have other concepts from other disciplines doing that, as well, so things like critical race theory or intersectionality or the one that was I having a discussion about this morning: the concept of proofreading is not what most people think it is from within the sort of formal editing perspective versus the common place

---

[1] Yusoff's (2018) eviscerating critique of White Geology's ongoing complicity and willful ignorance or intentional blindness toward communities who continue to live within its racist-colonial-capitalist-ecocidal wake (see Fig. 21.2) resonates. She argues:

> If the Anthropocene proclaims a sudden concern with the exposures of environmental harm to white liberal communities, it does so in the wake of histories in which these harms have been *knowingly* exported to black and brown communities under the rubric of civilization, progress, modernization and capitalism. The Anthropocene might seem to offer a dystopic future that laments the end of the world, but imperialism and ongoing (settler) colonialisms have been ending worlds for as long as they have been in existence. The Anthropocene as a politically infused geology and scientific/popular discourse is just now noticing the extinction it has chosen to continually overlook in the making of its modernity and freedom. (Yusoff, 2018, p. xiii, italics added for emphasis)

We too say this is not new; it is known, has been known, and is only finally of concern to White Imaginaries of exceptionality or difference due to the existential risk it poses to a particular way of being human that has masqueraded as the only way of Being Human—Man (Wynter, 2003).

understanding. Within the geological sciences, the term Anthropocene has a very specific and limited but important purpose for *that* discourse community. It is a tool for thinking and research and complicated conversation.

Our question is: what happens when we take it outside of that? I do agree that it functions outside of that as a mask. It's a discursive signal to see that we're doing something or thinking about something that's related to what the geo-scientists are thinking about. It is attention grabbing, which is important in today's limited attention economy and fragmented attention landscapes. It does that without really getting at the ways that particular systems are organized, and the logics that underpin those systems and how they came to be, so that historical dimension of our many planetary precarities that are more directly related to our day to day work in education than the geologic. It is simultaneously historical in the geological sense and ahistorical in the social sense. I feel as well that the term diffuses and generalizes responsibility for these global effects so that, ultimately, no one is accountable: not locally or not globally. That is where part of the issue is.

Thinking about the Anthropocene and all the differing variants of STEM,[2] how do you move that to the political need to organize and operate differently from how we are today but similar to the ways in which humans have operated in the past and continue to operate in the present in different ways? For me, that means less individual and more communal ways. How do we come together to do particular things, given the way that we've come together in the present, which is a result of all these movements and forces that move people organisms around the planet and also move the planet both literally and metaphorically.

What does it mean to do this work in the space of math education? There is already a strong socio-political, socio-ethical dimension in mathematics education—work from over the last four decades, at least, and perhaps even longer. It's always been aligned with this idea of opportunity, whether that's economic opportunity or other types of opportunities. In this case, it's being *for* flourishing. In the years leading up, the work of mathematician Francis Su struck a chord with lots of people. I entered the field through Martin Seligman's concepts from positive psychology and flourishing, even though I don't think Francis uses them. They are concepts that make sense to me. Psychology has very much focused on

---

[2] For example, STEM, STEAM, STREAM, STEMSS, SAMBA, etc.

pathology for quite a long time, but there is another side to being human which is being well, so how do we shift our attention to do that as well, to make that a focus or goal of ours in teacher education.

The fact is that the beginnings of focusing on the human haven't always forced us to act or be compelled to organize. Spending time in Alberta, in particular, with Indigenous, First Nation, and Metis colleagues (I don't think I had the good fortune to meet any Inuit colleagues), with scholars doing really important work, like yourself, Florence, Trudy, Sharla, and Brooke, we've been challenged in that way and in a very particular landscape that is Alberta, Edmonton, with the mountains and the plains. As well as being away from home, looking at what's happening in the Caribbean and to other small island nation states, I've been thinking again about the more-than-human or the other-than-human, not in a salvific way but in a partnering way. For example, in thinking with bees, we partner by altering the landscape or altering it back, to invite those relations that are also necessary for well-being and survival in ways that it's not just simply the number. It absolutely and very much is a continuation of ethics in math education; it is a move toward a multi-species, planetary-type ethics rather than a human-centric ethics of how we relate to other humans only through our disciplinary apparatuses.

Marc Higgins: I really appreciate the notion that focusing strictly on the human doesn't always leverage the kinds of openings that might be desired. Here, the notion of mathematics education as being actively shaped by colonialisms is not a new theme in your own scholarship, drawing and extending critiques of the ways in which ethnomathematics could at once be a radical challenge, but it could also be subsumed back into dominant interpretations where it's supplemental, an otherness to be consumed or exoticized. This takes a new shape in your recent work as you take up Sylvia Wynter's notion of Man and its co-constituting vectors of oppression. Following Wynter, the *After Man* that you call for is not a temporal after, but rather a working toward a structural otherwise. This takes on particular significance when you invite us to consider the ways that STEM skills were central to plantation economics:

> Land, and lots of it, was central to wealth generation at this early foundational stage of nascent state capitalism, and STEM skills were critical. Key aspects of the plantation model include the replacement of diverse native species by monocultures of economically productive ones; the transformation of diverse topographies into monographies of the grid allowing for

easier calculation rates related to yield and harvesting; and the use of cheap, replaceable, substitutable, or enslaved labour, including the labour of animals, micro-organisms, and machines and the study of any and all factors affecting yield/productivity in order to produce novelty and speed up generation time to market. (Khan, 2020, p. 237)

Further, you invite consideration of the ways that plantation logics continue to permeate and persist within mathematics education. Could you elaborate upon why this is significant, be it in terms of *who counts* as mathematics learner, *what counts* as mathematics, as well as *why* and *how* we learn mathematics?

Steven Khan: This is part of the affordances of doing a university degree in Canada and elsewhere. I stumbled into Sylvia Wynter's work while at University of British Columbia, through others' work. If I had stayed elsewhere, I probably would not have had that looking across, that looking back, that looking forward, I would not likely have had that space in mathematics education in the Caribbean itself. I would have done more traditional mathematics education and would not have had these influences necessarily at that time.

The idea of the plantation is certainly something that we talk a lot about in the Caribbean, and in a small place like Trinidad its presence is always felt (at least at the time I was growing up). Thinking back, my high school history education is mostly Caribbean history with almost no European history—European history is the one that is not heavily emphasized (this makes reading European theory harder at times) so mostly Caribbean, and with more North and Latin American history. I think back to the teachers that I had there who really took us through this step by step to understand how we came to be. We talked about the arrival of Europeans in the Caribbean and genocide, although not with that label at first though certainly with the language of population decimation: through slavery, through abuse, through labor, through disease. These changed or radically reduced the numbers of peoples in the region and then how that impacted the next part of history. This is the conversation we were having at 11–12 years old and throughout our high school years, which I think is very different from the types of conversations that perhaps happened here (in Canada). Again, it's a different society, different history. But the idea of the plantation, its logics and economics which are really significant in our region, comes out of the work of Sylvia Wynter and, in particular, economist-philosopher Lloyd Best who worked with Terry Levitt on plantation economics. The key idea is that in plantation or agricultural colonialism, as opposed to say, settler colonialism, you are using the land over and over in order to generate profits in order to export. It's not about meeting local demand beyond that of the planter class and what is to be provided for labor. As a result, there's always scarcity on the local front where there might be abundance on the consumer front and in the emerging and expanding metropolitan marketplaces.

All this is taking place in a context of an explosion in European science, technology, engineering, and mathematics—from building ships, to navigation, to the birth of modern accounting practices, to large-scale economies. All those "standard algorithms" for the basic mathematical operations that we still use and the ones that we continue to fight over as part of what used to be called the Math Wars are spreading during this period of time, such as keeping good ledgers, for example. At the same time, there's lots that's being pushed to the sides, there are things that are going underground, and new things that are emerging. Not everyone, for example, is being taught mathematics, and not everyone at this time is being taught the same type of mathematics.

In terms of plantation logics permeating and persisting in math education, you'll be familiar with it if you've been employed at some point in North America: overwork, uncompensated labor, and scarcities. This is a conversation I keep encountering in a number of books: the plantation system is definitely premised on scarcity, or rather, it's premised on if there's scarcity, then we need this, and our goal is really to keep increasing profits or markets, moving past satisfying needs to creating desires. Math is like that, as well, at times: there's math that we need and which is part of a culture and society meeting its members' needs for survival, transcendence, dignity, belonging, and equipping them to meet some challenges, then there's math that's enjoyable, and then there's math about creating desire and manipulating behavior.

In terms of Sylvia Wynter, there's an interview with David Scott, where she talks about genres of being human and alternative genres which resonated with me. It is not about looking for a singular notion of *After Man*, but it is about looking for different ways of being and doing. Because there is nothing that this swarm of beliefs and discourses and practices hasn't touched in education, and in particular math education, which happens to the place where I do my own work. The structural otherwise is about working with others and is about just starting or even noticing what is being pushed aside by dominant discourses and epistemic cultures. It is not a utopic "if this then this will come" so just get started. So, in terms of who counts as a mathematics learner or what counts in mathematics is definitely very much influenced by the ethnomathematics and critical mathematics programs. For me, mathematics, at its core, is really about patterns. Humans and other species do not survive and certainly do not complexify without attention to patterns and attaching significance to patterns in some way.

So *what* counts as mathematics is tied up with *who* counts in terms of mathematics is tied up with *what* mathematics *do we teach*, as well as *where* we learn, and *how* and *when* we go about doing it. The plantation system states, "here's what you need to learn and when." There's no real *why* or the why is endlessly deferred. That's changing in the last few decades with *why* being put forward around the STEM argument of economic competitiveness, around the environmental argument, or around understanding the political impacts of and participating in democracy and being a democratic citizen. However, the idea of flourishing only came about fairly recently as an explicit goal. There are lots of educators who've been doing that for many years differently. For example, Indigenous educators propose that we learn in order to become good ancestors, to become good relatives, and to live well in the world. This is different as well as more expansive than the Christian version of being good stewards. This distinction between ancestry and stewardship is a very different sort of relationship. Which is not to say that they do not have places of similarity and congruency, but they are grounded in different visions of the world (or creation) and our human place in it.

So, *who* counts as a mathematics learner is tied up with the history of mathematics around who's been excluded. We're at the point today where we now have two women who have won Fields medals, but several colleagues posted breakdown by country, showing a large number from the US and smaller numbers elsewhere. Who counts as a mathematics learner is also related to what counts: we can find mathematics, in terms of everyday, in things that are needed for survival and things that are needed for belonging, in practices that allow and promote this feeling of going beyond oneself, and in ways that are always challenging.

The key for me to mathematics is *challenge*: all species, I think, address the challenges that the environment that they find themselves in poses them in one way or another. Humans are a particularly interesting case on that tree. In the multi-species flourishing framework we place play with challenge as we think this is where it starts—the posing of challenges by the organism that results in pleasure and learning for its own sake rather than for someone else's sake. The set of animals that engage in play keeps increasing annually. As I revise this, for example, I just read about a study examining bees who exhibit behavior that they categorize as play. So, yes, there is mathematics related to survival, but there is also mathematics that is related to challenge through play and which I think is where its potential for flourishing emerges most forcefully as a rationale for learning mathematics.

Marc Higgins: These notions of the *who*, the *what*, the *why*, and the *how* being so intertwined really speak to your recent writing about over-representation being so present in math education (Khan, 2020).

Steven Khan: Informed by Sylvia Wynter's work, this goes back to when I was doing my doctoral work and noticing, doing that listing of the people that we (were required to) study and who they are, and seeing the real focus on mostly men. There is some diversity among those men, but they were almost all men, and they were mostly in the North American Academy, with the exception of Paulo Freire.[3] Then, having the opportunity to design a curriculum history course with my supervisor and being more intentional about different types of diversity in terms of our readings to bring that attention to representation and to open up to different perspectives. This is critical in this particular type of work.

Marc Higgins: This transitions us nicely. Importantly, as you consider decentering *Man*, there is an invitation to move beyond the human as "there is not and has never been human flourishing at community and population levels without—or independent of—multispecies' flourishing" (Khan, 2020, p. 239). What does it mean to address the paucity of work in STEM education which "mak[es] our more-than-human kin *central* to its theorizing and curriculum innovations" (Khan & Bowen, 2022, p. 4). Importantly, why is this significant?

Steven Khan: Regarding the first part, that there has never been human flourishing, I think about our mega-billionaires, our Jeff Bezos-es and Elon Musks, who are again mostly men; they've created systems that allow them to flourish at the expense of many others, or have capitalized on the

---

[3] Again, coming from a different place and being a bit of a prolific reader, I was very much confused by the limited engagement with educational thought and thinkers from other parts of the world apart from North America and England, France, and Germany, which I had already encountered through, for example, http://www.ibe.unesco.org/en/document/thinkers-education. The absence of African thinkers, for example, was glaring, as well the limited engagement at the time with Indigenous thinkers.

systems, logics, and principles that exist to do that. So it is possible for individual flourishing to exist without multi-species flourishing, without human flourishing. This throws us back to the fifteenth, sixteenth, and seventeenth centuries and the transatlantic slave trade, and the Indigenous slave trade in the Americas. This extends into the present with what's happened in the US with *Roe vs. Wade*, this idea of personal autonomy and responsibility to a collective. Extending this is not leaving behind the human being but stating that we are not the center of this conversation. We've forgotten how to do that; we need to re-learn it in the spaces that we now find ourselves in. There is good work in STEM education that does address it. For example, the work out of the University of Hawai'i System with their STEMS,[2] which includes sense of place and social sciences, fits squarely in that frame of keeping our more-than-human kin central and working toward thinking about them as partners and not as resources.

I think that's the other hard part: our language is very "resource"-oriented in teacher education and in education more generally. So, one of the things I always try to trouble is this idea of resource, working toward using it much less frequently, as well as really thinking about what it means and the baggage that comes with the word resource. *Partner*, for me, or kin, is that shift in language, theorizing, and innovation.

What does it mean to address the paucity of this work? It's significant that those who do this work are often not sufficiently acknowledged in the literature. I can see lots of people doing this work who aren't academics. Again, thinking about our responsibilities, how do we share the privileges and the rewards that come from the academy, while at the same time transforming those systems of privileging and rewarding? For me, it is a reorientation toward gratitude. Here, I'm really dependent on the work of Mohawk Mathematician Edward Doolittle and others who keep reminding us that a first move in anything is to recognize and to be thankful, to know what you have received and what you are a part of. Oriented toward gratitude and sufficiency is the idea of enough. This is again another difficult concept that goes along with the resource view and the scarcity that abundance holds. Scarcity is easy to see in our education system, abundance less so. But then there are also the places where there is abundance but also waste. Universities can sometimes be places like that as well. So, it's about bringing back ways for having respect for partners, in our learning, so they are not just another set of rules on paper but rather codes in our consciousness. This is what, in the old world, we used to call values and virtues, and still do, and bringing those back to life in our actual practice and not merely ethics.

Marc Higgins: In terms of phrases that really landed with me, as I was reading your work, is this notion of sensible and sense-able, to be able to sense, as they relate to quantitative literacy. One of the themes which permeates the first volume of *Reimagining Science Education in the Anthropocene* is that perhaps "the Anthropocene(s) need new ways to be felt" (Wallace et al., 2022, p. 6). As you state, "science and mathematics have their own poetics and construct powerful mythologies even if it might be difficult to conceive of [their] language ... as *also* poetic language" (Khan, 2011, p. 16, emphasis in original). In making the Anthropocene(s) sense-able, our responses sensible, we might need to calibrate our attunement otherwise: "mathematics, while able to describe and give a quantitative accounting of the magnitude of the planet's loss in its ledgers, perhaps has no language yet to audit such mourning" (Khan, 2020, p. 231). This might require differently attending to this moment's poetics, its silences, and rhythms. Can you elaborate upon the ways in which this quantifying must go beyond a more robust and reliable form of quantifying the world (e.g., inspire new ways of listening, mourn the innumerable and unquantifiable losses marked by the many end-of-the-worlds which co-constitute this one)?

Steven Khan: This question about new ways for the Anthropocene to be felt and sensibilities/sense-abilities is a conversation I've been having with Mike Bowen. One place we started together was Bernie Krause's soundscape ecologies, which is about recording the same landscapes at different times. One example is Sugarloaf Ridge State Park, where there's been intensive logging, but looks the same from the road. However, the bio-phonic signature between years is radically different: you hear this *silencing*. The way to knowing, noticing the pattern, and thinking about the implications all go together. It's good drama in that it doesn't give you all the details. It leaves enough space for you to have to fill in those gaps, or good horror as it doesn't show you everything. It leverages what our cognitive and emotional systems are meant to do well: fill in gaps. That led us into talking about people who work on sound science and bio-acoustics, and he shared a researcher's work on hearing rainfall patterns across different states in the US to hear how that's changing plant distress. I shared work on hearing sounds in the soil. Again, things that we don't really think about a lot.

Returning to the idea of merging, of synesthesia, of multi-modality: if we think of what we actually know right now, we know our brain functions best with multi-modalities. There are times where we want to focus only on one modality, but we are embodied creatures who are working with many modes of engaging with and experiencing the world. I remember reading years ago about work on nerve conduction: when we hear, the time for the signal to go from here to here is a lot faster than from the eye to the brain, so we hear a lot faster than we see, even though the light travels faster than sound. Even in the absence of hearing, most species, or maybe all species, respond to vibration. They have vibration sensation mechanisms. This is another thing that unites us. You can think about what we would classify as non-living within traditional biology, but even rocks resonate. They have vibration sensitivity, even if it is not the same response-ability toward those vibrations.

This type of work involves collaboration to really think through, as these are not places where most of us in education have expertise, be it in technology or in sound and visual engineering. For example, I think that sound engineers and other artists, I would put them within STEM as well. There's work, for example, with film and game studios, who have ways of affecting us and orienting our attention to these end-of-the-worlds. There's *WALL-E*, as one example where there's not a lot of human sound for a while, and the satire that came out earlier this year, *Don't Look Up*. I think our ancestors, our Indigenous partners, as well as our non-human partners have things to teach us about how to attune and what to prioritize in terms of what's necessary for well-being. Again, this requires us to slow down.

Another example here is whale song and how it literally moves the oceans, not just metaphorically. The water is the medium, similarly to how we don't see the air moving. Some whale sounds can travel thousands of kilometers, a great distance in the ocean. But what happens when that sound disappears? We have that silencing, that loss of movement, that stillness. This is what it means to go beyond a more robust and reliable form of quantifying the world.

Nora Bateson's work on warm data, for example, is where I think our work needs to go: making our data more relatable, partnering with our data, and partnering with our multi-species kin in the data-making process. Again, I think about some of our Indigenous colleagues and their traditions: for example, the winter count. This is very different from the annual report that we have, but I'm not really "partnering" with this computer to do that. The significance is not in this partnership. Whereas, in traditional winter counts, hides and other record-keeping devices are very much tied to animal partners, to plant partners, to other human partners, as well as quipu used by the Incan and other Andean cultures. Winter count involves a lot of plant partners, as well as time, slowing down and remembering. We can make quipu for these unquantifiable numbers: What would that look like? What would it feel like? How long would it take to actually represent that loss? I'm going to go with a billion for the number of sea creatures lost in the last year in British Columbia due to heat. How long would that physically stretch out in order to represent that?

I think that both our Indigenous colleagues and our colleagues who are the descendants of enslaved people have a lot to teach here, in particular, around that mourning of unimaginable loss while continuing to move forward, as well as responsibilities for reparation, for reconciliation, for new partnering.

This box plot is from *Reconsidering Reparations* (Táíwò, 2022). I very much now feel sadness for not appreciating just how under-represented and underappreciated box plots are. These two plots function as a visual poem: one is around how pollution is distributed based on whether a nation was colonized and the other is on how climate change vulnerability is distributed across colonized and colonizer countries.

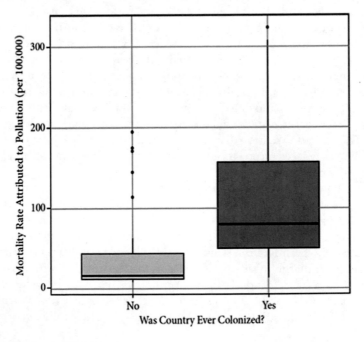

How pollution is distributed

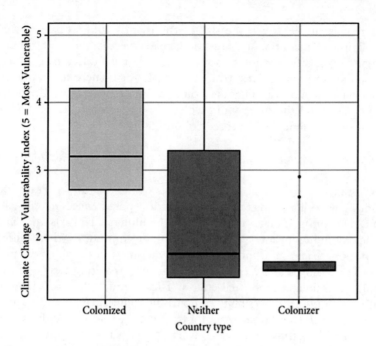

How climate vulnerability is distributed

Marc Higgins: When we talk about poetics, the meaning is made in the in-between space. Importantly, you are moving this work into your teacher education practice. Could you share some considerations that you bring for doing this work in teacher education? For example, in Khan et al. (2021) you speak to the importance of explicitly framing what might promote, limit, and work against multi-species flourishing, as well an attentiveness to the already present and existing ecologies of practice pre-service teachers bring with them.

Steven Khan: I will start by stating that I am operating from a position of a lot of ignorance (not intentional but just a humble recognition that there is more than one will ever know in a lifetime about most things); that's the place that I start from. Secondly, there things that I think I know very well and so there is appreciation for what I do know and the value of that. I don't fully understand the workings of teacher education here in Canada and the US and am constantly puzzled by its promotion, organization, administration, all those elements. The scale is different for me, here in Canada than from the Caribbean, where you typically meet teachers over and over again as there is less movement and the range of movement is smaller given the size of the islands, and so part of this is appreciating the continental-ness of Canada versus the island-ness of Trinidad. Even different institutions in Canada have different degrees of constraints in terms of what you can actually do. There are fewer constraints on what you can imagine doing. In reference to the Khan et al. (2021) piece, and the other work we've done when we started planning our course, we framed it in a three-column table:

flourishing-promoting, *flourishing-limiting*, and *flourishing-extinguishing*.

For example, social-emotional learning would be under flourishing-promoting as well as things that encourage joy, persistence, and challenge. What can we bring in under that first rubric? Things such as arts, movement activities, community involvement. Flourishing-limiting is where difficult conversations come in around things like racism, sexism, able-ism, pretty much all the -isms, scarcity, and other things that have long-term and wide-ranging systemic and individual effects. These reduce the probability but doesn't extinguish the possibility of flourishing; anticipating the argument of *What about this particular case?*, for example. How do we address those? How do we bring those into the conversation? It's not the only focus, but it is part of the conversation. For example, in my work, I typically don't do as much on gender diversity and sexual orientation as I do with respect to racism. However, that's an area that I can grow, and I add a bit more to that, over time; this is not my area of expertise and deep knowledge, as well as lived experience in that way. The third and last category of flourishing-extinguishing includes things like murder, genocide, genetic, or linguistic extinction, the unmaking of worlds. These are our Thanos-level events. One of the things that I do, for example, is lay out the characteristics of genocide: does what happened in place X or Y meet these criteria? It's an analogy to common geometry practices: okay, here are these characteristics of the rhombus, is this shape a rhombus based on its characteristics? It's about taking the idea classification to a different place and to recognize that the classifications skills we learn aren't only about mathematical objects. And it's not about thinking about genocide as a mathematical object but about thinking about genocide using skills and tools that we've learned in mathematics, through those habits of mind. These are not meant to be sharp categories. For example, war is hard to place. It is probably closer to flourishing-limiting than -extinguishing depending on the type of war. Climate change is probably closer to flourishing-extinguishing, but it's not completely in that category.

For teacher education, this means bringing in these conversations in ways that are meaningful, with pre- and in-service teachers as well as administrators, that we don't shy away from them.

I'm really excited to potentially be teaching an ethnomathematics course this year that really would be taking up some of these within that frame. [Well, only four students registered, and so the course did not run ... in the logic of our current academic system where it is almost impossible for something like this to grow or establish itself unless it already starts with some mass appeal. So even though this is something that teacher candidates and teachers have said they want more of ... for a variety of reasons when offered it can't even gain a foothold as yet. And if it fails to run the second year will be likely not to be offered for at least one year or more.]

Marc Higgins: Something that struck me as I was reading the Khan et al. (2021) piece was attending the ecology of practice that pre-service teachers already bring with them. I'm still learning a lot about teacher education as well, but there's a piece that stuck with me about aggressive and tender navigations: what does it mean to care for the individual while simultaneously critically and aggressively working the structural (Galman et al., 2010). I get that sense in reading your work that you recognize that they come from a particular place and inquire into what it means to care for teachers, where they come from, while still working with them toward something that might be more flourishing-promoting for themselves and their students, as well as their greater human and more-than-human communities.

Steven Khan: In the North American context, pre-service teachers have been learning about teaching and learning from pre-kindergarten: they have ideas of what that means, their communities have ideas about what that means. I don't have that same lived experience, though my daughter will; all I can do is really note how it is different and how it has really been radically different across the height of the pandemic. Teacher education is this potential space in which we could talk about those things, but we perhaps don't as much as we would like to or they get pushed off to smaller, more emerging-type spaces. In that frame, students have histories, they have things that they are experts in that they don't know they're experts in. For example, the practices of folding, packing, and putting on headscarves; any type of practice where you work with measurement with the body, such as sewing, cooking, etc.

Marc Higgins: I really appreciated how Hang's sharing about making *Bánh Chưng* generated a slew of student comments in which they came to seeing themselves or their own cultural or traditional practices as having always been mathematical but not registering as such because of the ways over-representation functions.

Steven Khan: I can't say that all students took that away, but the idea that "Okay, this is a not a practice that you can engage in in the classroom, this is not your tradition necessarily. If you want to do this, then you need to find a partner." That's where I want them to go with this, as opposed to, "Oh, this is a resource I could take into the classroom" because they might not have the intentionality and understanding that the person who is steeped in this culture can. That's not the place you want to end up, as this often ends in ways that are hurtful at the minimum and offensive or harmful at its worst.

There's a piece coming out that talks more about the caring part, which is work that I've done with Stephanie LaFrance around curating as a deep caring for. The period for which again Caribbean, and other migrations, when we think about all these migrations we have these things that people hold on to because they care very deeply about them. Like the practice of *Bánh Chưng* and all the other parts of the New Year celebration, not just the practice of *Bánh Chưng*.

When we think of one math education within the plantation system of teacher education, we offer resources rather than things to care about. How do we teach care? Recognizing that you don't have to care about the same things I care about, nor in exactly the same way. But we should know what each other cares about and how to care for those things well in this space, as well as little humans and other-than-humans. I have some teachers that I taught in the past, I have no responsibility for how good they become, but who do that type of work: for example, showing pumpkin decomposition by having it in a container in the classroom, naming it, and observing it decomposing over time, and reflecting on the process. In the spring, they took it outside and put it back into the ground. So, this is something this teacher cares about and is trying to communicate how you care about that: without faking it.

Marc Higgins: I think it's a fair statement to make that your recent work (e.g., Khan, 2020), which creatively weaves decolonial theory, Indigenous mathematics, and multi-species thinking to critically respond to mathematics education in this particular moment marked by a "we" of climate crisis that erases the ways in which multiple human and other-than-human communities are unevenly impacted (and the ways in which this particular Anthropocene is predicated on multiple others, in ways that are marked by [attempted] genocide), speaks to a long-standing commitment to fostering collectivities: allies and alloys. As you stated early on in your scholarship, "the field [of mathematics education] must find allies and alloy itself with disciplines and perspectives in which the imagination is central if it is to address or redress some of the inequities and injustices of the present" (Khan, 2011, p. 17). This is not only evident in the ways in which you've worked from and woven together multiple and often disparate fields, such as ethnomathematics (e.g., Khan, 2011) and environmental education (e.g., Karrow et al., 2017), but also the multiplicity of different collaborations that you've sustained and lifted (which more recently include working with folks in science education; e.g., Khan & Bowen, 2022). Can you speak to the importance of "allies and alloys" in your (collective) work?

Steven Khan: I was a secondary school biology teacher, so that remains a strong influence in terms of how I conceptualize allies: math and biology folks are those I chat with the most. Allies is pretty straightforward in that we all need communities of belonging and to feel a sense of belonging, for a variety of reasons. We need to feel part of communities. Part of this comes from this deep sense of academic loneliness that begins in different places and continues. I don't like to think by myself; I like chatting with students and colleagues. When I find people I like to write with and work well together, that is an extension of the talking with. When I wrote that piece (Khan, 2011), allyship was also about discipline. It was stating that curriculum studies and math education did not talk a whole lot to each other, but this argument is not exclusive to this context.

The idea of alloys comes from chemistry, not my favorite discipline: the calculation part great, the actual lab stuff less so—I don't think I was very good at titrations. It's also a variation on notions of identity, like creolization and intersectionality, but in the compound sense, rather than the cumulative, collective sense. We are not a collection of identities, which some versions of intersectionality frame it as. Rather, we are a *compound* of identities: that the thing we are now can't be disentangled, can't be re-decomposed into these individual constituent parts. We are all, and it is always, something new. This entanglement makes unitary locations and positionings impossible. The idea of alloys is also meant to bring to mind energy, forces, and spaces, and involves things like heat, light, sound, and crucibles. Although I don't think I've ever worked with a crucible, in terms of their real smelting processes, at least not yet. This is very much related to ideas, the history, and evolution of the steel pan movements, so working with alloys there. The description I offered in that piece, from Leroy Clarke, was about what it means to write poetry: to read words until they give off flavor, color, and scent. So he's also drawing on the idea of the steel drum: yes, you're hearing it but you're hearing it with your whole body, as there is a visual, gustatory, and olfactory element to the music. After becoming a parent, this gives us another example, although I don't think our partners would appreciate us calling our children alloys—so don't use that one! They are an admixture of two and more than two and that the result is one and more than one.

The idea of allies and alloys also comes out of some of the work of maroon theory and maroon mythopoetics in the Caribbean, with the idea of allies struggling against the plantation system. Welcoming not only those fleeing the plantation if they were enslaved or freeing others from the plantation but also welcoming those cast aside by the capitalist system. Often, if the sailors were injured, they'd be abandoned on one of these islands and left—or if they got old or sick. There are allies and alloys to be made with disability studies and other fields: these are the communities that are currently struggling against the plantation system, its logics, economics, and religion. So, how do we ally ourselves with them in terms of learning, organizing, and doing? It's also about speaking to the culture of mistrust, protectionism, and secrecy that's part of the DNA of the Academy that works against relationships in order to protect the integrity of your scholarly words.

Marc Higgins: Are there any closing thoughts that you would like to add to this particular interview at this juncture?

Steven Khan: Bringing together a number of things: for me, I think the multi-species flourishing idea resonates with people in ways that other discourses don't, where there is still that possibility for collapsing into some other discourse. [I recognize in it a variation of one of my initial concepts, that of inter-vulnerability, but I think flourishing seems to work better than vulnerability.] For now, I think it's managed to hold the conversation open a little to give pause: to go hmm, that's not something I've thought about or framed it that way. However, it is something that I think many people have thought about and are engaged with in different ways. It's continuing to expand as well.

I don't think I'm an organizer, but I think this is the next part: attending to others' writing and others who find themselves in particular histories and locations, making that choice to do that type of organizing work. Setting others up to take it up and do good things with it.

Marc Higgins: This also speaks to the ways in which we're all differently gifted. We all flourish when plurality is welcome rather than rendered a liability.

> Relationships matter; ... relationships with our more-than-human kin matter; and ... learning mathematics, science and technology for survival, transcendence, dignity, belonging and to meet challenges through studying the networks among land, language, lore (story), living, logics and (emergent) learning is a necessary first step in repairing relationships damaged through the various forms of colonialism (settler, extractive, plantation), ongoing colonialities and its attendants—racial capita lism and multispecies exploitation. (Khan & Bowen, 2022, p. 7)

References

Galman, S., Pica-Smith, C., & Rosenberger, C. (2010). Aggressive and tender navigations: Teacher educators confront whiteness in their practice. *Journal of Teacher Education, 61*(3), 225–236.

Karrow, D., Khan, S., & Fleener, J. (2017). Mathematics education's ethical relation with and response to climate change. *Philosophy of Mathematics Education Journal, 32*(November 2017), 1–27.

Khan, S. (2020). After the M in STEM: Towards multispecies' flourishing. *Canadian Journal of Science, Mathematics and Technology Education, 20*(2), 230–245.

Khan, S., & Bowen, G. M. (2022). Why multispecies' flourishing? *Journal of Research in Science Mathematics and Technology Education, 5*(1), 1–10.

Khan, S. K. (2011). Ethnomathematics as mythopoetic curriculum. *For the Learning of Mathematics, 31*(3), 14–18.

Khan, S. K., Tran, H. T. T., & LaFrance, S. (2021). Mathematics for multispecies' flourishing. In D. Kollosche (Ed.), *Exploring new ways to connect: Proceedings of the Eleventh International Mathematics Education and Society Conference* (Vol. 2, pp. 555–564). Tredition. https://doi.org/10.5281/zenodo.5414972

Táíwò, O. O. (2022). *Reconsidering reparations.* Oxford University Press.

Wallace, M. F., Bazzul, J., Higgins, M., & Tolbert, S. (2022). *Reimagining science education in the Anthropocene.* Springer Nature.

Wynter, S. (2003). Unsettling the coloniality of being/power/truth/freedom: Towards the human, after man, its overrepresentation—An argument. *CR: The New Centennial Review, 3*(3), 257–337.

Yusoff, K. (2018). *A billion black Anthropocenes or none.* University of Minnesota Press.

CHAPTER 22

Conclusion: Amplifying Science Education Research with(in) a Minor Key

Maria Wallace and Marc Higgins

How many people today live in a language that is not their own? Or no longer, or not yet, even know their own and know poorly the major language that they are forced to serve? (Deleuze & Guattari, 1975/1986, p. 17)

The chapters in this second volume collectively annunciate a refrain: another science education is not only necessary, but also possible. They demonstrate examples of what it might mean to enact science education research in a minor key: working within, against, and beyond a "major language" of science and science education that they are forced to serve but that no longer serves them. In the midst of this ecological reckoning without a roadmap, *majoritarian thinking* in science education that values (only) dominant discourses, epistemologies, and views of reality (i.e., what

M. Wallace (✉)
University of Southern Mississippi, Hattiesburg, MS, USA
e-mail: Maria.Wallace@usm.edu

M. Higgins
University of Alberta, Edmonton, AB, Canada
e-mail: marc1@ualberta.ca

© The Author(s) 2024
S. Tolbert et al. (eds.), *Reimagining Science Education in the Anthropocene, Volume 2*, Palgrave Studies in Education and the Environment, https://doi.org/10.1007/978-3-031-35430-4_22

409

students "ought" to think) cannot wholly account for and be accountable to the Anthropocene, this new geological epoch we live in. Within the Anthropocene, the planet is predominantly shaped by extractivism, the ever-accelerating project of extracting resources for energy production and in service of economic growth. As a result, it disproportionately threatens large swaths of the Global South, endangered animal and plant species, Indigenous peoples, and marginalized communities of color (both urban and rural). As science education makes possible and palatable such realities which render so many worlds within this world minor(itized), with both material and semiotic consequences, moving from a reliance on majority thinking as well as its prevailing onto-ethico-epistemological frameworks and methodological orientations toward thinking in a *minor key* is significant: "there is nothing that is major or revolutionary except the minor" (Deleuze & Guattari, 1975/1986, p. 26).

Drawing inspiration from Deleuze and Guattari's (1975/1986) notion of minor literature, we offer *science education research in a minor key* as a figuration to articulate the ways in which the chapters in this book, as well as the previous volume, depict images of what is possible for science education research within a new tenor. Of minor literature, Deleuze and Guattari (1975/1986) state that, "a minor literature doesn't come from a minor language; it is rather that which a minority constructs within a major language" (p. 16). Stated otherwise, this minor(itized) language, that which is rendered other by majoritarian language, becomes a minor literature when it is put in relation with majoritarian language in a way that makes it stutter, stumble, or stop in its tracks. In turn, science education research in a minor key is situated firmly in relation to both science education as well as ideas, literatures, disciplinary and beyond-disciplinary knowledges, voices, and beings made-to-be at the periphery of science education. Importantly, working the minor key is not necessarily an escape from major articulations of science education; rather, it is a means of creating new possibilities for a structure otherwise marked by impossibility, of reconfiguring what possibilities are possible, through a "minor utilization" (Deleuze & Guattari, 1975/1986) of major language.[1] In other words, it is to make science education "a sort of stranger *within* his own language" (p. 26, emphasis in original).[2]

---

[1] Importantly, we heed Spivak's (1988) critique that not all possibilities are possible, or desirable: whilst resistance to majority thinking can happen in innumerable locations and manners, it does not mean that all are equally significant in critical import and potentiality.

[2] Here, retaining the gendered pronoun *his* from Deleuze and Guattari's (1975/1986) *Kafka: Towards a Minor Literature* is an intentional nod to the ways in which Western Modern Science is often referred to as White Male Science.

In making meaning across these novel and necessary ways of being and knowing in science education research which register discourses and practices of anti-racism and -colonialism; ecopedagogy; speculative fiction; spatial, social, and ecological justice; and (post-)critical pedagogies, amidst other orientations, we turn to Deleuze and Guattari's (1975/1986) characterization of minor literature. As they articulate, the three most significant qualities of minor literature are deterritorialization, political immediacy, and collective enunciation.

Deterritorialization: Where territorialization is the metaphorical constraining and containing of a territory, deterritorialization asks us how we might (re)open spaces that are stuck or sedimented. In science education, this invites us to engage in questions beyond *How do we best teach science?* to engage with related and relevant questions of *What and who is science for?* and *What counts as science?* as a means of identifying and enacting potentialities beyond the major image of science education proposed.

Political immediacy: Political immediacy calls us to consider the ways in which the personal is political: it asks us to consider the ways in which individual learning and concern reverberate and resonate with questions of politics. For example, we might ask *Whose or which perspectives are included or excluded from science education?* Or, *If they are included, are they included in ways which refuse and resist the logics which excluded them in the first place?* in order to investigate what is at stake within the classroom and beyond.

Collective enunciation: When speaking with or from the margins, we may be in communication with different communities of practice beyond the classroom, the school, or the field of science education; when we speak in a minor key, we always speak with others. Collective enunciation elicits us to ask *With whom am I relationally in conversation?* and *What are the possibilities of and for a community of practice otherwise?* as well as the significance of these new or different constellations of relationships.

To conduct research in science education in a minor key is simultaneously an act that denaturalizes thought and illuminates political immediacy and the necessity for collective enunciation. We understand these three qualities as being sometimes implicitly infused and sometimes explicitly articulated throughout each section of this book, but always present. For example, the authors in this volume intentionally trouble the linguistic assumptions that produce normative and normalizing grids of intelligibility within science education. Considerably, we see these three qualities as provoking critical questions with respect to the means of research in science education in addition to its ends—that is, how might science education consider the ways in which methodologies of majority thinking further complicate what possibilities are possible for science education (Higgins et al., 2018)? For example, we might ask *How do we think about how we think without using the thing with which we think (when the thing with which we think is part of the problem)?* (see Higgins et al., 2019).

Enacting science education research in a minor key is to confront the co-constitutive trends in majority thinking which render thought circular, a circularity which produces and (fore)closes itself against its rendered "non-scientific" Otherness by rendering these supplementary ways of thinking inadmissible, unintelligible, and at times unimaginable. More specifically, minor thinking is *a refusal* to move directly to the center of such circular thought (i.e., majority thinking) through too-simple forms of inclusion while simultaneously not renouncing a potentially productive relation to this problematic center. This is done by moving along the circular contours of majoritarian thought while on the lookout for ethico-political lines of thought which move us away from what rigid majorities would have us think and embody which move us toward a yet-to-come that is with, from, or made-to-be-periphery (e.g., following the *mights* of science education rather than its *oughts*).

To animate this conversation, we revisit some of our own earlier work on *Thinking with Nature* (Wallace et al., 2018)[3] as *an* additional and explicit example of science education research in a minor key. Therein, we invited science educators to consider the always-already capacity of Nature to address some of the persistent dilemmas confronting our work as science educators in this contemporary moment. There is increasing awareness that the (re)production of "nature" is only in part a human meaning-making practice, and one that is often deeply territorialized within majority thinking. However, in deterritorializing "nature," we might take seriously the notion that it may not only be co-constructed with other humans, but also with other-than-humans (e.g., thinking with

---

[3] Thinking with Nature is a differential articulation of Jackson and Mazzei's (2012) work on *thinking with theory* developed by and for science education to stay with the trouble of science education.

lightning, with holobionts; Wallace et al., 2018; thinking with horseshoe crabs; Byers & Wallace, 2021), and more-than-humans as well (e.g., thinking with ghosts; Higgins, 2022)—those who would come to co-constitute the collective enunciation. The political immediacy of such a task takes many shapes in the Anthropocene as well having multiple bearings on how nature and those who are positioned as "closer to nature" are consequentially conceptualized (e.g., Higgins & Tolbert, 2018). One such example of lines of questioning that might be made possible through thinking with Nature, in the context of North America, which makes explicit the qualities of science education research in a minor key is as follows:

> How does sustainability science seriously contend with the genocides of large Indigenous populations (as a marker of the Anthropocene) and our more-than-kin (such as the disappearance of Buffalo herds and grass species)? How are practices of forgetting these disruptions, intentional or not, part of genocides-in-the-making? (Higgins et al., 2019, pp. 162–163)

We distinguish these tensions in science education with "N" versus "n": little-n nature and capital-N nature. Where n̲ature depicts majoritarian thinking, a (re)articulation of the language of science based in logics of control, representation, and dominion/domination, the expansive concept of N̲ature (beyond, but not oppositionally defined against nature) is synonymous with thinking in a minor key. Some further examples of departures of Nature from nature are depicted in Table 22.1.

With the advent of the ontological "turn" in education more broadly, and in science education specifically—a movement which, while not wholly unproblematic in and of itself, offers new possibilities for engaging science education research in a minor key—the role of Nature in the construction of "nature" is increasingly being considered (as well as the role of Nature in constructing Culture). From this view, Nature itself (as immanent totality) exceeds and continues to trouble our constructions of nature (e.g., as a simple opposite to culture). Furthermore, the ontological turn challenges researchers to deeply grapple with the ways nature and culture become entangled—rather than in binary opposition. The age-old discussion of nature versus nurture is no longer relevant, as the questions have now become *What does it mean for nature and nurture to be co-constitutive entities?* For example, there is much research coming out of the ecological and biological sciences depicting ways in which communication and knowledge making exceeds the human subject. One fairly mainstream

Table 22.1   Departures of Nature from nature

[With] **N**ature	[About] **n**ature
• A flow, intensity, and force in the making of knowledge • Knowledge generation that occurs within, against, and beyond traditions of inquiry • Includes other-than-humans, more-than-humans, and the not-yet as possible agents in phenomena	• Content or standards to be absorbed or mastered • Reality is strictly empirical (i.e., measurable observation through the senses—touch, feel, see, hear, or taste—in which some senses are valued over others) and within specific traditions of inquiry

example of this can be found in Peter Wohlleben's (2016) *Hidden Life of Trees*. It is becoming increasingly important in science education research to account for and be accountable to the other-than- and more-than-human actors whose meanings and practices are rendered unintelligible by majoritarian thinking. In its most succinct articulation, it is to think *with* rather than *about*.

As one might already sense, thinking with Nature is a non-normative and non-normalizing perspective in science education. Research, as it's typically produced and disseminated within the field, tends to methodologically function as a mirror. That is, it is an attempt to reproduce sameness, elsewhere, circuitously, in the interests of power. Alternatively, thinking with Nature invites a non-linear view which does not aim to reproduce thought or life as it is already conceptualized within the logics of representation, but instead tries to keep thought on the move. As an irritative and iterative movement, we understand thinking with Nature as an additional entry point into minor thinking like the chapters provided in this collection which reveals new questions rather than solving old ones by attending to Nature's molecular connectivities inherent to the work of science education.

In conclusion, we invite researchers of science education to explore and enact modes of minor thinking as methodological practice rather than the "common sense" logics[4] that permeate our field which reproduce majoritarian thinking (e.g., framing knowledge of nature as its own desirable and

---

[4] This is particularly important as appeals to "common sense" in science education rarely account for and are accountable to the power dynamics and structures inherent to how a particular sense is made-to-be common (see Higgins, 2021).

atomizable object of acquisition). Again, this is a task of particular significance in the Anthropocene when science education is caught up in the distributed responsibility toward making possible and palatable the extractivist practices leading up to this point as well as the systematic ongoing devaluation and erasure of peoples and their practices who are most negatively impacted by this era. Such questions and quandaries remind us that science education is at a critical juncture. Whereas the pendulum of science education cyclically swings between progressivism and conservatism as a function of majoritarian thinking, the work found in this (and the former) volume explores critical and creative ways of knowing and being in science education: science education in a minor key. In this contemporary moment in which science education is easily susceptible to further territorialization within majoritarian lines of thought (i.e., dominant, hegemonic, dogmatic), we see hope in the inseparable enactments of deterritorialization, political immediacy, and collective enunciation that are brought to life by the diverse scholars who have contributed here. They demonstrate the limits of anthropocentric ways-of-knowing and -being, creatively generate a proliferation of onto-epistemological and ethico-political possibilities to attune otherwise, and explore the potential and possibilities of a minor science education that, by design and with purpose, goes against the grain.

Once more, and louder for the folks in the back, another science education is not only necessary, but also possible.

References

Byers, C., & Wallace, M. (2021). A story of bodying in science education. *Cultural Studies of Science Education, 16*(2), 387–401.

Deleuze, G., & Guattari, F. (1975/1986). *Kafka: Toward a minor literature.* University of Minnesota Press.

Higgins, M. (2021). *Unsettling responsibility: Indigenous science, deconstruction, and the multicultural science education debate.* Palgrave Macmillan.

Higgins, M. (2022). Toward a hauntology of science education. In J. Beier & J. Jagodinzski (Eds.), *Ahuman pedagogy* (pp. 77–95). Palgrave Macmillan.

Higgins, M., & Tolbert, S. (2018). A syllabus for response-able inheritance in science education. *Parallax, 24*(3), 273–294.

Higgins, M., Wallace, M., & Bazzul, J. (2018). Disrupting and displacing methodologies in STEM education: From engineering to tinkering with theory for eco-social justice. Editorial for special issue in *Canadian Journal of Science, Mathematics and Technology Education, 18*(3), 187–192.

Higgins, M., Wallace, M., & Bazzul, J. (2019). Staying with the trouble in science education. In C. Taylor & A. Bayley (Eds.), *Posthumanism and higher education: Reimagining pedagogy, practice and research* (pp. 155–164). Palgrave Macmillan.

Jackson, A. Y., & Mazzei, L. (2012). *Thinking with theory in qualitative research: Viewing data across multiple perspectives.* Routledge.

Spivak, G. C. (1988). Can the subaltern speak? In C. Nelson & L. Grossberg (Eds.), *Marxism and the interpretation of culture* (pp. 271–313). University of Illinois Press.

Wallace, M., Higgins, M., & Bazul, J. (2018). Thinking with Nature: Following the contours of minor concepts for ethico-political response-ability in science education. *Canadian Journal of Science, Mathematics and Technology Education, 18*(3), 199–209.

Wohlleben, P. (2016). *The hidden life of trees: What they feel, how they communicate—Discoveries from a secret world* (Vol. 1). Greystone Books.

Index[1]

[1] Note: Page numbers followed by 'n' refer to notes.

418   INDEX

Printed in the United States
by Baker & Taylor Publisher Services